High-Security Mechanical Locks:
An Encyclopedic Reference

High-Security Mechanical Locks: An Encyclopedic Reference

Graham W. Pulford

AMSTERDAM • BOSTON • HEIDELBERG • LONDON
NEW YORK • OXFORD • PARIS • SAN DIEGO
SAN FRANCISCO • SINGAPORE • SYDNEY • TOKYO

Butterworth-Heinemann is an imprint of Elsevier

Elsevier Academic Press
30 Corporate Drive, Suite 400, Burlington, MA 01803, USA
525 B Street, Suite 1900, San Diego, California 92101-4495, USA
84 Theobald's Road, London WC1X 8RR, UK

This book is printed on acid-free paper. ∞

Library of Congress Cataloging-in-Publication Data

Pulford Graham W.
High-security mechanical locks : an encyclopedic reference/Graham W. Pulford.
    p. cm.
  Includes bibliographical references and index.
  ISBN-13: 978-0-7506-8437-8 (alk. paper)
  ISBN-10: 0-7506-8437-2 (alk. paper)
  1. Locks and keys–Encyclopedias.

  TS520.P854 2007
  683′.32–dc22

                                2007004202

British Library Cataloguing-in-Publication Data

A catalogue record for this book is available from the British Library

ISBN 13: 978-0-7506-8437-8
ISBN 10: 0-7506-8437-2

For all information on all Elsevier Academic Press publications
visit our Web site at *www.books.elsevier.com*

To Jill, Arthur, Lehene, HweeYing and Chloë

Contents

Acknowledgments

Following the release in 1994 of the "Catalogue of High Security Locks v1.00" [101], the author has received correspondence from many people who have offered their views and feedback on material that has helped to improve the content, coverage, and accuracy of the information in this book. Many of these people, being contributors to the alt.locksmithing newsgroup and other online forums, are known to the author only by their electronic mail addresses. For the sake of discreetness I cannot identify these people here, but my thanks go out to them nonetheless. Their helpful remarks were a regular reminder to me to continue working on this project.

In Australia, I had the good fortune to meet Norm Axford of Acorn Locksmiths and the Master Locksmiths Association (Australia), whose willingness to support this book was a great source of encouragement to me. I am particularly grateful to Ian McColl of Stockade Locksmiths, who greatly contributed to the completeness and coverage of the book and also assisted with the early stages of proofreading. I would also like to thank Rodney Loschiavo who generously shared with me his knowledge acquired during many years in the trade. I am grateful for the endorsement of Mark Wilson from the Sydney Institute of Technical and Further Education, who granted me access to the Ultimo workshop. Geoff Birkett of Olympic Keiler Security gave generously of his time to help ensure the accuracy of the material presented on several high-security locks. Meilin Yum's willing assistance in typing part of the manuscript is gratefully acknowledged.

In addition, I am fortunate to have had international collaborators. In this connection I would like to express my gratitude to Nathan Schlossman, Oliver Diederichsen, Tetsuya Ozawa, Han Fey, Paul Prescott, Tony Beck, and Alex Carter. I owe a special debt of gratitude to my friend and colleague, Jean-Marie Machefert, for many stimulating discussions and for imbuing me with some of his knowledge of the subtleties of the art, particularly in respect of French high-security locks. In the United Kingdom Michael Fincher, Richard Hopkins, and Jon Millington all provided valuable assistance with the proofreading of the manuscript. I am indebted to Michael Fincher for allowing me to photograph some of his lock collection; his in-depth knowledge, especially of German locks, spared me the embarrassment of committing some major historical errors. It would be remiss of me not to extend my sincere thanks to the editorial and production team at Elsevier for transforming the manuscript into a book.

A number of organizations are also deserving of mention for their assistance either with the supply of samples or in the provision of product information or services. In this context, the following businesses and companies are acknowledged: West Philadelphia Locksmith Co. (Philadelphia, U.S.A.), Lockmart (Toronto, Canada), Serrurerie Rapid-Securit (Paris, France), Quincaillerie d'Alembert (Paris, France), CISA S.p.A. (Faenza, Italy), Centro Sicurezza Minerva (Rome, Italy), Diesenhofener Schlüsseldienst Gabriel (Munich, Germany), Münchener Schlüsseldienst Willi Killian (Munich, Germany), Cire Electronics (Sydney, Australia), Rivers Locking Systems (Sydney, Australia), Eastwood Lock & Key service (Sydney, Australia), ABA Locks Manufacturer Co. (Taipei, Taiwan), Tover Security Systems (Banyoles, Spain), Helason SicherheitsTechnik GmbH (Vienna, Austria), ANTO Translation Services (Turin, Italy), Can-Am Door Hardware Inc. (Québec, Canada), Bramah Security Equipment (London, U.K.), Supra (UK) Ltd. (Worcestershire, U.K.), K. J. Ross Security Locks Pty. Ltd. (Melbourne, Australia), SEA Schliess-Systeme AG (Zollikofen, Switzerland), and Australian Lock Company (Unanderra, Australia).

A final note of thanks is due to the keepers of the alt.locksmithing newsgroup archives at ftp.indra.com, who kindly made a place on their system for the document that formed the basis for this book.

Preface

I have been interested in things mechanical and electrical since the age of two when, my father tells me, I unscrewed the back of the washing machine, having seen him do it. I also liked puzzles and used to infuriate my mother who, being of a more literary inclination, had considerably more trouble with them than I did. I had previously heard of things like lockpicking, skeleton keys, and Houdini, and had seen schoolmates open cheap padlocks with screwdrivers. But I had never stopped to think carefully about what made a lock tick. Like many other people I knew, I took locks for granted. My mindset regarding locks in those times can be summed up in this way, "if there is a lock on it, then it can only be opened by the person with the proper key."

My real interest in locks, which grew into a fascination, started when my family moved. As people tend to do, I thought we should have the locks changed. So in order to save money I unscrewed the front door lock, removed the rim cylinder, and took it to the local locksmith. I remember being totally amazed when, about five minutes after I'd handed it to him, he returned it to me recombinated with a new key. I realized there must be a trick to this and decided to find out some more about it. Over the course of the next few years, I learned about how pin-tumbler locks worked and how to pick them using homemade tools. Some of my friends who had contact with people in the locksmithing profession helped me along the way.

A pivotal point in my growing interest in high-security locks occurred while I was traveling in Europe. The locks I saw in France, Germany, Austria, and other countries were so different from those I was used to in my home country of Australia. I realized then that there was a whole world of ingenious locks out there to learn about, each with its own particular features. There were also many similarities in the operating principles.

Since that time, I have collected locks from Europe, Asia, and the United States, as well as from Australia. It was on the basis of these travels and experiences that in 1994 I compiled a document called the "Catalogue of High Security Locks v1.00." The present book contains the information in that previous work as well as many additional high-security lock descriptions. The obvious deficiency of that document— its lack of pictures—has been corrected in this book. This information is offered to readers in the hope that they can share in some of the excitement that I have had in discovering the amazing world of high-security locks.

Finally, a word of caution seems in order. This book will help you to understand the operating principles of a large number of high-security lock designs. It provides a rough estimate of the manipulation resistance provided by each lock. However, no detailed information on lock-defeating methods has been given. Thus, after reading this material, you may know if a given lock is susceptible to manipulation by impressioning or picking, but you will not be given instructions on how to do it. Nor is prescriptive information such as where to drill a lock or how to construct a decoder for a particular lock provided herein. Needless to say, the first step in picking or bypassing a lock is to have a detailed understanding of its design and how it works. The reader should be aware that specific tools and decoders are available not just for simple pin-tumbler locks, but for most mechanical high-security locks that enable them to be opened nondestructively without the key. However, such equipment is, for obvious reasons, not generally available to the public.

Chapter 1

Introduction

The construction of locks, is a subject on which many ingenious mechanics have employed their thought, and the art hath received many, and great improvements from their labours. *Joseph Bramah, c. 1784*

1.1   Prologue

The king did not wish to be disturbed. He had withdrawn to his atelier and bolted the door. The entire morning had been wasted on bothersome matters of state in which his wife showed considerably more interest than he did. Once inside his workshop he could concentrate on his favorite pastime: locks. For the last few days, he had been busy fashioning some intricate sash warding for an ornate lock he hoped to fit to his chambers. He was still engrossed in his work when there came a knock at the door. "Sire, the commoner Gamin requests an audience with you. He says that he is in possession of an item that you wished to see." The king put down the file with which he had been shaping the warding for the lock, his hands dirty from the work. "Bring him to me," he said. Some minutes later Gamin was brought to the now open door of the workshop. The king greeted him warmly, eager to see the article the man carried with him. Gamin beckoned the king over to a bench that seemed to be a little bit freer than the others. Pushing aside some tools and other clutter, he placed the article on the bench. He took a nearby screwdriver and proceeded to pry off the front of the case, revealing the interior mechanism. "It is a double-acting lever lock, sire. Come, see how it works!" The king had not seen this type of lever lock before, though he had heard about it. He picked up the lock in his sooty hands to examine it more closely. "Hand me the key!" he said.

The king was of course Louis XVI of France, his home the sumptuous palace of Versailles to the west of Paris. The particular episode described here is fictional, although given Louis' penchant for locksmithing and neglect of his state duties, it could quite easily have happened. Louis XVI did in fact associate with Gamin, a locksmith who taught Louis much of what he knew about the trade. There is no doubt that he was fascinated by locks and spent many hours in pursuit of his hobby. Perhaps if he had spent more of his energies responding to the cries of the French people for social reform he would not have come to such a sticky end.

But what does this speculation have to do with the subject of locks and their operating principles? Louis lived at a time of great change: not only in respect of the French Revolution, during which he was executed for treason, but the world of locks and locksmithing was equally undergoing a revolution. Until the late 18th century and for over 700 years, the only locks used in Western Europe were of the warded type (see Figs. 1.1 and 1.2). They were installed on every door, chest, and armoire and in almost all padlocks. It was the romantic era of the skeleton key and the wax imprint, and it was coming to an end.

Soon following Britain's lead, Western Europe would be transformed by the industrial revolution, and with it came the practices and reforms that we now take for

Figure 1.1: (Top to bottom) Late Renaissance "box-of-wards" key; examples of French "passe-partout" or skeleton keys from the 15th and 18th centuries.

Figure 1.2: French 17th-century steel "masterpiece" key.

granted in western countries. The tireless repeatability and accuracy of machine tools powered by the steam engine gradually supplanted manual labor, first in the cotton and wool industries and later in heavy manufacturing. Already, before the reign of Louis XVI was brutally ended, men like Robert Barron and Joseph Bramah were overturning the status quo of the lock industry. The new designs were often backed up by advances in manufacturing techniques that resulted in increases not only in security and reliability, but also in the rate of production. Industrialization also brought with it urban growth and rising rates of crime, which further increased the demand for security products such as locks. Increased demand put pressure on workers in lock factories such as Chubb's in the United Kingdom. This in turn led to the formation of unions in the late 19th century, such as the National Union of Lock and Metal Workers [114], to campaign for improved working conditions on behalf of their members.

We will have more to say about the history of locks since the late 1700s, but now let us return to the present. Two hundred years ago, while Western societies were coming to grips with industrialization, no one could have foreseen the profound changes that we today face in the information revolution. In the space of one generation we have experienced the birth of the Internet and the so-called information super highway. We also live in an era of unprecedented corporate globalization.

The impact of these phenomena on the traditional art of locksmithing has been immense. Since the late 1980s, the Internet has increased the availability of specialized knowledge to the general public by orders of magnitude. Teenagers can download articles on lockpicking and key impressioning, not to mention more arcane subjects. Technical information about these subjects is now freely exchanged between "nontrade" people from the comfort of their own homes. This situation would have been unthinkable 50 years ago.

Moreover, the modern-day epidemic of globalization is leading to the disappearance of small companies producing locks inside a "family business" setting. Any local company sporting a commercially successful product is liable to be acquired by a global corporate entity seeking to increase its market share. The future of the product is then determined by the directors of the global company and not by the people who originally developed it.

Another force is also shaping the industry—the "digitization" of locks. The proliferation of traditional mechanical locks is fast giving way to locks that combine electronic and mechanical elements to fully electronic ones. In the motor vehicle industry the key-top transponder is becoming increasingly prevalent, while for hotels and other large complexes, magnetic and proximity card systems are gaining the upper hand owing to their lower cost and better flexibility. It is against this backdrop of technological change that this treatise on mechanical high-security locks has been compiled.

Before getting down to business, I should tell you about the motivation for this book. Locksmithing is an ancient art that has been practiced in every industrialized country for centuries. There are many good treatments of locks and locksmithing, and the ones that have been consulted in the writing of this book are listed in the bibliography. However, it is not my intention to offer the reader a course in practical locksmithing techniques. Accordingly, many aspects of the trade such as tools and equipment, manufacture, assembly, installation, servicing, key cutting, and key duplication are not covered. The important subject of master-keying is mentioned inasmuch as it affects the design of a lock, but the design of master-keyed systems is outside the scope of the book.

The raison d'être of this book is the desire to encapsulate and share my fascination for high-security locks and their operating principles with locksmiths and lock enthusiasts alike. This knowledge comes from many sources, including books, product catalogues, marketing brochures, the World Wide Web, as well as from face-to-face and online discussions with locksmiths, lock collectors, and security consultants, both locally and overseas. A large part of the information was obtained through inspection of the actual locks that are presented.

While there are many books and resources that cover a few high-security locks, few provide a comprehensive coverage of many different types and brands. Moreover, the treatments tend to be aimed at people in the locksmithing trade or industry. Thus there seems to be a lack of resources that do justice to the enormous variety of high-security locks that exist today while remaining accessible to the public, who, after all, are paying for all of this. This book is an attempt at redressing this apparent deficiency. Its content represents a fusion of information from a wide variety of sources and more particularly from sources that are widely separated geographically. With all the changes confronting the worldwide lock industry today, it is an opportune moment to take stock of the multitude of mechanical high-security locks without regard to their country of origin or their commercial viability, celebrating their diversity.

On a didactic note, the material presented is at times quite specialized. Thus, although an attempt has been made to set the scene at the start of each chapter, the reader may find it useful to consult an introductory locksmithing textbook or online resource (e.g., [47]) beforehand in order to become acquainted with the ideas and vocabulary used in this treatise. A good degree of familiarity with basic lock-operating principles is assumed in some of the descriptions.

The level of presentation is suited to an apprentice locksmith or intermediate hobby-ist wishing to gain a more complete understanding of high-security lock principles, including the similarities in the designs as well as what sets them apart. People with an engineering, electronics, software, or information technology background should have little trouble digesting the material. Even more experienced readers, both professionals and enthusiasts, may discover some types of locks of which they were previously unaware, due to the inclusion of locks from many different countries around the world.

1.2 Security Versus Obscurity

In a book on a potentially sensitive subject like high-security locks, a discussion of the topic of security versus obscurity is warranted. I have deliberately modified the usual term of *security through obscurity* to imply that security does not arise simply through secrecy, as will be argued in this section.

While attitudes toward the dissemination of information from the locksmithing pro-fession have fluctuated, there has never been a consensus on what represents an acceptable level of disclosure. The problem has always been to balance the legiti-mate right of the public to know about the product it is purchasing with the risk of the information being used for nefarious purposes.

The existence of the Internet has had a tremendous impact on the availability of detailed information on locks and manipulation techniques. With little skill or ded-ication, patents and other documentation can be located via Web search engines. This is not to say that the information was not available before (patents in their modern form, i.e., with a specification, have existed in the United Kingdom since the early 1700s), but it is now much cheaper and easier to find and redistribute. Despite copyright, it is not uncommon for entire books to be made available, albeit illegally, online in electronic format.

Because almost everybody buys and uses locks, the situation is somewhat less clear-cut than the protection of sensitive information in national defense, where secrecy is paramount. In this arena it is clearly unwise to divulge to a third party detailed information about, for instance, signaling codes or actual weaknesses in defense systems. Sensitive information is shared only on a "need to know" basis in order to minimize the occurrence of security breaches.

At the start of the industrial revolution, as noted by Ian McNeil in his biography of Joseph Bramah [82], the interchange of technical information about locks was very much restricted. Besides the cost of printing, the need for secrecy was no doubt justified by the locksmith's legitimate concern for the public's protection from criminals on the one hand and his fear of divulging trade secrets on the other. Nonetheless, the thieves and pick-locks of the time did not seem to suffer greatly from this lack of information disclosure. It seems fair to say that the main casualties were the public, which was starved of affordable products offering adequate security, and the lock industry itself, which initially suffered from slow progress due to duplication of work and an inability to derive benefit through the sharing of ideas.

By the mid-1800s, the situation had changed considerably. John Chubb [22] saw fit to publish at a meeting of the Institution of Civil Engineers in April 1850 the details of a warded lock from a London banking house to emphasize its weaknesses:

> . . . and to prove its utter insecurity, a drawing has been made of a lock and key, with picklocks.

Chubb went on to disclose in great detail the working principles of Bramah's lock and the Chubb detector lever lock, both of which are covered later in this book. A detailed drawing of a tool for prying open iron safe doors, called a "Jack in the box," was exhibited. Chubb added in an appendix a detailed sketch of a quadruple 24-lever lock for strong rooms, designed only four years earlier. A picture of this lock is featured in Chapter 5. The appendix of Chubb's paper also contains a complete list of U.K. lock patents from 1774 to 1849 (the year prior to the meeting). This signaled a clear departure from the tradition of secrecy in locksmithing.

Further argument in favor of disclosure is provided by A. C. Hobbs, the legendary lockpicker of the Day & Newell Company who picked Chubb's detector lock at the Great Exhibition of 1851. He wrote in 1854 [51]:

> Many well-meaning persons suppose that the discussion respecting the means for baffling the supposed safety of locks offers a premium for dishonesty, by shewing others how to be dishonest. This is a fallacy.

Hobbs followed this remark with the observation that:

> the spread of the knowledge [of the vulnerability of locks] is necessary to give fair play to those who might suffer by ignorance. It cannot be too earnestly urged, that an acquaintance with real facts will, in the end, be better for all parties.

It is clear that both Chubb and Hobbs intended this disclosure of frank and accurate information to allow people to make an educated choice on what security devices were worthy of their consideration and which devices should be avoided. It is also true that in publicizing the weaknesses of competing products, both of them were eager for the public to adopt their own respective brands of locks and safes.

The same is no less true of the American locksmith Linus Yale Junior. Only two years after Hobbs's book appeared in print, Yale published a 40-page book whose ostentatious subtitle included a sales pitch and product endorsement:

A Dissertation on Locks and Lockpicking, and the Principles of Burglar Proofing: showing the Advantages Attending the Use of the Magic Infallible Bank Lock, the Infallible Safe Lock and the Patent Door Lock, Invented by Linus Yale, Jr., and his Patent Chilled Iron Burglar-Proof Bank Doors, Vaults, and Safes, which are Adopted by the U.S. Treasury Department for All the New Mints, Custom-Houses, and Sub-Treasuries in the United States.

In his book, Yale Junior, a highly skilful lockpicker in his own right, made the following statement about a recently discovered soft-key impressioning technique that he had successfully applied to Hobbs's locks [135]:

> It was not at first intended to give the modus operandi of the new methods of lockpicking, lest a knowledge of the fatal facility with which a lock can be picked by any one of average ability, might tempt to depredation—but the constantly recurring remarks made to us that we are the only ones who know these processes, have decided us to publish our methods in self-defence; for we do not doubt that now the possibility of so doing is demonstrated, the method will soon be rediscovered by those who wish to do so for nefarious purposes: whilst those most interested in knowing whether that in which they place their reliance is secure, are still ignorant of the fact.

If Yale hesitated to make his method immediately known, George Price, a noteworthy English locksmith and author of a monumental work, entitled *Treatise on Fire and Thief-Proof Depositories and Locks & Keys*, was more forthright. Price's book [99], also published in 1856, pointed out that the technique, known at the time as mapping the lock, was merely an adaptation of the age-old method of smoking a key blank. Price had liberal views on the dissemination of knowledge about the weaknesses of locks, quoting, as we do, a translation from French of the 18th-century scientist Réaumur.[1]

> But is there not a danger that at the same time we shall be giving lessons to theives? It is not very likely that they will seek instruction from us, or that they have any need of it; they are greater masters in the art of opening doors than we. So let us learn the art of opening [locked] doors, so that we may acquire [the art] of securing them in such a way as to leave little or nothing to fear.

---

[1]René Antoine Ferchault de Réaumur was responsible from 1709 to 1757 for the compilation of a 27-volume dossier on Arts and Trades commissioned by the French Royal Academy of Sciences, which was published after his death by Henri Louis Duhamel du Monceau [28]. A large part of volume 6 of this work, published in 1776 and entitled *l'Art du Serrurier*, is devoted to locksmithing.

It is true in areas other than locksmithing that commercial success does not necessarily reflect technical superiority or ingenuity. In locksmithing, this is evidenced by the proliferation of cheaper products on the market that only serve to provide their owners with a false sense of security. Still it seems reasonable to assume that no harm can come from informing the customer of the pros and cons of the various products available. This idea is seen in almost all areas of consumer goods and services (e.g., *Choice*, *Which*, and *Que Choisir* magazines). The cost saving in buying a cheaper lock should be balanced by the increased risk of it being compromised by a thief or burglar.

The author's own view on the matter is that sufficient information should be publicly available to allow an informed choice to be made, but this information should be disclosed in a responsible manner that, as far as possible, respects the business interests of the manufacturer. Thus, to use a colloquialism, it is not really justifiable to "go the whole hog" and release very detailed accounts of how to pick, impression, drill, and bypass high-security locks, simply for the purpose of educating the public. This line of thought applies even more strongly where locks are used to protect the assets of many people (e.g., in bank safes and vaults). It stands to reason that the detailed plans and specifications for such equipment should be closely controlled, while the presence and nature of security features and the level of protection they provide should be made known.

Wherever possible, objective and factual information should be provided as this contributes to the state of the art without making life easy for unscrupulous individuals. On the other hand, if a serious flaw is found that comprises the security level provided by an existing product, then this information should be brought to the attention first of the manufacturer and subsequently the public. In this way, the manufacturer is given the opportunity to rectify the problem, with the eventual benefit being passed on to the consumer. It is important to realize that no product is perfect, and improvement is only possible through recognition of shortcomings combined with an iterative process of development and testing.

1.3   Innovation in the Lock Industry

Numerous factors contribute to the need for innovation in the design of high-security locks. Since the market for locks is huge, consisting of residential, commercial, industrial, and government sectors, there has always been a great deal of competition in the development of commercially successful technologies. This variety of end-users leads to a large spectrum of customer requirements in terms of function, price, size, convenience, finish, durability, safety, and security that reflect the different environments and uses to which the lock will be put. These requirements continue to evolve with time and have led to demand for more affordable systems providing better levels of security. In particular, as organizations grow in size, there is a need to supply larger, more complex suites or master-keyed systems, with tighter control over the supply and reproduction of keys and key blanks.

The need for continued development of high-security locks is also driven from within the industry itself. Many manufacturers patent or register their lock and key designs, and this provides a time window for the production and marketing of the product. Once this time window has elapsed, competitors can move in and copy the design. A further motivating factor is exposing weaknesses in existing designs. This may be due to ongoing testing by the manufacturers, locksmiths, independent labs, or even from people outside the profession. Progress in the design of new and improved locks would indeed be slow were it not for the feedback of information on the deficiencies of the product. As we mentioned before, for hundreds of years people used warded locks that could be defeated in a matter of minutes by a skilled thief without leaving any trace.

As it is in science, the exchange of accurate, up-to-date information leads to rapid progress. In the commercial arena, however, it is often not the core ideas that need to be protected, but rather their method of implementation. Thus, while the basic principles may well be explained in a patent specification or working model, a commercial edge can be maintained by safeguarding the actual processes used to manufacture the product reliably and economically. It is often more important to maintain the continuity of development of a product than to worry about competitors stealing the idea from a patent or other publication: "strike while the iron is hot," so to speak.

All of these factors provide a constant impetus for innovation in the lock industry. As new designs are introduced to satisfy evolving requirements, the state of the art advances incrementally. A further aspect of innovation is the capturing of the expected or achievable performance of existing systems in industry standards, which must be regularly updated. In the next few sections we review some of these ideas in more detail.

Patents and Registered Designs

Patents, or utility patents as they are known in the United States, are widely used in the lock industry for new designs because they provide up to 20 years of protection against unauthorized production and importation of copycat products, often of inferior quality. A patent typically reviews the state of the art, identifies one or more problems to be addressed, and then specifies the design (at least at a theoretical level) of a novel apparatus or method to solve the problem that involves an "inventive step" and is capable of industrial application. Since the design process is iterative, patents often relate to improvements in established designs or in the processes required to manufacture the product. There are tens of thousands of lock-related patents whose title includes the word "improvements."

The patenting process starts with the preparation of a provisional specification that is assessed for patentability and originality. This is followed by the submission of a complete design specification that is reviewed in the light of the existing body of patented and public-domain information. A properly researched patent should contain references to previous relevant patents or other public-domain sources. Each patent also contains a list of claims that precisely characterizes the design. During this time a provisional patent may be granted pending the award of a full patent. Worldwide patents require the submission of the specification to the patenting authorities in each country where patent protection is being sought.

As is often the case with legal work, the patenting process can be very slow, resulting in delays of several years between the submission of the provisional specification and the granting of the full patent. In this book, when we refer to the date of a patent, we take the *filing date*, which more accurately reflects when the work was actually done, rather than the *issue date*, which may be several years later. (In this text, patents are generally referred to by their reference number preceded by a country code.)

A brief but fascinating account of patents, as they apply to locks in the English-speaking world, is contained in an article by Millington [87] from which we cite a few facts. The first English patent for a locking device was issued to George Black in 1774. Patents have been used in the United States since 1790, although they were not numbered until 1836. One of the earliest U.S. patents relating to locks and keys is number 7,917 for a swivel-nibbed key invented by J. Hanley in 1851. During the researching of this book, one of the most recent patents for a lock was European patent EP 1,518,979, published in 2005, which describes the electromechanical control of a cylinder lock developed by the French company Dény-Fontaine. We mention numerous other lock patents in subsequent chapters of this book. Not all of these locks went on to achieve commercial success; indeed, a large percentage of patents never get past the prototype stage.

The registration of a design or trademark is also a popular method of protecting certain aspects of high-security locking systems.[2] This approach is typically used to prevent third parties from making after-market key blanks. Before the original registration has expired, the keyway broaching of a given high-security lock is modified and registered as a new design. Old systems are then progressively upgraded to use the new design.

Registration covers aspects of the system that would not normally be covered by a patent: for instance, the shape or other visual aspects of a particular key or keyway. The registration process is simpler, cheaper, and generally quicker than the patenting process (e.g., months rather than years) and covers the appearance and external qualities of the product rather than its internal design and functionality. Registration provides up to 25 years of protection against breach of industrial copyright. According to the Designs Registry of the United Kingdom Patent Office [94], the industrial design copyright was first enacted in 1787 in connection with the textiles industry.

Most manufacturers of high-security locks offer patented locking systems with registered key and plug broachings. This formula brings peace of mind to the end-user

---

[2] In the United States designs are covered by "design patents."

as well as to the lock companies, since they can ensure that no one will be able to make and supply unauthorized copies of their keys. When the original patent expires, it is often the case that the company that holds the patent will make sufficient modifications to the design so that a new patent can be taken out and, with it, a new lease of protection can be acquired.

Customer Requirements

There are reasons other than expiry of patent and registration that motivate the design of new high-security locks. Most innovations in high-security locks can be traced to customer requirements. We already mentioned the need for protection against unauthorized duplication of keys, which can be addressed through design registration and control of the distribution network from the manufacturer to the locksmith. Since the 1980s, locks have been produced that have a movable element in the key, which renders copying impractical. Another example is customer convenience, which has led to smaller-sized and reversible (symmetric) keys. The problem of key breakage, a great inconvenience to the customer, has led to a number of design refinements in terms of strength of materials, key section, and key-bitting patterns.

One of the principal motivating factors is the exhausting of key codes, which can happen when a lock has been in production for a long time or with the increasing scale of master-keyed (MK) systems[3] for large building complexes. Particularly with inline pin-tumbler locks, it is true that the more the system is master-keyed, the less secure individual locks tend to be. Thus there is a move toward lock designs that support very large numbers of master-keying options and retain as much system integrity as possible even when master-keyed. The traditional solution for large MK systems has been to use sectional or multiplex keyways together with more pin-tumblers. Multiplex systems allow expansion of a MK system using a hierarchy of different key profiles or broachings. Another approach is based on passive profile pins. Both of these concepts are described in Chapter 2. A more recent and much more secure solution is furnished by the class of dual-action side-bar locks, covered in Chapter 4.

The requirement of easy reconfigurability—that is, the ability to change or recombinate the lock with a minimum of effort—has led to the design of a number of key-changeable locks, some of which are covered herein. A common situation arises following the construction of a building when the key is to be handed over to the owner: a method called construction keying is applied to ensure that the key is different from that used to access the site during the construction phase. For some time the idea of an interchangeable-core (IC) lock has been popular for large centralized installations. A control key is all that is needed to remove the core of the lock so that it can be replaced, for example, following the loss of a key (see Chapter 2). Requirements from the hotel sector have resulted in a number of interesting designs that focus on easily rekeyed locks with low-cost keys (e.g., Vingcard).

---

[3]Further abbreviations are listed in Appendix C.

Naturally, a prime factor in the design of high-security locks is the requirement of better security against surreptitious and forced entry. These factors have greatly influenced the design of locks over the centuries and more particularly the locks presented in this book. Modern high-security locks often contain hardened, drill-resistant inserts and saw-proof collars to protect the cylinder. Manufacturers of lock cylinders must also guard against attacks by prying and wrenching. Although customers tend to be more aware of the consequences of forced entry, the nondestructive methods of lock opening have been no less of a driving force in the design of more secure locks. We encounter some of these further on.

Industry Standards

Real high-security locks are designed to industry standards that ensure quality, reliability, and fitness for purpose. Each country has its own industry standards. Different standards are applied to products with different end-user requirements. In particular, there are separate standards for cylinder and door locks, mechanical safe locks (keyed and keyless), and electronic safe locks, although some of these may be covered by the same standard. For instance, the European standard EN 1300 (2004) defines a high-security lock as:

> an independent assembly normally fitted to doors of secure storage units, into which codes can be entered for comparison with memorized codes (processing unit); a correct match of an opening code allows movement of a blocking feature.

A mechanical high-security lock (as opposed to an electronic one) is secured by means of mechanical elements only.

In Australia the standard for cylinder locks is AS 4145. In the United Kingdom the applicable standard is BS 3621. U.K. Product certification is carried out by the Loss Prevention Certification Board (LPCB) and Building Research Establishment (BRE) Certification. In France, the CNPP (Centre National de Prévention et de Protection) oversees the A2P rating system. For cylinder locks the A2P rating is one, two, or three stars depending on the level of security afforded by the product. Each star corresponds to an increment of five minutes in resistance time to various burlargy methods. In Germany the appropriate standard for profile-cylinder locks is DIN 18252 (classes P1–P3), with class P3 offering the highest level of security (e.g., resistance to drilling and forced extraction). Certification testing is carried out by the organization VdS Schadenverhüg. VdS stands for *Vertrauen durch Sicherheit*, which translates as "confidence through safety and security," and *Schadenverhüg* means "loss prevention."

The standards set by a country's industry standards bureau are not, in general, the same as those set out by its insurance and accreditation agencies. A case in point is the U.K. Loss Prevention Certification Board, whose security requirements for lock cylinders are set out in LPS 1242 [72]. This standard sets out requirements and testing procedures that allow locks to be sorted into eight categories, whereas the national standard specifies only five grades. Additional categories include design patenting, key registration, key cutting, and whether the lock can form part of a MK system. The testing procedures specify the types of tools and conduct of tests for the grading operation. Tools are sorted into six categories as specified in LPS 1175 [73], which deals with standards and tests for security enclosures. LPCB's standards for safes and strong rooms are set out in LPS 1183 [74]. A summary of the LPS 1242 security gradings is contained in Appendix G.

Locks in European countries are generally standardized for Europrofile cylinders, although a number of oval and larger format lock cylinders are still in use. The standard for European locks is set out in EN 1303 [16], which specifies a 7-digit code containing the grading of the lock against requirements such as security, temperature, durability, fire, and corrosion resistance. Durability refers to the number of cycles of use the lock has been tested for (e.g., grade 5 is 50,000). Security is measured in five grades that consider the number of effective key differs (key changes), the minimum number of movable elements (pins, levers, discs, etc.), and resistance to attacks by drill, chisel, forced extraction, and torque. A summary of the security rating (from [4]) is provided in Table 1.1. In this table, Direct Coding on Key means that the actual bitting code for the key cuts is imprinted on the key rather than an indirect or blind code that must be translated via a code book. For double cylinders, or locks with cylinders on either side of a door, the security rating may be different on either side. Fairly obviously, grades 1 to 3 should not be referred to as security locks.

Locks for safes and security containers are covered in EN 1143-2 and EN 1300 [18, 19]. For mechanical safe locks there are four classes (A, B, C and D) that follow the European standard EN 1300. These are summarized in Table 1.2. Note that the coding referred to in the table is the number of usable combinations, also referred to as the material coding, defined by the physical features of the key. This is distinct

Grade	1	2	3	4	5
Minimum Effective Differs	100	300	15,000	30,000	100,000
Minimum Levers/Pins/Discs	2	3	5	6	6
Direct Coding on Key	Yes	Yes	No	No	No
Maximum Net Drill Time	-	-	-	3	5
Total Drill Test Time	-	-	-	5	10
Number of Blows (Chisel)	-	-	-	30	40
Extraction Force	-	-	-	15 kN	15 kN
Torque Resistance	2.5 Nm	5 Nm	15 Nm	20 Nm	30 Nm

Table 1.1: EN 1303 security ratings. All test times in minutes and force and torque tests may be applied to either plug or cylinder. Reproduced with permission of BSI [17].

Class	Minimum Usable Codes	Manipulation Resistance	Destructive Burglary Resistance
A	25,000	30	80
B	100,000	60	135
C	1,000,000	100	250
D	3,000,000	620	500

Table 1.2: Security requirements for high-security safe locks (from [19]). Reproduced with permission of BSI [17].

from the mnemonic coding consisting of numbers and/or letters, which is the code assigned to the keys; this may be either direct or indirect, and there may be more mnemonic codes than actual key combinations. The manipulation resistance and destructive burglary resistance are measured in specially defined resistance units (RUs). For manipulation resistance, the units are the time taken and offset by the class of tools used (more advanced tools have a higher offset value). Burglary resistance units are a weighted index that takes into account the operating time for the attack and the attack coefficients of the tools used (more destructive tools have a higher attack coefficient).

In the United States, the American National Standards Institute (ANSI/BHMA A156.5) classifies security locks in three grades depending on loading and cycle testing. Due to its widespread acceptance, the de facto U.S. standard for keyed locks is Underwriters Laboratories Standard 437, with products satisfying the various levels of security referred to as UL-rated. UL Standard 437 sets out performance criteria for key-operated door locks, locking cylinders, and security containers (safes). UL Standard 768 applies to keyless combination locks. Lock cylinders are tested for resistance to (1) picking; (2) impressioning techniques; (3) forcing methods; (4) pulling; and (5) drilling. Door locks are additionally tested against jimmying, driving the lock assembly, sawing the bolt, and the use of small handtools. Other applicable tests are for the number of differs, which should exceed 1,000 for door locks and cylinders, and for endurance, under which locks must complete 10,000 cycles of operation at up to 50 cycles per minute. Attack resistance to specified tools and techniques is measured as a time in minutes. UL-rated door locks and cylinders should withstand ten minutes of picking and impressioning attempts and five minutes of destructive methods. Corrosion resistance (salt-spray) testing is also carried out.

Lockpicking

The impetus provided by independent testing of a product remains an important factor in standardizing and advancing the technology. This is true in fields other than locksmithing, such as software development and computer security, and in general in any area where R&D is undertaken. It would therefore be remiss not to mention what has been one of the major driving forces in the design of new locks: namely, the contribution of the "lockpickers." As Linus Yale Junior very aptly wrote in 1856 [135]:

The art of Locksmithing has become almost a science; and a review of the ingenuity and labor displayed in endeavoring to fill this great want of the community, would show to the inquiring mind the most ingenious system of attack and defence ever witnessed; difficulties and obstacles, instead of daunting, have only stimulated new effort.

Literally dozens of patents claim to have invented the "pick-proof" lock, but according to F. S. Holmes [56], one of the most famous lock makers of the world once said:

No lock having a key hole has ever been made or invented which is absolutely proof against picking, nor is it probable that one will ever be or can be made.

Despite these claims and counterclaims, efforts on the part of the lockpickers continue apace, as well as the development and testing of more destructive techniques brought to bear on the opening of locks. When a method by which a lock can be opened or bypassed without its correct key is brought to the attention of a lock manufacturer, it often results in modifications to the product to counter the attack.

Methods that have been applied to open or bypass mechanical locks include (together with references, where appropriate):

1. Picking the lock manually with flat, tubular, or Hobbs picks (see Figs. 1.10 and 1.11) [10, 65].

2. Manipulating the pins using specially designed tools such as wire lifters, "Sputniks" (Fig. 1.12), and comb picks (Fig. 1.8).

3. Impressioning the lock using blanks, foil, wood, or other "soft keys" in order to make a working key (Fig. 1.9) [129].

4. Decoding the combination of the lock optically, acoustically, electrically, magnetically, electromagnetically, or by mechanical measurement of the pins (or other elements) and hence cutting or assembling a correct key [120].

5. Impact-based methods such as pick guns (Figs. 1.4 and 1.5), vibrator picks (Fig. 1.6), bump-keys (Fig. 1.7), and rapping [130].

6. Partially destructive methods such as grinding and shimming the front of the plug, or bypass methods requiring drilling small holes or other minor damage.

7. Somewhat destructive or forced entry methods, including drilling the pins, side-bar or bolt stump, forced extraction of the cylinder, forced rotation of the plug, breaking of the coupling on profile cylinders, punching, and driving of the lock cylinder.

8. Totally destructive methods: chisels, pry-bars, sledge hammer, power tools (carbide-tipped drills, angle grinders, saws, and cutting wheels), oxy-acetylene torch, thermic lance (also known as a "burning bar"), hydraulic and scaffolding jacks, and explosives.

The list has deliberately been arranged in order of decreasing subtlety and should convince the reader of the lengths to which some people are prepared to go to defeat a lock, especially when there may be money behind it. Nitroglycerin, a liquid-form high explosive invented in 1864 by Alfred Nobel, was used to blow the rear casing off safe locks. Gunpowder was also a popular choice [100]. These tactics became less effective with the invention of "powder-proof" lever locks in the mid-1800s (see Fig. 1.3) and safe relocking devices in the early 1920s [104]. Contrary to popular belief, hand-guns are not very effective for opening locks and padlocks.

Most of these techniques are discussed in textbooks on locksmithing [95, 105, 106] and in Tobias's two books [121, 122]. Many of the less destructive techniques are used by locksmiths in perfectly legitimate circumstances (e.g., lock-outs). All of these techniques are known and employed by security and specialized personnel in state and federal government agencies. Less fortunately, but inevitably, they are also available to criminals, who tend to prefer the more rapid and often more destructive methods.

If we have implied that this information on "opening techniques" is due to dishonest people, then a correction is in order. Many of the techniques have been developed either by locksmiths, lock designers, or in laboratories where the security level of locks is tested (such as Underwriters Labs in the United States).

Figure 1.3: Milner's "double-patent" solid powder-proof 6-lever lock (patented in 1854) minimized the free space around the tumblers that could be filled with gunpowder.

Both the maker and the user of a lock have a right to know and are often fascinated by the question of how hard a given lock is to defeat. In much the same way that some people like puzzles, some people spend countless hours practicing their picking techniques. These days, lockpicking competitions, sometimes called "lock sports," are held regularly by people both inside and outside the trade. There are a number of historic instances where manufacturers have offered large sums of money to the first person who could open a particular lock (nondestructively) without its proper key. Examples of where such rewards have been offered may be found in the sections on Chubb, Bramah, Medeco, Newell, Parsons, and Yale locks later in this book.

The fact that some people, whether or not in the locksmithing trade, devote their time to defeating locks and developing methods for surreptitious entry should not necessarily be seen as a bad thing. Indeed, it is a major driving force behind the development of new lock designs, which may not have arisen if the weaknesses of previous designs had not been uncovered. One should not conclude that the people involved in these activities are motivated by dishonesty.

To further pique the reader's curiosity we have provided in Figs. 1.4–1.12 some drawings from publicly available patents depicting some of the curious instruments that have been applied to the picking, decoding, and impressioning of locks.

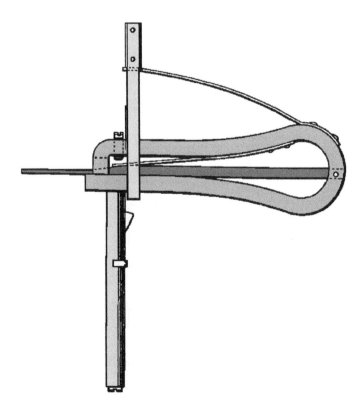

Figure 1.4: Impact pick gun from US patent 1,403,753 (1922) by N. Epstein.

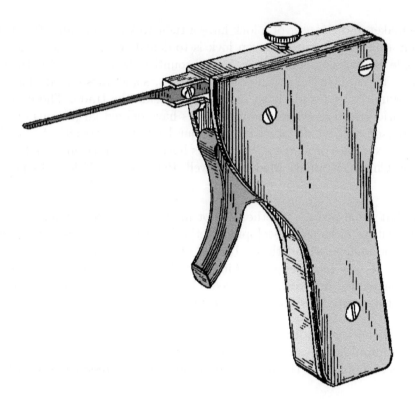

Figure 1.5: S. Segal's 1939 pick gun from US patent 2,309,677.

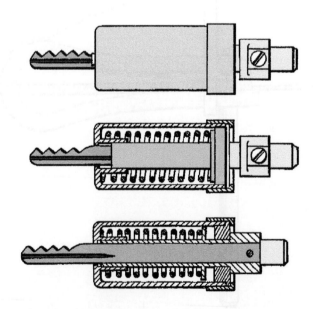

Figure 1.6: G. J. Barron's vibratory lock-pick from US patent 1,639,919 (1925).

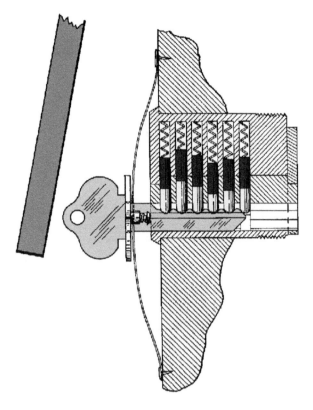

Figure 1.7: H. R. Simpson's rapping- or bump-key from US patent 1,667,223 (1928).

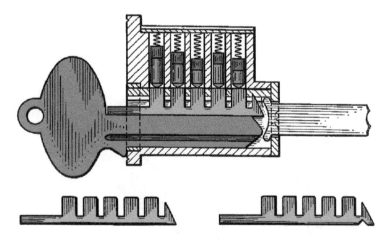

Figure 1.8: Use of a blank key to raise a comb pick from F. Buday's 1934 patent (US 2,064,818).

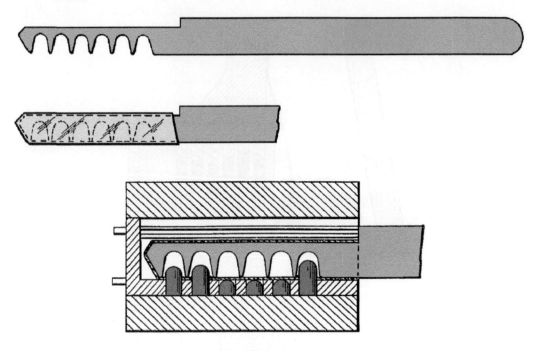

Figure 1.9: A 1955 patent by M. L. Tampke describes how to impression a lock using a foil-coated "soft key" (US patent 2,763,027).

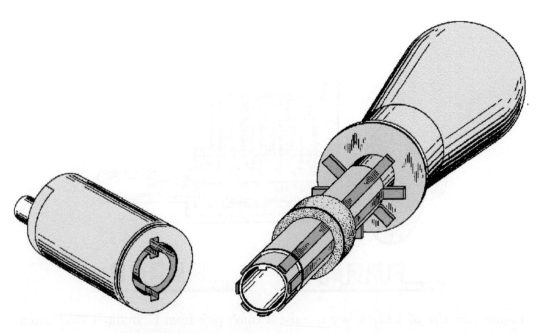

Figure 1.10: A pick for axial locks from US patent 3,251,206 (1963) by R. Gruber.

Figure 1.11: A two-in-one or Hobbs pick for manipulating lever locks.

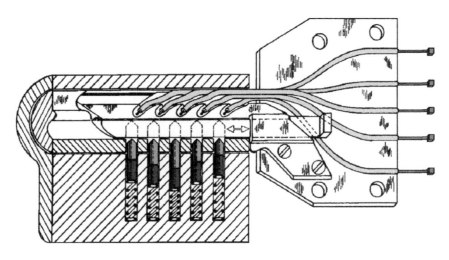

Figure 1.12: S. A. Bitzios's 1991 design for a "Sputnik" decoder-pick for pin-tumbler locks (US patent 5,172,578).

The problem of unauthorized access by manipulation, including lockpicking, impressioning, and decoding, has led to a great many design modifications in the field of high-security locks. In Chapter 2 on pin-tumbler locks, we encounter highly paracentric keyway designs, spooled and mushroom driver pins, active profile pins, multiple inline pin-tumblers, twist-and-lift pins, blocking pins, trap pins, and rockers. Disc or wafer-tumbler locks (Chapter 3) have been improved through the use of serrated tumblers and multiple lines of action. The side-bar design, covered in Chapter 4, is in itself an answer to increased security against manipulation. Lever locks (Chapter 5) have had many design modifications, including false and serrated gates, floating cams, balance levers, detector levers, and gears. Magnetic locks, dealt with in Chapter 6, have been produced in various arrangements to increase the number of codes and also to provide enhanced security to decoding and picking. Car locks, the subject of Chapter 7, have benefited from many of the innovations bestowed on pin-tumbler, disc-tumbler, and side-bar locks.

1.4   Administrative Matters

A number of administrative issues need to be dealt with before we launch into the main fare. The first of these is a discussion of the scope of the material covered in the book. This is followed by a brief description of the conventions and terminology used in the following chapters. The reader will also find a section explaining the difference between the theoretical number of key combinations and the practical or usable number. There is also a discussion on the grading of manipulation resistance as it has been interpreted in this book. Finally, the organization of the sections is presented.

Scope

Current locks can loosely be classed as electronic or mechanical, although quite a few use both principles. It would be ambitious to attempt to cover both electronic and mechanical locks in a single book, and the skills required to describe and understand both types are quite different. This book is restricted to mechanical locks, and in particular to *key-operated* mechanical locks. All the same, we will occasionally mention whether a given design has electronic enhancements, as these now seem to be gaining popularity at the high end of the market.

Electronic locks include electromagnetic and fully electronic locks. Electromagnetic locks employ such devices as magnetic card readers that are read by the lock or resonant circuits (coils) that "talk to" a receiver in the lock by RF electromagnetic induction. Another type relies on an array of Hall effect sensors to read a magnetic signature in the key. Electronic locks use radio frequency transponders, opto-electronic or mechanical switches (that can be either on or off), or numeric keypads to detect whether the correct code is being presented. An important class of electronic systems is that of smart cards [126], where the code is held digitally in a silicon chip carried in the card key. Other fully electronic methods include biometric techniques like retinal and fingerprint scanning and voice recognition [35, 61, 62]. Electromechanical principles are also used in electric strike plates, solenoid-operated bolts, and electromagnetic induction door fasteners. For safety reasons, many of these designs must also be teamed with a mechanical lock or override in the event of an electricity failure.

Mechanical locks, which all have moving parts, can loosely be divided into three classes: (i) conventional; (ii) keyless combination; and (iii) magnetic. The class of conventional mechanical locks includes pin- and wafer-tumbler, side-bar, and lever locks. Examples of keyless combination locks are wheel pack and push-button "digital" locks. Some locks are situated in between conventional and combination types (e.g., the Vingcard lock, which uses a matrix of "binary" pins). Magnetic mechanical locks use the attraction or repulsion between pairs of permanent magnets to actuate their tumblers. A further subdivision of mechanical locks is based on their application, and in this connection we distinguish between architectural (domestic, commercial, industrial) and automotive (car) locks. The vast majority of locks covered in this document are conventional mechanical locks of the rim-, mortice-, or profile-cylinder variety. The highly specialized area of keyless combination locks (typically used for safes, vaults, and strong rooms) is not dealt with herein, although we describe numerous keyed locks for these applications.

Conventions for Lock Descriptions

Each lock section is arranged according to the plan shown in Table 1.3. The country codes shown in Table 1.4 have been adopted. This is followed by a description of the mechanism, mode of operation, security features, and other comments relevant to the lock(s) in question. Some descriptions include patent references and information on the development history of the lock. Where several brands of locks are considered to be very closely related or equivalent in operating principle, these are listed together.

The conventions and terminology used to describe particular classes of locks, such as pin-tumbler, wafer, and lever, are given in the introductory section of the relevant chapters. Since so many of the lock descriptions we present conform to the conventions for pin-tumbler locks, we give some basic definitions later in this section.

The reader will appreciate that there is no absolutely correct terminology for locks. Differences arise between various countries, companies, and schools of locksmithing. Since this book is intended for an international readership, we have taken the liberty of mixing our terminology, employing both U.K. and U.S. terms. Equivalences between U.K. and U.S. terminology for locks may be found in Appendix C. For readers whose native language is not English, Appendix B contains listings of lock-specific vocabulary in French, German, and Italian.

country	brand	type	picking difficulty

Table 1.3: Key for lock description headings.

AT	Austria	AU	Australia
CA	Canada	CH	Switzerland
CN	China/Taiwan	DE	Germany
ES	Spain	FI	Finland
FR	France	HK	Hong Kong
HU	Hungary	IL	Israel
IT	Italy	JP	Japan
NL	Netherlands	NO	Norway
PT	Portugal	SE	Sweden
UK	United Kingdom	US	USA

Table 1.4: Country naming conventions.

In many cases the distinction in terminology is unimportant since there is no possible ambiguity. Thus we may equally well say "plug" for "core" or "barrel" for "cylinder." Other examples of words that have been used interchangeably include bitting, cut; section, profile, broaching; and change, code, differ, permutation, combination. When speaking of key or keying permutations, we have used the word "code" to mean the actual key bittings corresponding to the permutation. The reader should be aware that these are equivalent for so-called direct codes but not for indirect or blind codes.

In some cases, particularly for the more complex locks, the usual terminology does not provide enough differentiation, and ambiguities arise. In these cases, we have used both U.S. and U.K. terms to mean *different* things. For example, in the Citroën (Simplex) lock, which has several levels of "housing," we refer to the innermost part as the core, and then in order of increasing diameter, plug, barrel, and cylinder body. The table of equivalences in Appendix C should help readers to equate the more commonly used terms with those familiar to them.

We have attempted to keep to a consistent set of terminology for all locks in a given category. Thus for Bell-type locks we use the term "bar-wafer;" for Bramah and Ava-type locks we use "slider;" for Abloy type locks we use "disc." In some sections we have had to adopt other words like "rocker" since none of the conventional ones seemed to fit the purpose.

The convention used for describing the position of features or motion in a pin-tumbler lock cylinder, whether of the rim/mortice or profile variety, is that of a clock-face. The assumed viewing orientation for cylinders in this book is with the pin chambers vertical and above the keyway, as shown in Fig. 1.13. (An apology is in

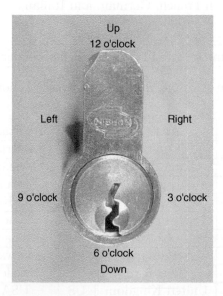

Figure 1.13: Terminology and positional conventions used for pin-tumbler lock cylinders.

order for European readers, who are more accustomed to viewing the cylinder with the keyway at the top.) In this orientation, the bottom of the keyway is at 6 o'clock, the right side of the cylinder at 3 o'clock, and the left side at 9 o'clock. The 12 o'clock direction is referred to as up and the 6 o'clock as down. Directions toward or away from the central axis of rotation of the plug are called radially inward and outward, respectively. Directions along the axis are called longitudinal or axial. Along the longitudinal axis, the direction toward the front of the lock is called forward, and the direction toward the rear is called aft. Directions perpendicular to a given axis (usually the major or longest one) are called transverse. The term "lateral" is used to describe longitudinal warding in a plug or milling on a key blade.

For locks with a single row of pins, pin position or "space" numbering starts from the front of the cylinder or the shoulder of the key. (Note that some manufacturers number pins in the opposite way: from the tip of the key back toward the shoulder.) For other locks the description is tailored to the particular geometry in question. Pin-tumbler sizes (depths) for bottom or key pins are generally numbered from 0 upward, corresponding to the required depth of cut. Thus a size 0 pin requires the minimum cut to the key blank. Note again that some manufacturers use the convention that size 1 is the minimum depth and size 0 is actually size 10 (i.e., greater than size 9).

For certain locks we have provided a theoretical analysis of the number of keying combinations supported by the system. Several points should be noted in connection with this. First, this information is of a theoretical nature and provided only for illustrative purposes. In practice, the lock manufacturer and/or locksmith who combinates the lock determines which key codes are supported and which are not. The net result is that the practical or usable number of combinations is always less than the theoretical number, usually significantly so. The estimates that we provide by combinatorial analysis are based purely on simple factors like the number of pin sizes (or depths of cut) and the number of pin positions. This type of analysis is common in the literature on locks (e.g., see [20]).

In some cases we also account for the maximum adjacent cut specification (MACS). In a few cases we have also provided estimates that account for bitting rules such as the exclusion of repeated cuts of the same size. Such estimates must in general be worked out with the aid of a computer program (see Appendix F). We make the point that even though the estimates for usable combinations may not coincide with the actual figures obtained in real keying systems due to the use of different bitting rules, the inclusion of MACS and other constraints makes the figures considerably more realistic.

For dual-action side-bar locks and other locks where two independent locking mechanisms coexist, the theoretical number of combinations is taken as the product of the combinations provided by each mechanism (see Appendix A). In all cases, the number of combinations assumes that the same keyway profile is in use. Thus we do not consider that different key sections, warding, or multiplex master-keying provide additional key combinations (see Chapter 2), since these necessitate a change to the physical design of the lock.

Grading of Manipulation Resistance

A rather subjective estimate of the manipulation resistance of each lock has been provided. This shows up as a number on a scale of 1 to 5 according to Table 1.5. We have appended an indicative, net time range in minutes for the manipulation exercise. This time range is to be interpreted as the median time to pick or impression the lock in "laboratory conditions" (i.e., when firmly mounted, properly illuminated, and using appropriate tools) by an experienced person who has practiced on several locks of the same brand (keyed differently).

For most wafer and pin-tumbler locks, picking is quicker than impressioning, so the grading refers to the time required to pick the lock. Some locks, depending on their construction, may be easier to impression than to pick. In general, however, the grading relates to the method, whether picking or impressioning, that is quicker.

The grading only refers to the difficulty of picking or impressioning the lock manually and is based on (i) theoretical considerations such as the design and construction of the lock; (ii) reported evidence of picking/impressioning difficulty in locksmithing forums; (iii) reported picking times; and (iv) the author's own experience. It does not refer to the lock's resistance to impact-based methods (pick guns and bump-keys), decoder-based picking, overlifting, or opening by any other nondestructive means. As such, it must *not* be taken as an indicator of the overall security rating or the time taken to defeat the lock by the quickest possible means.

Note that in ruling out decoder-picks from the assessment of picking difficulty, we have effectively excluded from consideration a large number of specialized tools and tool sets that are available to appropriately qualified individuals. We have ruled out this class of methods since (i) although decoding tools are hand-held, they give such an advantage to the lockpicker as to void comparison with other manual techniques; (ii) decoder-based picking tools allow each locking element of the lock to be manipulated independently while maintaining elements that have already been decoded/picked in their correct positions; and (iii) data on decoding of high-security locks is not in the public domain. In much the same way as a combination safe lock can be opened by an "auto-dialer," conceptually at least, any

Rating	Meaning	Picking Time
1	easy	< 1
2	moderate	1–5
3	hard/special tools required	5–10
4	very hard	10–60
5	impossible	> 60

Table 1.5: Picking difficulty is graded from 1 to 5 with corresponding (median) time taken in minutes.

mechanical lock can be defeated by exhaustively trying every possible combination. In this respect, decoding a lock, especially through exhaustive search for the correct key, is fundamentally different from manual picking—a point we discuss further in Chapter 2.

The assessment of manipulation difficulty is highly error-prone. The actual degree of difficulty depends in practice on variables such as (pinning) combination, wear, lubrication, and even on the location of the lock (whether on a bench or in the field). A lock from a master-keyed system will generally be easier to pick than one that is not, in as much as it has been designed to be operated by more than a single key. More importantly the difficulty depends on the tools being used and the skills of the lockpicker. It is well known that even among locks with the same combination of pins, picking difficulty may vary enormously due to tiny differences within the range of manufacturing tolerances. This is particularly true for locks in the higher grades (3 and above). The same comments apply to impressioning. It is therefore to be expected that there is a high degree of statistical variation in the time taken to open a given lock in a given situation. For example, the author was provided with some actual figures for Ingersoll 10-lever padlocks by someone skilled in the art. The picking time on 11 trials for different padlocks varied from 5 to 900 minutes with a median of 30 minutes. In two further trials, the padlock was unable to be opened (in a single session).

We have reserved the category "impossible" for locks that would take an expert in lockpicking and impressioning a "long time" (e.g., several hours, or even days, to open or for which there have been no reported openings). Locks in this category include the Fichet-Bauche 787, the EVVA MCS, and a number of German safe locks. We also class dual-control safe deposit locks (with separate renter and guard keyways[4]) as unpickable since they require the action of two keys.

Security is a relative term. The reader will appreciate the need for different standards for door locks compared with locks designed for safes and vaults. Although locks for the latter category are usually defined as high-security locks, we employ rather looser terminology here. For commercial and residential purposes, any lock with a rating of 3 or above could be classed as a high-security lock in terms of the time required to manipulate the mechanism. It should be remembered that once a lock requires more than a few minutes to open by manipulation, it is an effective deterrent to a would-be thief, who will either go elsewhere or look for a quicker way in (e.g., a window or the roof).

Destructive methods (drilling, extraction, forcing, bypassing, etc.) are not considered in this assessment, although many of the high-security locks featured in this book can include hardened inserts and/or a cylinder-guard to counter attacks of this sort. Further information on ratings for attack resistance of locks may be found in the section on industry standards.

---

[4]For safe deposit locks, the term *guardian key* is used in the United Kingdom whereas *guard key* is used in the United States.

Chapter Organization

The rest of the book has been organized under six main chapters, with a concluding section in Chapter 8:

Chapter 2: Pin-Tumbler Locks

Chapter 3: Wafer Locks

Chapter 4: Lever Locks

Chapter 5: Side-bar Locks

Chapter 6: Magnetic Locks

Chapter 7: Car Locks

Each chapter contains an introductory section giving a historical perspective and a description of the basic mode of operation for the locks presented. In order to balance the presentation, in some cases, I have chosen to defer an account of the history of a lock or lock-making company to the section pertaining to the lock in question.

Following this the reader will find a classification table that sorts the locks into categories according to their operating principle. The remainder of each chapter presents the different locks in more detail. Patent diagrams and references are used occasionally to illustrate concepts both in the introductory and descriptive sections, especially for locks with historical significance.[5] The patents that are referenced in the text are listed in Appendix D. At various points the reader may wish to consult the index to help with cross-referencing.

Some of the locks cited in the text are described without illustrations. The reader may find it helpful in these cases to employ a Web search engine to obtain further information. As Web sites constantly shift and change, only a few of the major ones are actually provided as references. Although it is easy to compile, I decided to omit a detailed listing of Web sites, including those of lock manufacturers, because in the current climate of company mergers and acquisitions it would rapidly become obsolete.

The chapter on pin-tumbler locks is longer than the other chapters for two reasons. First, there is a large number of different types of pin-tumbler locks to cover. Second, the class of pin-tumbler locks has been used to illustrate concepts such as key codes, master-keying, and key-bitting constraints. These concepts recur in later chapters.

---

[5]Patent diagrams used in this book were obtained from the European Patent Office esp@cenet Web site at http://ep.espacenet.com.

Appendices are also provided that cover the following topics:

Appendix A1: Permutations and Combinations

Appendix A2: Lock Permutations and Fractals

Appendix B: Translations of Lock Vocabulary

Appendix C: Terminology and Abbreviations

Appendix D: Lock Patents

Appendix E: Brief History of the Bramah Lock

Appendix F: Computer Code for Key Computations

Appendix G: Security Gradings for Cylinder Locks

The discussion in Appendix A2 on lock permutations points out a connection with a new type of fractal image.

Chapter 2

Pin-Tumbler Locks

> It would seem that there is a limit to the number of distinctive ways which are purely mechanical of providing security and differing in locks and that the limit has been reached, or nearly so. *F. J. Butter, c. 1958*

2.1 Introduction

The pin-tumbler lock is loosely based on a locking principle employed in ancient Egypt as early as 2000 B.C. [91] where a wooden key with pegs or prongs was required to retract a bolt fastening a door. Although not widely recognized, the modern pin-tumbler principle was enunciated as early as 1805 in a British patent (UK 2,851) by A. O. Stansbury, an American who emigrated to England [99]. Stansbury's idea involved a pronged, axial, or bit key that operated a pair of circular plates, one fixed and one rotating. Inserting the key brought the ends of the pins to the interface between the two plates, allowing the front plate to turn. The pins were all of the same length, however, with variation achieved through the use of wards or different positioning of the holes for the pins [31]. Another invention relating to pin-tumbler locks of the axial variety was proposed in 1839 by W. M. Williams in England, although it was not commercialized [22]. The conventional inline pin-tumbler cylinder lock, which we recognize today, has its origins in the mid-to-late 18th century with inventions by the North American locksmith Linus Yale Senior and his son Linus Yale Junior.

Professionally, the Yales were not a father-and-son team. By the time he was in his mid-30s, Yale Junior was at pains to distance himself from his father in regard to a pin-tumbler lock that Yale Senior had invented some years earlier and subsequently sold to a lock producer called Bacon on the grounds this lock had been repeatedly picked [135]. Yale Senior produced a number of security locks for banks including a "Quadruplex" lock, patented in 1844 (US 3,630) with four rows of radial

Figure 2.1: Linus Yale Senior's Quadruplex 4×2 pin cylinder lock and key.

pin-tumblers having a key with a round section (Fig. 2.1). Although this mechanism used eight or more pin-tumblers, it was bulky and lacking in modularity since the pins were chambered in the case of the lock [21, 66]. In 1857, Yale Senior produced a 5-pin padlock with a sliding, rather than rotating, mechanism similar in principle to the ancient Egyptian lock (refer to US patent 18,169).

Like his father, Linus Yale Junior also designed a number of ingenious key-operated bank locks including the "Infallible," the "Magic Infallible," and the "Double Treasury" [31, 57]. We review the Magic Infallible lock, which was a lever lock, in Chapter 5. From about 1863, these key-operated locks were followed by a number of keyless combination locks, which Yale believed held the answer to true security [134].

Figure 2.2: (Top) Early Yale flat pin-tumbler key. (Bottom) Yale corrugated key.

Whatever misgivings Linus Yale Junior may have had for Yale Senior's earlier locks, the son built on the work of his father, developing an inline pin-tumbler cylinder lock with a rotating plug. By 1865, he had developed a mortice cylinder lock with five pin-tumblers operated by a flat "feather" key as shown in Fig. 2.2 (US patent 48,475). This lock is clearly identifiable as the forerunner of the modern cylinder lock, on which almost all pin-tumbler locks are now based. He subsequently co-founded the Yale Lock Manufacturing Company with Henry Towne in 1868. Following Yale's untimely death in December of the same year, the firm was taken forward by Towne, becoming the Yale & Towne Manufacturing Company in 1883. The pin-tumbler cylinder design was refined throughout the 1870s. The corrugated keyway, taking a key with a wavy profile as in Fig. 2.2, was introduced in 1883 and the paracentric keyway in the late 1890s. The familiar Yale & Towne oval cylinder was patented in 1923. Unlike Yale Senior and Yale Junior, who preferred to work on complicated and expensive bank locks, Towne recognized the commercial potential of the pin-tumbler lock.

Like many other security locks, the basic idea of the pin-tumbler lock, though simple in essence, requires a high degree of mechanical precision to implement effectively. Early pin-tumbler locks were not affordable except to commercial customers and only became so with the advent of mass production. The pin-tumbler lock is now the most widespread lock in the Western world. In its simplest form, it offers a reasonable level of security for a moderate cost. The pin-tumbler cylinder is manufactured in various shapes and sizes: most commonly the rim, mortice, and Europrofile varieties.[1] It is the central component in rim and mortice cylinder locks, a modular

---

[1]DIN Standard European profile cylinder.

design that allows the combination of the lock to be changed simply by the removal and repinning of the lock cylinder followed by the cutting of a new key based on the same key blank. It is also adapted for key-in-knob entrance sets and deadbolts. Unlike lever locks, the same format cylinder can be fitted to many different locks, thus providing many different locking functions. The basic principle of the pin-tumbler lock, while probably already familiar to the reader, is explained next.

Construction and Operating Principles

The pin-tumbler lock, illustrated by the Corbin Europrofile cylinder in Figs. 2.3–2.7, comprises a plug or core fitted to a barrel or cylinder. Whereas the barrel is stationary with respect to the lock, the plug can be turned when the lock is operated by the correct key. There is a close analogy with electric motors where the plug is like the rotor and the barrel acts as the stator. The barrel is normally part of a larger locking mechanism comprising bolts, latches, and so on. A cam or tail-piece attached to the plug actuates the mechanism.

The plug contains a broaching called the keyway (Fig. 2.4), through which the key is inserted. The keyway is normally of irregular shape or paracentric, containing fixed lateral obstructions called warding.[2] The broaching of the plug matches the profile or section of the key blank. Both the plug and barrel possess a set of vertical borings called pin chambers (Fig. 2.5). The borings in the top of the barrel are either capped or sealed with brass plugs or a slide and penetrate about halfway through the plug's diameter. Each pin chamber contains a pin-tumbler pair consisting of a bottom pin or key pin and an upper or driver pin (Fig. 2.6). The combined length of the pin pair exceeds the length of the plug bore. A phosphor-bronze spring atop

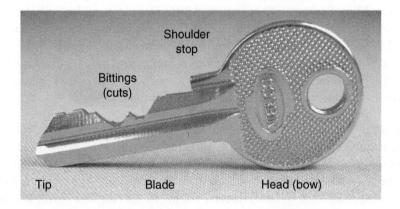

Figure 2.3: Naming conventions for pin-tumbler lock keys.

---

[2]This terminology is borrowed from medieval times prior to the invention of the lever lock, when wards were the only barrier to opening the lock.

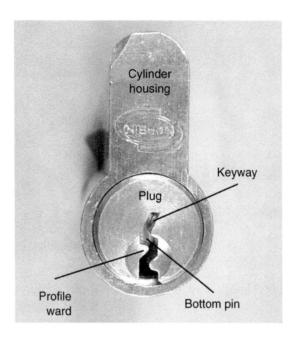

Figure 2.4: Terminology for pin-tumbler cylinders.

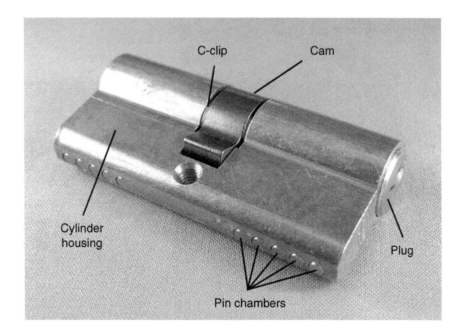

Figure 2.5: Terminology for pin-tumbler cylinders (continued).

each driver ensures that the pin pair is maintained at its lowest position. The pin stack may also contain additional pins for master-keying. In the locked position the plug is prevented from turning by the presence of the upper pins that straddle the interface between the plug and barrel.

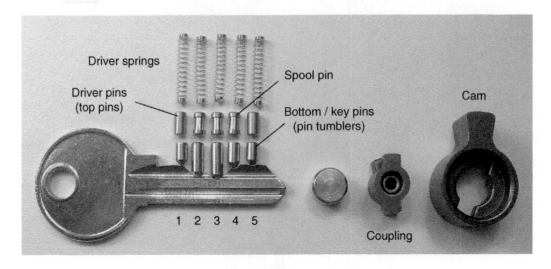

Figure 2.6: Component naming for a 5-pin profile cylinder.

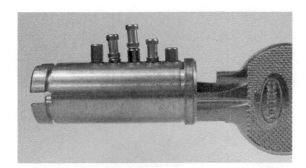

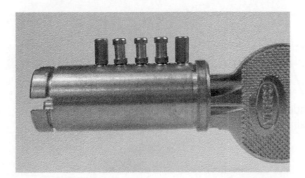

Figure 2.7: (Top) Insertion of key into plug. (Bottom) Key fully inserted, aligning pins at shear line.

The key (Fig. 2.3) has a series of V-shaped cuts with a spacing that matches that of the pin chambers. As the tip of the key blade is inserted into the plug, its angled leading edge lifts the lower pins. The key blade continues its passage under the pins until its shoulder contacts the front of the plug or until its tip contacts the backstop of the plug. The centers of the key cuts or bittings are then in alignment with the tips of the lower pins, each of which is raised to a height determined by the depth of the key cut. If the cut is such that it raises the junction between the lower pin and driver to be flush with the rim of the plug, then this pin pair can offer no resistance to the rotation of the plug. When this occurs, the lower pin is said to be at the *shear line* (see Fig. 2.7). Of course, it is not sufficient to raise only one or two pins to the shear line to open the lock; the key bittings must be such that all pins are raised simultaneously to the shear line, at which point the key can operate the lock in either direction. As the plug is turned, the lower pins remain at the shear line and capture the key until it returns to the 12 o'clock position. On removal of the key, the pins are returned by spring biasing to the bottom of the pin chambers, locking the cylinder.

Pin-tumbler locks typically have five or more lower pins, each one available in several lengths, corresponding to different cut depths on the key. Miniature pin-tumbler locks may have fewer than five pins and correspondingly fewer sizes. For instance, Lockwood locks have 10 pin lengths[3] ranging from 0.150″ (size 0—shallowest cut) to 0.300″ (size 9—deepest cut) in depth increments of 0.015″ and with a spacing of 0.156″. The driver pins are usually 0.220″ long, although shorter pins are used in key-in-knob cylinders due to the limited height of the pin chambers. The drivers may also be compensated or balanced: i.e., shorter drivers are matched with longer bottom pins. This ensures that the spring tension is roughly the same in each pin chamber, avoiding key-insertion difficulties. In interchangeable-core locks, the drivers are adjusted so that all pin stacks have the same overall height.

For master-keying, or when one key is required to operate two or more locks that are keyed to differ, master pins (also called chips or differ bits) are inserted in the pin stack. A pin stack containing master pins is said to be "segmented" since it has more than one shear line. A less common method utilizes a master ring around the plug to create a second, independent shear line. In a master ring cylinder, the change keys[4] raise the pins to the inner shear line, while the master key or keys raise them to the outer shear line. This idea, described in [21] and illustrated in Fig. 2.8, dates from the last quarter of the 19th century and is analogous to the control shear line in an interchangeable-core cylinder, which we cover in the next section.

Master pins come in various sizes, for example 1–9 ranging from 0.015″ to 0.135″. In general, a difference of two sizes or 0.030″ is preferable to avoid jamming due to wear and tear or tilting of the pin. This also ensures that jiggling an adjacently coded key in the lock is less likely to open it. The inclusion of each master pin in a chamber introduces an additional shear line. In a lock having only one segmented pin stack with N master pins, the number of keys that can operate the lock is equal to $N+1$, corresponding to the number of different shear lines. Whenever more than a single pin stack is segmented, which typically occurs in master-keyed systems,

---

[3]The symbol ″ is used to denote inches.

[4]For ordinary locks, a change key, as opposed to a master key, operates a unique lock or set of locks that are keyed alike. For recombinatable locks, a change key is the tool required to reset the lock to a new combination.

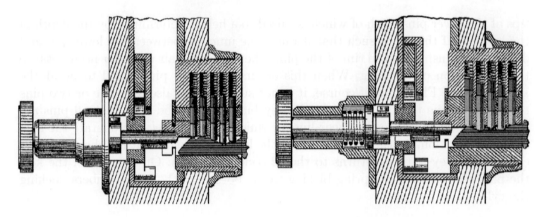

Figure 2.8: Master-ring cylinder from E. J. O'Keefe's 1889 US patent 414,720. Pins aligned at outer shear line (left) and at inner shear line (right).

a multiplicity of operating keys results. The effect on the number of operating key combinations is multiplicative. Thus a 5-pin cylinder with a single master pin in chambers 1, 2, and 3 yields $2 \times 2 \times 2 = 8$ possible keys, all of which operate the lock. This may be good or bad, depending on the requirements of the MK system in terms of the number of levels and the number of change keys required at each level. In general, the more complicated the MK system, the larger the number of different keys that will unintentionally operate the lock. The presence of unintended operating keys is known as key interchange. The interested reader is referred to [102, 105] and to a recent online article [11] for further discussion of master-keying.

The reverse of master-keying is called maison-keying. This situation occurs when all change keys are required to operate the *same* lock (e.g., the building entrance door). Usually accomplished by removing one or more of the pin stacks, it results in a severe loss of security in the lock that has been maison-keyed. In a multilevel MK system, where more than one level of master key exists (that is, some master keys operate only a subset of differently keyed locks), it may be necessary to employ other methods, such as multiplex master-keying, to implement the system without resorting to maison-keying. Multiplex master-keying is dealt with further on.

Interchangeable-Core Locks

In any master-keyed system, the loss or theft of a key necessitates recombinating that part of the system operated by the lost key. Depending on the size of the system and the security compromise incurred by the lost key, the owner must make a decision on whether or not to have the locks changed in the affected part of the system. In large multilevel systems, the cost of restoring the integrity of the system can be prohibitive. For this reason, locking systems that are easily reconfigurable are an attractive option. For hotels and other temporary accommodations where occupants

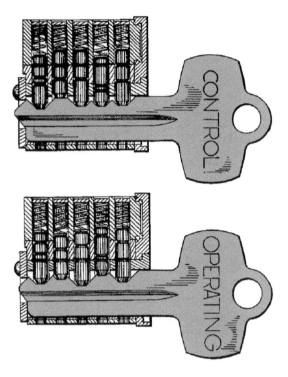

Figure 2.9: Interchangeable-core cylinder from F. E. Best's 1963 US patent 3,206,958.

may only need access for short periods of time, it makes sense to use electronically reprogrammable locks or card-in-slot locks (such as VingCard). When occupancy is more stable, such as in office buildings and apartment complexes, a key-operated lock may be a more appropriate choice.

Replacing a rim or mortice cylinder requires, at a minimum, unscrewing the fastening screws followed by the recombination of the cylinder. Recombination itself may require the removal of a cam and the insertion of a follower to remove the plug. Some locks may even require drilling, for instance, older style Europrofile cylinders and particularly padlocks. In a master-keyed system, the work associated with a rekeying job can be very significant. The interchangeable-core (IC) cylinder, produced since the late 1930s, is a cost-effective solution to the problem of reconfigurability for pin-tumbler locks (see Fig. 2.9). The function of an interchangeable core should not be confused with that of a master-ring cylinder, where an inner and outer cylinder provide different combinations (Fig. 2.8).

Small-format interchangeable-core locks, as supplied by Best, Falcon, Arrow, and Corbin, are designed primarily to be easy to change over. First, a control key is inserted and turned, allowing the entire lock cylinder to be extracted. The pins can then be accessed by removing a slide cover to expose the pin chambers. Although the locksmith has the option of first removing the core, recombinating, and replacing it, it is preferable to prepare new cores off-site so that they can be swapped over rapidly. The actual changeover process is accomplished without the need for tools.

Another removeable-core system is produced by Schlage, and this enables a key-in-knob cylinder to be adapted to a padlock, for instance. In this system the cylinder is loaded into an adaptor fastened to the padlock by a screw that is only accessible when the lock is open.

Codes, Permutations, and MACS

The question naturally arises regarding the total number of permutations, key codes, or pinning arrangements that are supported by a given pin-tumbler system. Note that we are equating a key code with the bittings or cut depths on the key—this is only the case for direct codes. (Other codes, called indirect or blind codes, are deliberately made to be different from the direct code and are used to hide the actual cut depths. The direct code and indirect code are then cross-referenced in a code book.) In a 5-pin system with 10 pin sizes, there are theoretically $10^5 = 100,000$ different keys[5] in the series that can be made to operate the cylinder. In practice, this is a gross overestimate for a number of reasons: the most important is that in most systems the spacing between the pin chambers makes it infeasible to place a very shallow cut (for a short pin) next to a very deep cut (for a long pin). This can be better understood by reference to the diagram in Fig. 2.10 where a shallow cut at position 4 on the key blade is adjacent to a deep cut at position 5. The geometry is such that any further increase in the cut depth at position 5 would cause the bitting at position 4 to be undercut.

A constraint called the *maximum adjacent cut specification* (MACS), or adjacent cut difference, is therefore imposed on the sequence of bittings. The MACS is specific to the particular type of lock and depends on the pin spacing (D), the cut depth increment (d), the cut angle (θ), and the cut root (or pin footprint) (δ).

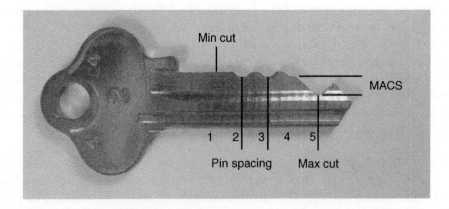

Figure 2.10: Five-cut key with maximum difference of cut from position 4 to 5.

---

[5] Ten to the power of 5, or $10 \times 10 \times 10 \times 10 \times 10$.

The relationship between these quantities is defined in Fig. 2.11. For key cutting, the important parameters are the cut depths for each pin size and the spacing of cuts along the blade of the key, which may be measured from the shoulder or from the tip. Manufacturers provide this information in the form of depth and space charts for each type of lock. In some pin-tumbler locks, V-shaped cuts are used that have an effective cut root of zero: the bottom pin rests on the slopes of the cut.

According to Fig. 2.11 the distance D_{MACS} is the maximum depth difference between adjacent cuts and is given by $D_{MACS} = MACS \times d$ where the MACS is measured in terms of the depth increment. It is straightforward to show from the diagram that $\tan(\theta/2) = \frac{D-\delta}{D_{MACS}}$, from which it follows that the MACS is given by the formula:

$$MACS = \lfloor \frac{D - \delta}{d \tan(\theta/2)} \rfloor = \lfloor \frac{\text{(pin spacing)} - \text{(cut root)}}{\text{(depth increment)} \tan(\frac{1}{2} \text{cut angle})} \rfloor$$

which must be rounded down to the nearest whole number (denoted by the outer brackets or "floor" function).

For example, lock cylinders of a type commonly used in Australia have the following approximate dimensions: pin spacing 0.156″, depth increment 0.015″, cut angle 96°, and cut root 0.045″. The above formula implies that $MACS = \lfloor (0.156 - 0.045)/(0.015 \times \tan(48 \deg)) \rfloor = \lfloor 6.6 \rfloor = 6$. In practice, it is allowable to undercut the shallower cut slightly to increase the MACS. If we allow an undercut of u units, such as on the cut at position 4 in Fig. 2.10, the cut root becomes $\delta - u$ and the previous formula must be modified to

$$MACS = \lfloor \frac{D - \delta + u}{d \tan(\theta/2)} \rfloor$$

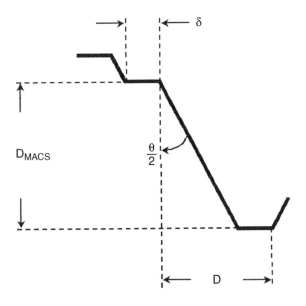

Figure 2.11: Notation for adjacent pins with maximum difference in depth of cut.

For example, allowing a minimum cut root of 0.030″, instead of 0.045″, gives a MACS of $\lfloor 7.6 \rfloor$ or 7. This value of MACS is typical for 5-pin locks produced by Lockwood and Schlage. Further details on MACS are contained in [102].

The MACS has a significant effect on the number of available system permutations. For instance in a 5-pin system with a MACS of 7, the bitting code (2 7 1 9 3) would be ruled out since placing a depth 9 cut next to a depth 1 cut violates the MACS constraint. The computation of the number of permutations that satisfy the MACS is, in general, difficult. Nevertheless, for locks with six pins or fewer, one can write a computer program that simply generates all the theoretical permutations and excludes those that do not satisfy the MACS. (For locks with seven or more pins, this approach can lead to computational problems.) We have listed in Table 2.1 the number of permutations for various values of the MACS for both 5- and 6-pin cylinders, assuming 10 available pin sizes. The number of permutations for 5- and 6-pin systems using only 9 instead of 10 pins sizes is also tabulated. The figures were obtained using the computer program check_macs.m given in Appendix F, written in the Matlab^{TM} programming language [115].

Recall that for a 5-pin system with 10 pin lengths there are theoretically 100,000 codes, rising to one million for a 6-pin system. One can see that even with a MACS of 7 (the maximum for most pin-tumbler locks), the number of codes is reduced by 20 to 25 percent from its theoretical value. For lower values of the MACS, the effective number of codes can be less than 40 percent of its theoretical value. This, however, is not the final word on the number of key codes, as there are a number of further constraints imposed in practice. These constraints, called coding or bitting rules, ensure that trivial codes that would be easy to duplicate by sighting the key or too easy to manipulate are excluded. The following four additional constraints are typically applied to generate real key series.

[C1] Only two adjacent cuts can be the same depth.

[C2] A total of three or fewer cuts may be the same depth.

[C3] Three or more cuts must be of different depth.

[C4] A number 8 or 9 cut is not allowed in position 1.

MACS	9 sizes 5-pin	9 sizes 6-pin	10 sizes 5-pin	10 sizes 6-pin
3	10,619	63,111	12,990	79,258
4	21,141	148,433	27,142	198,034
5	33,101	258,953	44,692	367,826
6	44,573	374,641	62,948	562,670
7	53,769	472,943	79,666	753,754
8	59,049	531,441	92,674	909,602

Table 2.1: Number of permutations for 5- and 6-pin locks with 9 and 10 pin sizes satisfying a given MACS constraint.

The first three constraints (C1–C3) ensure that pinning sequences such as (5 5 5 5 5), (3 3 3 4 5), and (3 2 3 2 3) are not permitted. Constraint C4 is needed to minimize key breakage due to a maximum-depth or near maximum-depth cut in the position nearest to the key shoulder. When we account for the MACS and conditions (C1–C4), the number of different codes is reduced further, though not massively (refer to Table 2.2). These results were obtained with the help of the program perms_macs.m in Appendix F.

The above figures are for the set of bitting rules C1–C4 as given. If we change the bitting rules, then a different key series is generated with a different number of codes. For instance, in order to obtain a larger series, some manufacturers relax the first constraint, allowing up to three adjacent cuts to be of the same depth. Many other sets of bitting rules can be applied, each generating its own key series. The programs in Appendix F may be used as a starting point for creating code series according to different bitting rules. For instance, another bitting rule called a "pull-out" rule is sometimes enforced. This stipulates that the sequence of cut depths must not increase monotonically along the blade, for example, as with (5 5 6 7 9). For codes like this, the key could potentially be pulled out in the unlocked position if the lock and key were sufficiently worn.

The security advantages of a 6-pin over a 5-pin cylinder can be readily appreciated from Table 2.2. The multiplication in the effective number of differs due to the single extra pin is better than 8 for practical values of the MACS (greater than or equal to 5). If constraint number C4 is ignored, about 20 percent more differs are available than those indicated in the table, although many of these weaken the key blade near the shoulder, leaving it prone to breakage. We have deferred the presentation of results for 7-pin locks until a later section.

In a single key series for a master-keyed system, an even smaller number of codes is achievable since, in order to avoid key interchange (two different keys unintentionally operating the same lock), a minimum difference of two cut depths is usually required between codes. Instead some systems require a difference of one depth increment in at least two positions. The minimum difference between successive codes is referred to as the *progression step*. In more secure systems, key codes may be required to

MACS	9 sizes 5-pin	9 sizes 6-pin	10 sizes 5-pin	10 sizes 6-pin
3	8,799	51,401	10,010	60,072
4	17,967	124,457	21,364	153,896
5	28,381	219,519	35,341	287,817
6	38,287	318,571	49,703	440,225
7	46,046	401,240	62,514	586,584
8	50,176	447,552	71,928	700,528

Table 2.2: Number of codes for 5- and 6-pin locks with 9 and 10 pin sizes, taking both MACS and additional constraints into account.

differ by three depths in at least one position. As we will see, the effect of such restrictions on the available number of system codes is significant and somewhat counterintuitive when applied to the entire key series.[6] One would surmise that the requirement that all codes must differ by at least two sizes would approximately divide the number of codes by two. However, this is only the case if we consider differences in a single bitting position.

As an example of the effect of the preceding requirement, we take a 5-pin system with a MACS of 3. From Table 2.2 we know that such a system provides around 10,000 unique codes, taking MACS and constraints C1–C4 into account. Now let us eliminate every code that differs from its preceding code by less than two depths of cut in position 5. This is approximately the same as removing every odd combination from the series. It can be shown that we are left with 5750 key codes starting with (0 0 1 0 2), (0 0 1 1 2), (0 0 1 1 4), (0 0 1 2 0), ... and ending in (7 9 9 8 9). However, we have left many other codes that differ from existing codes in this series by a single depth in one of the other positions (1–4). For instance, the first code can be paired with (1 0 1 0 2), which differs by only one cut in position 1.

We must therefore remove all codes that are equivalent to an existing code in the sense that they differ by one depth in only one position. Many of the codes are equivalent in this sense to more than one other code. For 5-pin systems with larger MACS, we can get up to 10 equivalent codes for each code. For example: (7 8 7 5 1) differs by one depth from each of the following 10 codes: (6 8 7 5 1), (7 7 7 5 1), (7 8 6 5 1), (7 8 7 4 1), (7 8 7 5 0), (8 8 7 5 1), (7 9 7 5 1), (7 8 8 5 1), (7 8 7 6 1), (7 8 7 5 2). The program differ_by_2.m in Appendix F allows us to compute a set of codes that are unique and differ by two or more increments. The result for the MACS = 3 example is surprisingly low: only 1,533 out of the 10,000 codes satisfy all the constraints! Furthermore, the resulting key series is not unique; we can obtain other series by changing the order in which we verify the constraints. For example, checking position 1 first, 2 second, up to position 5, which is done last, results in a series with 1571 codes. Some consequences of the equivalence of codes are explored further in Appendix A, where we demonstrate a link to a new class of discrete fractal image.

Table 2.3 summarizes the results for 5- and 6-pin locks for a range of MACS values when the progression constraint is applied starting from the last position. It can be seen that a drastic reduction in the number of codes in the series results from using a progression step of two between codes. Usually this is acceptable for master-keying, since the system is essentially broken down into subsystems that have their own set of codes, each subsystem having its own master key or set of master keys. Note that increasing the MACS does not necessarily increase the number of codes satisfying all constraints. For 5-pin locks, the net effect is similar to halving the number of pin depths from 10 to 5, which yields $5^5 = 3,125$ permutations.

---

[6]If the progression step is only applied between adjacent codes in the series, then more codes will result, but there will be residual codes that do not satisfy the differing constraints.

MACS	5-pin	6-pin
3	1,533	7,462
4	2,608	12,544
5	3,793	22,412
6	4,831	28,326
7	4,404	34,660
8	4,386	39,409

Table 2.3: Number of codes for 5- and 6-pin locks satisfying all constraints, including differing by at least two depths of cut as a function of the MACS.

A significant aspect of 6-pin systems is the increased number of possibilities for master-keying, although this invariably decreases the manipulation resistance of the lock as more shear lines are introduced in the pin stacks. We can see from Table 2.3 that a 6-pin cylinder offers around 20,000 to 30,000 codes for a progression step of 2 compared with only 4,000 to 5,000 codes for a 5-pin lock with the same MACS. Additional access control may be achieved by varying the keyway profile (or section), with blanks of one profile being incompatible with blanks of other profiles. Each new profile is a multiplier in the overall number of distinct key codes in the system. Thus a system with 10 key profiles would support 10 times the number of codes compared with a single-profile system.

Multiplex Master-Keying and Profile Control

Profile variations are also used to supplement master-keying. This is achieved by a hierarchical design of the set of key blank profiles and matching keyways called *multiplex master-keying* or *multiple broaching*. Key blanks in the upper levels of the hierarchy can enter or pass the profiles at the lower levels, but not vice versa. This idea, invented in 1896 by W. F. Donovan (US 567,305) of the Yale and Towne Manufacturing Company, is analogous to the way in which a skeleton key can be cut to pass a warded lock. A given MK system can be replicated on each different key profile, providing a multiplier for the number of system codes.

As an aid to understanding this method of profile control, consider the generic key profile in Fig. 2.12. This particular profile is loosely based on a system used by DOM [110], but simplified for illustrative purposes.[7] The labels P1 to P6 in the diagram are points on the key section at which longitudinal profile-control grooves can be produced on the key blade. A multiplex system would use this as its top-level profile because it can pass any keyway generated by the omission of one or more of the profile grooves. Since a profile groove may either be present or absent, this generic system supports $2^6 = 64$ possible key sections and up to six levels. In keeping with

---

[7]The DOM system has up to 10 profile variation points; refer to US patent 5,287,712 (1992) by G. Sieg.

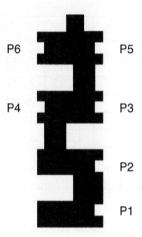

Figure 2.12: Generic key profile for multiplex MK illustrations.

the terminology used for master-keying,[8] we refer to the sections at different levels in the hierarchy as (1) C (change); (2) M (master); (3) GM (grand master); and (4) TM (top-level master).

A key blank with grooves at points P1, P2, and P4 can be referred to as a (124) section. The TM section, or all-section blank, is (123456) in this hierarchy. It can be seen that a key blank with a (124) section cannot enter, say, a keyway with a (125) section since it does not have a profile groove at point P5. A key blank with section (1245) would pass both the preceding keyways. It is important to remember that the profiles are not the same as the actual key codes, of which there may be thousands for each profile.

An example of a simple two-level multiplex or doubly-broached system is given in Fig. 2.13. A level 2 profile with a (1234) section is compatible with the six level 1 sections in the diagram. Examples of real two-level multiplex MK systems include Russwin and Yale, with 14 and 22 different key sections, respectively [110]. In these systems, all the keyways on the level 1 are incompatible, while the single level 2 profile fits all the lower level profiles.

As soon as we graduate to higher levels of profile-controlled master-keying, we encounter more than one architecture for the profile hierarchy. For instance, ASSA's multiplex system for 5-, 6-, and 7-pin locks, illustrated generically in Fig. 2.14, is a three-level system with a single level 1 profile, three level 2 profiles, and a single level 3 profile that fits all of the other profiles.

A more complicated three-level multiplex system appears in Fig. 2.15. This is a system of 13 key profiles on a tripartite tree: a single TM section at level 3; three level 2 sections, each one compatible with three sections at level 1. All nine of the level 1 sections are incompatible. Schlage 6-pin cylinders use a similar three-level

---

[8]Strictly speaking, this terminology should only be used for the master keys themselves and not for the key profiles.

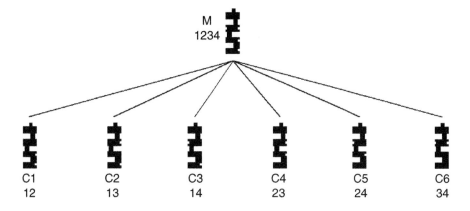

Figure 2.13: Two-level multiplex profile hierarchy.

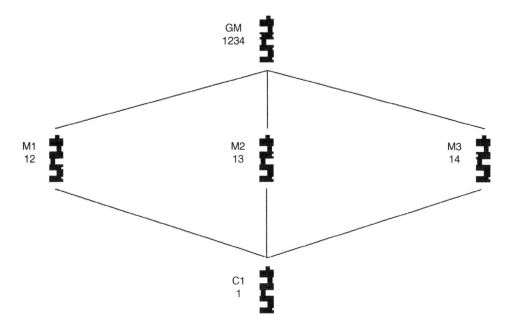

Figure 2.14: One possible three-level multiplex profile system.

multiplex system with 11 different key sections. Another popular three-level system is used variously by Meroni, Yale, and Dexter.

The final example we present is a four-level multiplex system with 16 key profiles. This system has been adopted by Sargent in the United States and by Vachette in France. The system, shown in Fig. 2.16, has a single level 4 profile that fits all of the other 15 profiles. There are three level 3 profiles, each of which passes only two of the three level 2 profiles below it. The level 2 profiles each pass three different profiles at level 1.

It can be appreciated that, although there is in principle no limit to the complexity of a multiplex profile hierarchy, there are practical limitations on what can

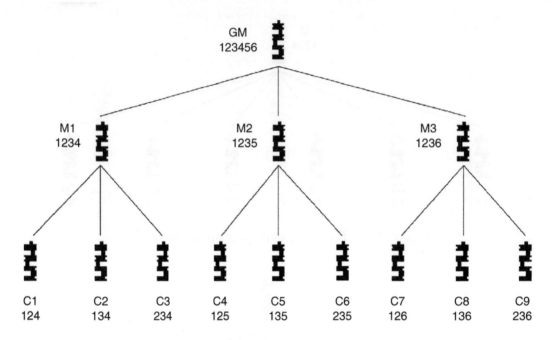

Figure 2.15: A more complicated three-level multiplex profile system.

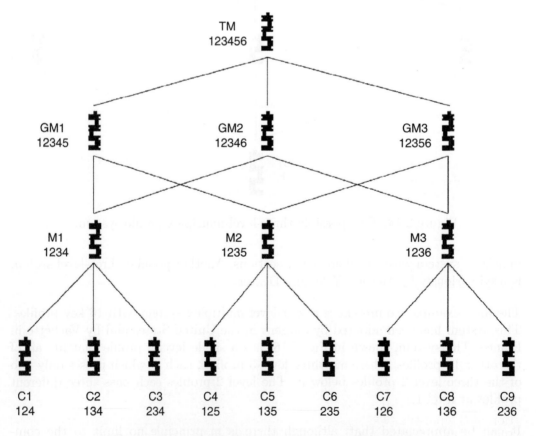

Figure 2.16: A four-level multiplex profile system.

be achieved through variation of the keyway broaching. The larger the number of profiles, the less tolerance there is to wear and tear through frequent use. Moreover, the design has an inherent security flaw in that it is frequently possible to convert a low-level profile so that it fits all the other profiles simply by grinding down the sides of the key. This can obviously lead to key interchange or unintended operability. In the days of warded locks, a similar situation existed that led to the appearance of skeleton keys that were created by filing off the inessential parts of the key bit.

In an effort to avoid the shortcomings of key control through profile variations and also to reduce production costs, numerous manufacturers have moved toward locks with *profile-control pins*. We will see many examples of such locks later in this chapter. The fact still remains that whenever the profile-control mechanism is passive, the key blank can be doctored so that it will pass all levels in the system. A better method is the use of active profile control as evidenced by locks such as ABUS, ISEO, Vachette, MLA, Winkhaus, and Schlage.

Increasing the Level of Security

In this section we consider a number of aspects that influence the security of a pin-tumbler lock. These include the design of the keyway, which is important to prevent the insertion of picking and bypass tools. We also discuss physical security aspects such as drill resistance. Security can be interpreted in terms of the number of combinations or differs, which is greater when more pins are included. The subject of key control, via the restriction of key blanks and requiring authorization for key copying, is also discussed.

A major step toward the modern pin-tumbler lock was the development of the corrugated keyway by Linus Yale Junior in 1883 [134]. Prior to this invention, it was possible to insert a crude, flat blade into the keyway of the lock in order to pick it. This was followed in the late 1890s by Yale's so-called paracentric keyway, which is still used in Yale cylinder locks. The paracentric keyway was first suggested in an 1891 patent (US 457,753) by W. H. Taylor, a long-time employee and the principal inventor at the Yale and Towne Manufacturing Company [50]. This lock included beveled pins with locating lugs to ensure vertical lift by the (very wavy) key blade. Despite its improvements, such a lock could not be economically manufactured at the time.

Since then many other refinements have been made, to the point where today there exist tens of thousands of different keyway sections from hundreds of different manufacturers, each with its own distinct key blank. The Silca range of catalogues [110, 111] is a good reference on this subject. Only a relatively small number of these profiles contain severe enough warding to hamper significantly the manipulation of the pin-tumblers using modern tools, and some of these locks are covered in this chapter.

An early method for enhancing the security of a pin-tumbler lock was devised by V. J. M. Eras of the Lips lock factory (Lips Brandkasten Sloten) during a 1903 visit to John Mossman in New York [30]. The method involved inserting a pair of ball bearings in the first pin chamber. In a 5-pin lock, a sixth chamber was added, and all the pin stacks were shifted back by one chamber, as shown in Fig. 2.17. The first chamber was modified such that the depth of the hole in the plug equaled the diameter of the lower ball bearing, which remained at the shear line. The second ball was located above the first and was held in place by a hardened rod. The arrangement was such that the interface between the two ball bearings coincided with the shear line. The presence of the ball bearings did not impact on the normal operation of the lock, but it made it considerably more difficult to drill the plug at the shear line. A slight modification of the idea is described in the 1912 UK patent 27,511 in which a hardened rod was mounted in the front pin chamber and a hardened ring surrounded the front of the plug.

In terms of drill resistance, many of the locks featured in this book contain hardened balls, rollers, crescents, and/or other inserts at various locations in the plug and housing. The pins themselves can be made of a hard material, such as stainless steel, or contain hardened inserts. Further protection, especially against sawing and wrenching, is available in the form of guard plates and sleeves made of toughened steel. The sleeve may be fixed or rotating, which makes it impossible to get a purchase on the lock in order to saw off the barrel. Some locks are more prone to attack by wrenching than others: the older style solid-brass profile cylinders, popular in Europe, are a case in point. These are now being produced in a composite format that is both easier to rekey and also more resistant to being snapped off.

Alternatively, a secondary lock or locking shield can be installed to guard the keyway of the primary lock. An example of such a system is the Drumm Geminy

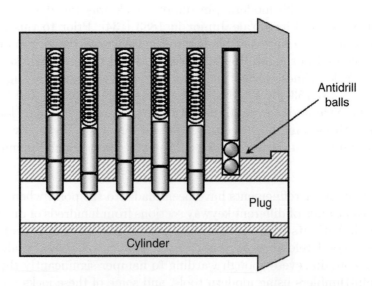

Figure 2.17: Eras's drill-resistant pin-tumbler cylinder.

shield, which has a lockable sliding cover operated by a 10-pin lock with concentric pin-tumblers [36]. Protection is also available for lever locks (Chapter 5), which often include steel hardplate in strategic locations and captive rollers in the bolt to resist sawing. Naturally, there is a limit to the level of protection that these measures can afford, but a well-designed lock is still an effective deterrent to a would-be thief since it increases the time required to gain access.

The vast majority of pin-tumbler locks are of the 5- or 6-pin variety. Five-pin locks have traditionally been used for residential premises as they are low-cost and provide a level of security that is considered adequate, given the ease of unauthorized access by other means (e.g., via the windows or roof). With only five pins, we saw before that the number of differs, taking a MACS of 7, is around 60,000 to 70,000, depending on the set of bitting rules used to generate the key codes. For houses equipped with locks that use the same key blank, there is the possibility that one person's key may unintentionally open someone else's lock due either to duplication of a code, loose tolerances, or wear in the lock. Shortcuts such as using rounded-end pins and maison-keying (leaving out some of the pins) also lessen the security of the pin-tumbler cylinder and leave it vulnerable to key interchange (i.e., unintended operability).

Six-pin locks are an option that provides a higher level of security, and these are primarily used in the commercial, industrial, and public sectors. The increased level of security results from a higher level of manipulation resistance due to the presence of a sixth pin, as well as a lower probability of code duplication in uncontrolled systems since most 6-pin systems with 10 pin sizes support between 600,000 and 700,000 differs.

For completeness we have tabulated in Table 2.4 the number of system codes for 7-pin locks as a function of the MACS and the various bitting rules C1–C4 mentioned on page 42. The table shows how the theoretical number of codes is reduced from 10 million to a generally much smaller number of usable differs. In particular, the requirement to differ by at least two depths of cut from any other code in the series

MACS	Codes with MACS	Codes with MACS and Rules	Progression Step of Two
2	90,790	57,431	5,700
3	483,646	347,500	34,252
4	1,444,904	1,082,065	62,424
5	3,027,314	2,302,060	133,256
6	5,029,530	3,842,520	176,223
7	7,131,596	5,435,049	237,208
8	8,927,810	6,745,186	280,393

Table 2.4: Number of codes for 7-pin locks with 10 depths of cut as a function of the MACS when indicated bitting rules are accounted for.

reduces the number of codes by a factor of between 15 and 30 compared with the number of codes satisfying the MACS.

This chapter contains several examples of inline pin-tumbler locks with seven pins, including a number of locks produced by Lockwood (Australia). A 7-pin security cylinder manufactured by ASSA is illustrated in Fig. 2.18. We add in passing that conventional pin-tumbler locks having more than seven pins are also produced. The Spanish company FAC makes an 8-pin cylinder, while the Italian firm Wally produces a 9-pin cylinder [110].

As one might expect, increasing the number of pins in a cylinder lock is not the only way to increase the number of available system codes or the security offered by the lock. In practice, while increasing the pin count of an inline cylinder reduces the chances of key interchange, it does not greatly increase the overall level of security. A 6- or even 7-pin lock remains relatively easy to manipulate or drill unless further security features are added or the inline design is modified. The trend in high-security locks, however, is to move away from conventional inline designs to other systems such as side-bar and dimple-key locks (e.g., Kaba, KESO). These systems, which provide vastly increased security and key control without the need for a longer key, are discussed later in this chapter.

The risk of code duplication and unauthorized access can be further reduced by controlling the availability of key blanks. Whereas patenting may be applied to protect the design of a lock, it is an expensive and time-consuming process: it is inherently not well suited to minor variations of a well known concept such as changing a key profile. To ensure that third parties are discouraged from reproducing the design, the key profile is usually registered with a national agency. In this way control is established over the production and supply of the registered key blanks, which are only legally available to authorized locksmiths. An added level of security results when the manufacturer issues key codes centrally to prevent two end-users being assigned the same code. In this process each key code is assigned to a registered owner, with copies of the key requiring a signature or an ownership card that can be compared with a record held on file in order to authorize the making of a duplicate.

Figure 2.18: ASSA 7-pin cylinder and core with antipicking sleeve and pins with hardened inserts.

In practice, once a registered key system has been in service for a sufficiently long time, the number of available codes may be inadequate to cover future orders. Moreover, design registrations and patents have a finite lifetime, following which aftermarket blanks can be made by third parties. At this point it becomes necessary to move to a new key profile, which requires changes in the manufacturing process and a new design registration or patent.

The examples we have mentioned so far represent only a small fraction of the large number of modifications that have been proposed to improve the security of pin-tumbler locks. We will encounter more of these subsequently once we have given a brief overview of picking, impressioning, and decoding as it applies to pin-tumbler locks.

Lockpicking, Impressioning, and Decoding

Ever since locks were invented, people have sought to open them by means other than using the correct key. Techniques such as lockpicking, decoding, and impressioning are well established and evolve continually as new lock designs are put into service. Although it is not the purpose of this book to discuss these techniques in detail, it is necessary to give a brief coverage in order to appreciate the security features of various locks. We also provide a few pointers to reference materials that the reader may be interested in pursuing. Needless to say, the success of all of these techniques depends on a thorough understanding of the mechanism of the lock one is trying to open.

The manipulation of an inline pin-tumbler lock depends in large measure on the ability of the lockpicker to "set" each pin at the shear line. A tensioning tool is used to apply torque to the plug of the lock while a lock-pick is inserted in the keyway to lift the pins. The torque causes the pin-tumblers to "bind," that is, to prevent the plug from turning since they straddle the interface between the plug and the cylinder. Since the borings for the pin chambers are never in perfect mechanical alignment, as the pins are lifted, some pins will tend to set before others when light turning tension is applied to the plug. The art in lockpicking is to determine which pins are correctly set and which are not, and to proceed in an order that does not unset the correct ones. This task, while requiring considerable practice to perfect, is not overly difficult to accomplish when standard cylindrical pins are used and the keyway is accessible.

The lockpicker is at times aided by the observation that the tops of very short pins are visible in the keyway if the intervening pins are pushed up by inserting a probe. This property, referred to as shear line vulnerability, gives important information on the overall shape of the key.

A number of manufacturers (e.g., EVVA, DOM, and ABUS) have developed keyway sections that are highly paracentric to guard against manipulation attempts. An example is the CISA/ABUS "Top Security Profile" cylinder pictured in Fig. 2.19.

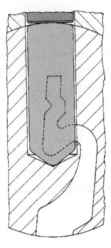

Figure 2.19: D. Errani's 1985 design of the CISA/ABUS Top Security pick-proof keyway profile (US patent 4,683,740).

The keyway profile of this lock is very difficult to navigate with conventional lock-picks, which will inevitably overraise some of the pins.

As well as manual picking, another technique, called raking, is less methodical: a rake or snake pick is used to lift and drop the pins simultaneously while tension is applied to the plug. This method is usually applied first, since if it works it is more rapid than manual picking. If raking is unsuccessful, the lockpicker must resort to manual picking. Any lock that can be opened by raking should not be referred to as a high-security lock. A further method involves the use of an impact-producing implement called a pick gun. This is briefly described later on.

Other kinds of locks, such as tubular, wafer, disc, and lever locks, can also be picked. For instance, lever locks require picks with an L-shaped end, which may be used in conjunction with a similarly shaped tension wrench. Alternatively, both pick and tensioner can be combined into a coaxial tool called a 2-in-1 or Hobbs pick, illustrated in Fig. 1.11. While the tools required for the job may be different, the principle is the same: tension is applied while the active elements of the lock are being manipulated in an effort to set them in the correct positions for the lock to open.

Impressioning refers to the process of fashioning a working key for a lock while it is in the locked state, without dismantling it. This may at first seem impossible, but in the imperfect world of mechanics it turns out not to be. When a blank is inserted in a pin-tumbler cylinder lock, turned to bind the pins, and then wiggled, tiny marks are left on the top surface of the key blade. It may take a considerable amount of force to make the marks visible to the naked eye. The remarkable thing is that marks are only left by pin-tumblers that are *not* at the shear line (i.e., they are binding). This is because, in this case, the lower pin is held by the chambers in both the plug and the core; when lateral force is applied to the plug, the leverage on the pin causes its tip to skew, leaving a faint mark or "impression" where it contacts the key.

The trick in impressioning a key is to remove only a small amount of the blank, by filing or cutting, from the pin positions where impressions have been left. If a key-cutting tool is available, the mark may be deepened to the next depth of cut. The process of wiggling the key blank and incrementally filing the key in the required locations is continued until no further impressions are left by any of the pins, at which point the lock should open. During this process the order of filing may change; thus some pins may not make a noticeable mark until other pin positions have been filed.

Like picking, the impressioning process requires skill and considerable time to learn. With practice, basic pin-tumbler locks can be impressioned in a matter of 10 minutes. Other impressioning techniques involve turning the key to bind the pins and then either tapping it or pulling it to take an impression. Impressioning can also be applied to other types of locks including wafer and lever locks.

The primary requirement for impressioning, apart from a set of files and a grip tool, is a blank that fits the lock's keyway. Makers of high-security locks exercise control over the distribution of registered blank keys by supplying only through authorized agents and locksmiths who duplicate keys on proof of ownership. Many security keys are restricted in this sense, and this is a first step in preventing unauthorized keys from being made by impressioning. The reader is referred to [122, 129] for more detail on the art of impressioning locks.

Decoding, as the name suggests, is the process of determining the code of the key from measurements taken on the lock. The measurements may be taken by any physical means, including mechanical, acoustic, optical, electrical, electromagnetic, or even X-ray. Some locks, for instance, wafer locks and low-security lever locks, can be decoded by inspection of the locking elements visible in the keyway. As far as mechanical decoding is concerned, any tool may be used that allows the shear line of each pin to be detected and the corresponding height of the pin to be measured. Examples include shims, thin wires, and calibrated pin-lifting devices. In pin-tumbler locks with a constant driver length, the compression of the spring is proportional to the length of the lower pin. Thus it is possible to estimate the pin length by measuring the force on each pin stack. For this reason many high-security pin-tumbler locks use compensated drivers to ensure that the pin stacks have roughly the same overall length.

The decoding idea can be applied after a lock is opened by picking, or it can be used to assemble or cut a working key. A key made by assembling a key from components of preset lengths is referred to as a "make-up key" or "pin-and-cam tool," and this is a popular method for opening lever locks. Once a pin-tumbler lock is picked open, a gauge consisting of a thin pivoting arm can be inserted into the keyway; the distances from the bottom of the keyway to the tip of each lower pin are then simply related to the angle of the pivot arm.

Decoding and picking can be combined into a single process whereby the individual locking elements are maintained at the heights determined via the decoding process.

The net result is that the tool both measures the active elements in the lock and opens it. An implementation of this idea for conventional pin-tumbler locks is the so-called Sputnik decoder-pick pictured in Fig. 1.12. Another example of a combined decoder and pick is furnished by the tubular lock-pick in Fig. 1.10. In this instrument, the cut depths may be read off a calibrated scale on each of the sliders once the lock is picked.

Using a combined decoder-pick is fundamentally different from manual picking. As explained above, manual picking relies on imperfections in the lock to bind pins at the shear line under light turning force: if picking proceeds out of sequence, pins that were previously set return to their locked positions. This contrasts with decoder picking in which the instrument effectively decouples the functions of the locking elements by allowing them to be manipulated independently. Once the correct height is determined for one pin, it can be maintained at that height and left alone while decoding proceeds on the other pins. The technique is amazingly powerful, and there are few locks in this book that are immune from this type of approach, given an appropriately fashioned tool (refer to [122] for examples, particularly those produced by Falle Safe Securities).

Spool and Mushroom Pins

A simple way to enhance a cylinder's manipulation resistance, and one that has been employed for many years, is to replace some of the straight driver pins with spool or mushroom-shaped pins, as in Fig. 2.20. These are a standard feature in many high-security locks such as Yale and DOM, and can be incorporated at minimal extra cost as they do not entail a change in the overall design of the lock. The effect of a spooled driver is to bind in the pin chamber if not properly set (see Fig. 2.21). This thwarts a picking attempt since it is then not possible to raise the spooled driver to the shear line without relaxing tension on the plug—which allows pins that were set to drop back down to their rest positions.

Although picking a lock with spooled drivers is still feasible, it requires much more finesse with the instruments. One reported technique is to raise all pins to the top of the keyway and set them from above the shear line by gradually relaxing the tensioner. This technique works on the assumption that the lower pins are not of

Figure 2.20: Mushroom drivers increase a lock's resistance to picking.

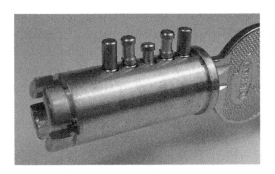

Figure 2.21: Tension applied to a spooled driver not at the shear line causes it to wedge in the pin chamber.

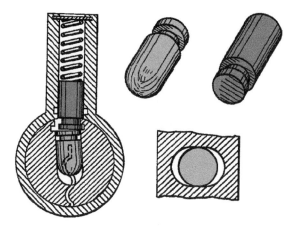

Figure 2.22: R. P. Crousore's 1940 patent called for spooled upper and lower pins (US 2,283,489).

the ribbed or spool-type. Often the front pins are spooled while the last pin-tumbler is regular (straight-sided) so as to avoid the inconvenience of the key sticking before it is fully inserted.

In some locks (e.g., American and Laperche), antipicking lower pins are also used. This idea was mooted around 1940 in a patent by Crousore (Fig. 2.22), which called for grooved upper and lower pins cooperating with a channel in the plug. An even earlier method that involved modifications to the lock cylinder was presented in a 1928 patent (US 1,739,964). This described a tamper-resistant sleeve similar to the one shown in Fig. 2.18. German manufacturers in particular have gone to great lengths to devise tamper-proof pin varieties such as stacked spool and rolling-pin drivers (see sections on Winkhaus and DOM in this chapter). The presence of several driver types in the same cylinder makes both picking and impressioning much more difficult because the dynamics of the mechanism can vary from one pin to the next.

It should be noted that, unless teamed with a very restrictive keyway section, the presence of spooled drivers does not make the lock less susceptible to pick gun attack. This method works by imparting a sharp blow across the tips of the bottom pins.

The impact is sufficient to create a gap between the two halves of each pin against spring tension. When properly administered, the gaps thus created momentarily straddle the shear line, and applying tension will cause the plug to rotate. As one might expect, modifications have been proposed to counter attacks of this kind, one of which is depicted in Fig. 2.23. Another impact-based method involving a bump-key is covered in the section on M&C locks later in this chapter.

Methods employing spooled pins are again only one way of thwarting picking and impressioning attempts. Many other, more exotic, examples have been put forward, although the majority of these proved too expensive to produce or had undesirable side-effects, such as lack of reliability or rendering the lock more difficult to open with its correct key. Among the more ambitious schemes we find Hucknall's patents from 1971 and 1980 shown in Figs. 2.24 and 2.25. The 1971 design was produced under the name "BHI Huck pin" [71].

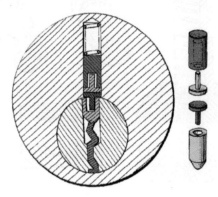

Figure 2.23: Independent Lock Company's 1938 impact- and pick-resistant pin-tumbler design by L. Gutman (US patent 2,158,501).

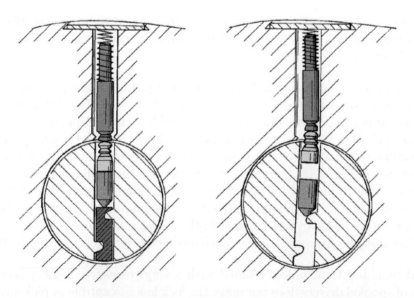

Figure 2.24: R. Hucknall's pick-proof pin (1971 US patent 3,762,193).

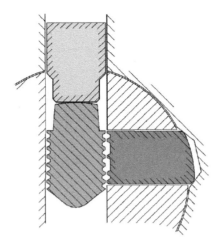

Figure 2.25: Cross-section of Hucknall's 1980 impression-resistant pin-tumbler lock (US patent 4,377,940).

Profile Pins, Pin Design, and Geometry

We have so far considered the influence of keyway design and the inclusion of hardened inserts and variously shaped pins and drivers. Further levels of security require more substantial modifications to the basic pin-tumbler cylinder design. We have grouped these modifications into three broad classes: pin design, profile pins, and pin-tumbler geometry. The additional category of pin chamber design, which has been suggested in some old patents, for example, US patents 1,860,712 (1930) and 2,043,205 (1932), is rarely used in practice for reasons of cost and complexity and has therefore been omitted.[9]

Pin design encompasses, as well as conventional spooled drivers, modifications to the function of a pin-tumbler pair. In this category we include both rotating and interlocking pins, as found in locks like Medeco and Emhart (note that Medeco is a side-bar lock, dealt with in Chapter 4).

Profile pins differ from conventional pin-tumbler pairs in respect of not having a spring-biased driver moving in a pin chamber, although some types are paired with a ball bearing. Profile pins are generally operated by cuts or dimples on the side(s) of the key that are supplementary to the main bittings for the pin-tumblers. The cut for a given profile pin is either present or absent on the key blade. An example of profile pin design is furnished by Kerr's 1966 patent (US 3,418,833) which was for wafer locks but applies equally well to pin-tumbler locks (see Fig. 2.26).

Two basic types of profile pins are discernible: *passive* and *active*. Passive profile pins are not related to the locking function of the cylinder, but instead provide a level of key control above that afforded by the keyway broaching itself. Profile pins

---

[9]An exception is the ASSA V-10 side-bar lock covered in Chapter 4.

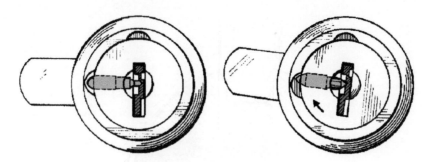

Figure 2.26: Inclusion of a passive profile pin in a wafer lock (US patent 3,418,833 by W. J. Kerr).

are usually mushroom-shaped (i.e., wider at the top end than at the tip) and are seated in borings in the plug at 3, 6, or 9 o'clock. There is a corresponding cavity or groove milled into the plug housing that accepts the head of the profile pin. As the plug is rotated, the profile pin rides out of the groove and its end impinges on the keyway. A key blade that matches the keyway section will be blocked unless it has both the correct pin-tumbler bittings and profile pin dimples. Pin-tumbler locks with passive profile pins that we cover in this chapter include DOM iX-10, EVVA DPS, Winkhaus VS, Azbe, Alpha, and Laperche Diam.

Passive profile pins can be put to great effect in increasing the number of key codes in a system. Since the operation of each profile pin is binary (i.e., it is either present or absent), each one multiplies the available number of key codes by two. Thus a system with 10 profile pins has $2^{10} = 1,024$ times as many possible codes as the same system without the profile pins. While this may sound impressive in marketing brochures, the passive profile pin has the drawback of being passive. That is to say: a key with the correct pin-tumbler bittings that is also cut in all possible profile pin positions will be able to open the lock regardless of the profile pins that are loaded in the plug. In the same way, the presence of passive profile pins does not impede the manipulation of the lock. These last two points lead us to the difference between passive and active profile pins.

As the name suggests, an active profile pin has a role to play in the actual locking function. As such, the active profile pin must be teamed with a device that will block the rotation of the plug if the key is not appropriately cut. Some systems use several sizes of active profile pins, although a single size is already a vast improvement over purely passive profile pins. The ambiguity of whether a profile pin is active or passive presents a dilemma to anyone seeking unauthorized access: a passive pin requires a key dimple, but an active pin usually does not. Thus it is not enough to bit the key in all profile positions in order to circumvent the profile pins. There is quite a deal of scope for innovation in active profile pins, and the locks we cover in this category include ABUS TS 5000, ISEO R11, Vachette 2000, Winkhaus Titan, Schlage Everest, and MLA Binary Plus. The idea is illustrated in Fig. 2.27, taken from a 1980 EVVA patent that was used in the ABUS TS 5000.

One of the obvious problems with key-operated mechanical locks is that, regardless of the level of physical security and manipulation resistance of the lock, a skilled person,

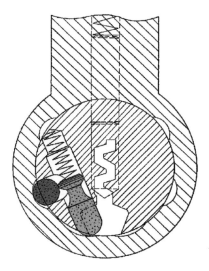

Figure 2.27: EVVA's 1980 active profile pin design (US patent 4,434,636 by K. Prunbauer).

given time and tools, can duplicate the key. One workaround for this situation involves the placing of *active elements* in the key blade itself. These elements can be either mechanical or electronic. The latter case includes key-top transponders and silicon chips, but is beyond the scope of this book (see instead [112, 126]). However, there are numerous examples where the key incorporates an active or mobile mechanical element. As with active profile pins, the active key element provides an additional blocking function. The element may consist of a floating ball (as in the DOM-iX KG lock) or pivoting member (as in the Bricard Chifral and BiLock NewGen) or one or more floating or sprung pins embedded in the key blade (as in KESO Omega, Vachette Radial Si, Laperche Diam, Pollux Interactive, Mul-T-Lock Interactive).

Whatever the form of the active element, the principal idea is that the element can be made to protrude below or above the surface of the key blade to miss fixed wards and actuate a blocking pin-tumbler. This is illustrated in Fig. 2.28 by DOM Sicherheitstechnik's floating-ball system from 1980, in which a captive ball in the key blade surmounts a fixed obstruction and then returns to the plane of the key blade to actuate the blocking pin. A correctly cut key without the active element simply cannot imitate this effect. Naturally the presence of the active element makes the key much harder to duplicate, assuming that access to the key blanks is controlled in an effective manner.

The final area of categorization of pin-tumbler locks relates to the geometrical arrangement of the pins. It is this factor that gives great variety to the range of pin-tumbler locks. The original inline pin-tumbler design, which still dominates the low end of the commercial and residential market, underwent many transformations in the second half of the 20th century. Since the 1960s new designs were made possible through advances in manufacturing engineering such as computer numerical control (CNC), which allows complicated components to be made reliably to much tighter tolerances than before. Nowadays, components for high-security locks are

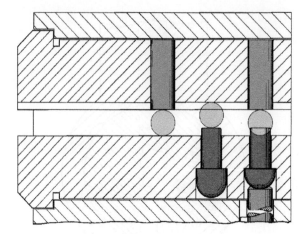

Figure 2.28: Longitudinal section of DOM's 1980 "floating ball" blocking pin (US patent 4,377,082 by H. Wolter).

routinely manufactured to a linear tolerance of 1 to 2 thousandths of an inch (less than 0.05 mm), and cut angles are made to within half a degree. Since the 1980s, Flexible Manufacturing Systems (FMS) [92] have led to even more rapid and reliable manufacturing capabilities.

Pin-Tumbler Lock Classification

The pin-tumbler range now includes many unusual and unconventional designs, which we have attempted to classify according to their geometrical features, as follows.

1. Inline: locks with a single row of pin-tumblers. Examples: DOM, EVVA, IKON, Lockwood, Rivers, Best interchangeable core, M&C.

2. Inline + Passive Profile Pins: locks with a single row of pin-tumblers having passive profile pins on one or both sides of the keyway. Examples: Winkhaus VS & VS6, EVVA DPS, Gege AP 3000, Vachette VIP.

3. Inline Horizontal Keyway: locks with a single row of pin-tumblers operated by the wide side of the key, possibly including passive profile pins and/or active element in key. Examples: Alpha, Azbe, Codem, Laperche Diam.

4. Twin Inline: locks with two rows of pin-tumblers with either a vertical or horizontal keyway, possibly including passive profile pins and an active element in the key. Examples: DOM iX-10, Lockwood V7, Head, Tover 2F30, Lancia (car lock), Renault TS (car lock).

5. Inline + Active Profile Pins: locks with a vertical or horizontal keyway and a single row of pin-tumblers, containing active and (optionally) passive profile

pins. Examples: ABUS TS 5000, ISEO R11, Vachette 2000, MLA Binary Plus, Schlage Everest, Winkhaus Titan.

6. Cruciform: locks with three or four rows of pin-tumblers arranged on axes at 90 degrees; keyway usually cross-shaped and key cut on three or four sides. Examples: IKON, Moreaux, Helason, Papaiz.

7. Multiple Inline: locks with three or more rows of pins arranged radially and operated by a "dimple key," which may also contain active elements. Examples: Kaba Gemini, Kaba Quattro, KESO, KESO Omega, Kaba ExperT, Bricard Chifral, Vachette Radial, YBU.

8. Tubular: axial pin-tumbler locks with pins arranged on a circle or other shape (e.g., an "E"); pins may be recessed or flush, in which case the key is pronged; key may be end-bitted, solid, tubular, or with radial fins; key may contain active elements. Examples: ACE, GEM, Apex, Central, Izis Arnov, Zenith Cavith, JPM, Pollux, Van Lock, ISEO R6, Tover 27A, Bramah (wafer lock), Picard (wafer lock).

9. Concentric Pin: locks with one or more concentric or coaxial pins (i.e., pins inside pins). Examples: Mul-T-Lock, Mul-T-Lock Interactive, Age, Wiselock.

10. Rotating Pin: locks whose pins must be rotated as well as lifted; key bittings are not all at 90° to the key blade. Examples: Emhart, Medeco (side-bar lock).

11. Pin Matrix: card-operated locks with binary pins arranged in a rectangular grid; pins can take only one of two positions ("in" or "out"). Examples: Vingcard 1050 and Japanese card-operated padlocks such as the Saiko CardLock.

12. Key-Changeable: locks that are capable of being code-changed by the insertion of one or a pair of keys (does not refer to keyless combination locks or construction-keyed locks). Examples: Code, Rielda (wafer lock), Winfield (bicentric wafer lock).

Given the astounding variety of pin-tumbler locks, other classification schemes are certainly possible. Some locks straddle the boundaries between two or more categories. For instance the Zenith Cavith is a tubular lock with additional radial pin-tumblers on one fin of the key. A variant of the ACE tubular lock by the American Locker Company has a combined tubular and flat key. One of the strangest locks was proposed in a 1975 U.S. patent by N. Epstein of the Norman Lock Company. The lock transmitted linear force to a hinged tail-piece via a chain of balls that were brought into alignment by the correct key. For interest's sake, we have included in Fig. 2.29 a picture of this bizarre concept, although it does not seem to fit in any of the categories we have specified.

The remainder of this chapter is devoted to detailed descriptions and modus operandi of a number of interesting pin-tumbler locks organized according to the preceding categories. As explained in Chapter 1, the locks in each section are listed

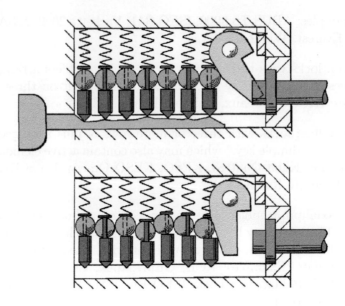

Figure 2.29: N. Epstein's 1974 "chain of balls" concept for force transmission (US patent 3,928,993).

along with their country of origin (refer to Table 1.4), the type of lock mechanism, and a grading of their manipulation resistance on a scale of 1 (low) to 5 (high). This structure recurs in later chapters.

2.2   Inline

EVVA, DOM, IKON

EVVA (AT) 5-pin (3)
DOM-S (DE) 5-pin (3)
IKON (DE) 5-pin (2–3)

The modern high-security inline pin-tumbler cylinder lock is exemplified by locks made by companies such as EVVA-Werk (founded in Austria in 1919), IKON AG (originating in 1926 in Germany), and DOM Sicherheitstechnik (founded in Germany by Joseph Voss in 1936). Locks from each of these companies are displayed in Figs. 2.30–2.32. IKON, which built its reputation on precision optics and mechanics for cameras under Carl Zeiss, merged with Abloy in 1989 and since 1994 has been part of the ASSA Abloy group.

Although the EVVA and DOM-S are conventional pin-tumbler locks, they incorporate some serious security features. These include very severe keyway broachings with multiple ribbings that make the insertion of all but the thinnest lock-picks

Figure 2.30: Inline cylinders from EVVA, DOM, and IKON.

Figure 2.31: Keys for DOM-S and EVVA GPI locks.

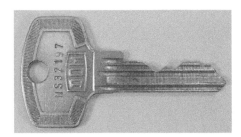
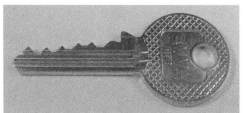

Figure 2.32: DOM-S (left) and EVVA GPI (right) cores, with antipicking and drill-resistant features.

difficult. Hardened inserts in the plug and cylinder body and spooled drivers are standard issue. Access control is achieved using a broaching with high-precision multiple side-wards. The EVVA GPI, introduced in 1976, is a de facto industry standard in this respect: the system offers 32,000 different keyway profiles, allowing a very high level of multiplex master-keying and access control.

The DOM-S 5-pin profile cylinder uses torpedo-shaped (tapered) antipicking lower pins as well as stacked-disc spooled drivers (Fig. 2.33). These drivers consist of a support pin that houses four independent discs. Each of these can wedge across the

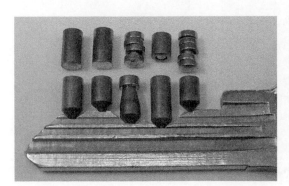
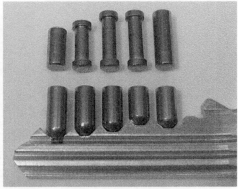

Figure 2.33: (Left) Pin set from DOM-S with hardened pins, torpedo, and multi-segment spooled drivers. (Right) Pin set from EVVA GPI.

shear line in the same way as an ordinary spooled driver. There are a total of 1,024 keyway profile variations.

Lockwood 7-pin

(AU) 7-pin (2)

The Lockwood 7-pin cylinder, shown in Figs. 2.34 and 2.35, is used exclusively by the Australian Postal Services for Post Office boxes. The cylinder is made of brass with a nickel silver key. The presence of seven pins gives an increased number of key codes (around a factor of 10 more than 6-pin locks), which is the primary requirement for this system to ensure privacy of people's mail. No further security features are present since the lock is not intended to be used for general architectural purposes. The cylinder is designed for easy recombination: the locking cam at the front of the cylinder is first removed, and the plug is rotated to 6 o'clock, at which point the lower pins may be unloaded through a row of access holes in the bottom of the barrel.

American Lock Company

(US) 5-pin (2)

The American Lock Company (Junkunc Brothers), founded by J. Junkunc around 1912 [50], makes high-quality pin-tumbler and wafer locks. A picture of an American removable-core padlock is shown in Fig. 2.36. The lock cylinder includes multiple

Figure 2.34: Lockwood postal services 7-pin cylinder and key.

Figure 2.35: Lockwood 7-pin plug with key partially and fully inserted.

Figure 2.36: American removable-core 5-pin padlock and key.

grooves on both the driver and lower pins to increase picking difficulty (see Fig. 2.37). A similar idea to enhance a lock's manipulation resistance by adding grooves to both the top and bottom pins was discussed in Crousore's 1940 patent (Fig. 2.22). The American padlock also has a spring-loaded ball-locking shackle. The mechanism captures the key in the open position so that the user is obliged to relock the padlock in order to remove the key.

Maxis

(CN) 5-pin (1)

"If you can't tension it, you can't pick it." This is the theory behind this unusual pin-tumbler lock. The Maxis lock, shown in Figs. 2.38–2.40, has a 5-pin cylinder with a perfectly circular keyway section. A single row of five pins extends from 12 o'clock to the midpoint of the keyway. The key is made from a nickel-plated flat brass blank 0.1″ in width, sandwiched inside a slotted tube to make it round in cross-section. The key cuts are thus nested in between the round edges of the tube. The key design is quite similar to Linus Yale Senior's pin-tumbler padlock, patented

Figure 2.37: (Left) Key and cylinder from American padlock. (Right) Pick-proofing grooves on upper and lower pins.

Figure 2.38: Maxis 5-pin round-keyway cylinder and key.

Figure 2.39: Maxis plug with pins at shear line.

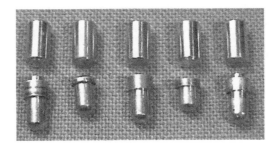

Figure 2.40: Set of lower pins and drivers from Maxis cylinder.

in 1857 (US 18,169). The pin-tumblers are conventional except that the lower pins are flanged to sit in a countersunk bore in the plug, suspending them above the keyway. Some of the lower pins may also have a reduced-diameter shank above the flange. The lock also has hardened inserts in the driver pins.

The keyway is blind, having a round brass plug at its end. The key has a conical tip and is supposed to be inserted right way up, but since the keyway is round, the key can be inserted in any orientation. Once the key is correctly inserted, its bittings raise the lower pins to the shear line and the key can then operate the lock. It is quite difficult to tension the plug without interfering with the pins, although a tensioner could be fastened to the front of the plug with adhesive. If tension can be applied either to the cam or plug, the lock is easy to pick open with a flat blade.

Interchangeable-Core Locks

(US) **Best, Arrow, Falcon** 5–7 pins (1–2)

The motivation for interchangeable-core locks was set out in the chapter introduction. In this section we focus on the workings of the small-format interchangeable-core (SFIC) system, illustrated by the Falcon IC cylinder in Figs. 2.41–2.45. The key codes in Fig. 2.41 are: change key (6 7 4 5 2 4); master key (8 9 8 5 2 4); and

Figure 2.41: (Left) Interchangeable-core cylinder by Falcon. (Right) Change, master, and control keys.

Figure 2.42: (Left) Core showing locking lug. (Right) Nomenclature for interchangeable-core cylinders.

control key (4 1 4 5 2 4). Note that cut numbering proceeds from tip to bow in conventional interchangeable-core locks.

A SFIC assembly comprises a plug, control sleeve, and shell, as shown in Fig. 2.42. The plug slots into the control sleeve, and this in turn slots into the shell. The control sleeve has an extended top portion that forms part of the pin chambers. The effect of the control sleeve is to create a second shear line (much as in a master-ring cylinder). The normal (lower) shear line is called the *operating* shear line, while the upper one is called the *control* shear line. The difference in height between the operating and control shear lines is 0.125″.

Figure 2.43: (Left) Control key aligns pins at control shear line. (Right) Retracting the locking lug for removal of core.

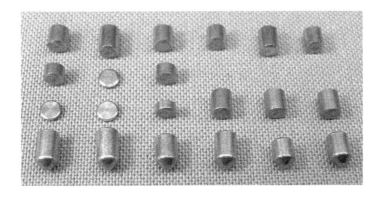

Figure 2.44: Pin set from a master-keyed interchangeable-core lock.

The rear section of the control sleeve has a locking lug that normally protrudes into a cavity created for it in the lock housing to which the core has been fitted. The control sleeve must be turned about 15° to the right for the core to be removed from the cylinder housing. The plug and sleeve are secured in the shell by a retaining plate that also acts as a key stop. The plug has two holes at the rear, one on either side of the keyway, to accept two prongs on the tail-piece of the cylinder. This provides mechanical coupling to the lock.

In general, each pin chamber in the lock contains a lower pin, master pin, and control pin, as well as a driver and spring (see Fig. 2.44). The actual dimensions of the pin stacks are available from Best's 1968 patent (US 3,603,123). There are nine standard sizes for bottom pins, ranging from 0.110″ to 0.222″ in increments of 0.0125″. A further 18 sizes are reserved for master, control, and driver pins, ranging from 0.025″ to 0.237″ in length [102]. The length of the overall pin stack in each chamber is maintained at 0.397″ by compensating the driver pins.

In addition to the change and master keys, as mentioned earlier, the IC also possesses a control key that differs from the former two. The operation of the change and

Figure 2.45: (Top to bottom): Change, master, and control keys align pins at either operating or control shear line.

master keys is the same as in a conventional pin-tumbler lock. The cuts on the change key bring the lower pins to the operating shear line, allowing the plug to be rotated (Fig. 2.45). Since at this point the control shear line is straddled by one or more control pins, the control sleeve cannot be turned. Similarly, a master key unlocks the lock by aligning the lower and master pins with the operating shear line, but again, will not set the pins at the control shear line. The control key creates a shear line at the interface of the control sleeve and barrel, allowing the control sleeve to be rotated to retract the locking lug that normally prevents removal of the core. The control key does not align all of the pins at the operating shear line since it has some shallower cuts than either the master or change keys.

IC locks are not especially difficult to manipulate, particularly in master-keyed systems. However, the control shear line is more difficult to pick since tension must be applied to the control sleeve. The lock is designed so that it is difficult to apply force to the control sleeve from the front of the lock, although comb-type tension tools are available for this purpose [121]. A patent was filed in 1963 for a method of deadlocking the plug in the case of a successful picking attempt (US 3,181,320); this is undesirable, however, since it cannot easily be undone. Interchangeable-core locks are also produced for high-security locks by Mul-T-Lock, Medeco, and Schlage Primus, among others.

Rivers

(AU) 6-pin (4)

What is the logical opposite of a conventional key-operated lock? The Rivers lock, shown in Figs. 2.46–2.49, must come close to this idea. It appears to be open by default, turning freely when no key is inserted! It was invented in Australia by

Figure 2.46: (Left) Rivers key-drive lock. (Right) Handle removed and key inserted.

 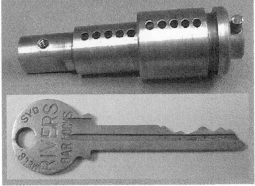

Figure 2.47: (Left) Rivers lock driver pins are captive in upper chambers. (Right) Core comprises inner and outer plugs, shown alongside key.

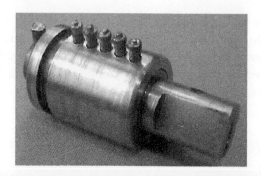

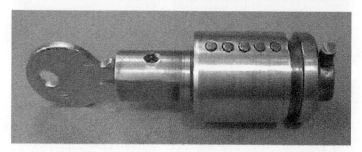

Figure 2.48: (Top) Suspended drivers leave inner core free to turn. (Middle) Incorrect key blocks outer core. (Bottom) Correct key positions pins at shear line of outer plug.

Figure 2.49: Correct key connects inner and outer plugs while freeing inner plug to turn.

Hector Rivers around 1922 and although based on a very simple principle, it is highly effective and has consequently met with commercial success. The patent for the design was filed in 1928 and published as US patent 1,770,864 in 1930 (see Fig. 2.50). The Rivers lock is distributed by Rivers Locking systems (Australia), which is now part of the Austral Monsoon Group of companies. It is a key-drive pin-tumbler lock for use with high-security bar lock and two-point locking systems, and as such is aimed primarily at the commercial and industrial sector. It is typically fitted to sheeted steel doors and shutters.

The lock has a cast brass body and round, polished brass front. The plug protrudes about 16 mm past the front of the lock and has affixed to it a knurled brass handle (Fig. 2.46). In the absence of a key, the handle spins freely, so there is no point taking a wrench to it to force it open. To protect against drilling, two hardened pins are installed in the front plate of the lock body, in line with the pin chambers. The key is long for a pin-tumbler lock: the blade length is 46 mm with cuts starting at 24 mm from the shoulder. The lock has five or six pin-tumblers.

Inside the lock body there is not one plug but two (see Fig. 2.47). We shall refer to these as the inner and outer plugs as they are concentric. The inner plug has a diameter of about 16 mm and houses the lower pins. The outer plug has an outside diameter of approximately 22 mm. The construction is not dissimilar to a Corbin master ring cylinder, except that the outer cylinder is blind at the back end.

The first five pin chambers house the driver pins and springs, with the rear chamber reserved for a retaining pin that rides in a groove around the outer core. Since the inner plug has a reduced-diameter front end, it is captive between the front plate of the lock body and the blind end of the outer plug.

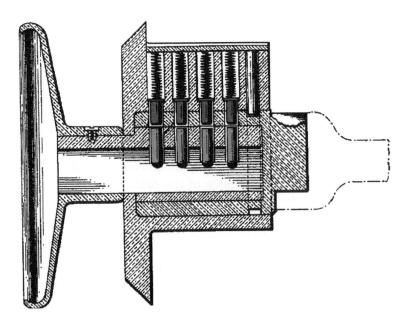

Figure 2.50: Diagram from a 1930 patent for the Rivers lock (US 1,770,864).

The lock would be easily defeated, despite the recessing of the core, if the driver pins sat atop the lower pins; but this is not the case. The driver pins are flanged at the top, and the pin chambers in the body are correspondingly fashioned so that only the bottom 1/8″ of the drivers protrudes into the core. This distance is enough to penetrate the outer plug down to the interface with the inner plug. In other words, the top pins remain suspended just above the inner plug while blocking the outer plug.

In addition, in their rest positions, the lower pins are wholly contained in the inner plug. There is thus a gap in each pin stack between the lower and upper pins that spans the shear line between the inner and outer plug. To operate the lock, it is the outer plug, to which is attached a drive cam, that must be turned. Turning the inner plug achieves nothing until a key is inserted.

The correct key raises the bottom pins past the rim of the inner plug and sets them at the shear line between the outer plug and the lock body (Figs. 2.48 and 2.49). With the inner plug thus coupled to the outer plug which is no longer impeded by the driver pins, one is able to operate the outer plug and drive the lock mechanism using the handle.

This brilliantly simple mechanism is remarkably difficult to manipulate with ordinary lock-picks. The standard trick of tensioning the plug while lifting the pins will not work since the plug just continues to turn. It would be necessary to raise one or more pins so that they reach the outer plug before applying tension, but setting all the pins at the outer shear line in this manner would require substantial dexterity.

M&C

(NL) 5-pin + 8 trap-pin (3)

The bump-key technique, or *slagmethode* as it is called in Dutch, has seen a recent resurgence in the Netherlands and Germany [131]. The method is simplicity itself. A suitably doctored key, similar to a rapping or "999" key, is first prepared (see Fig. 1.7). The bump-key is then inserted into the lock and, under light turning tension, "bumped" with a suitable impact-producing implement. The angled surfaces of the bump-key transfer the impact to the pin-tumbler pairs. If the impact is sharp enough, then despite the downward force of the driver springs, a gap is formed by the transfer of energy to the top pins that momentarily separates them from the driver pins. The bump-key is designed to cause this to happen simultaneously across all pin chambers, at which point the plug is free to turn.

Although a number of high-security cylinder locks are resistant to bump-keys [130, 132], the M&C cylinder (Fig. 2.51) is, as far as the author is aware, the only currently made lock specifically designed to resist attack by bump-keys. This is the reason for the symbol of the crossed-out hammer on the front face of the

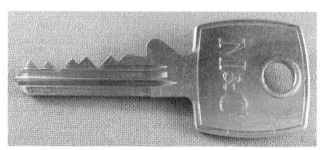

Figure 2.51: M&C 5-pin cylinder and key.

Figure 2.52: Conventional construction of M&C plug.

cylinder—since bump-keys are normally used in conjunction with a small hammer to provide the necessary impact.

Apart from the anti-bumping icon on the front, the cylinder (Figs. 2.51 and 2.52) looks much like a conventional pin-tumbler Europrofile cylinder. In terms of its basic specifications, it is a conventional 5- or 6-pin security cylinder with nickel-plated brass key. There are drill-proofing pins flanking the pin chambers, which is a requirement for the Dutch SKG security standard. The keyway is not highly para-centric in comparison with other locks like Winkhaus, DOM, and EVVA. However, the reader should not be lulled into thinking that the M&C is easy to defeat because there is more to opening this lock than just raising the pins to the shear line.

The lock's secret lies in the presence of four pairs of auxiliary pin chambers sit-uated to the immediate left and right of the regular chambers for pins 2–5 (see Figs. 2.53 and 2.54). These eight auxiliary chambers, which are of smaller diame-ter than the regular ones, contain trap pins with their own driver springs. The trap pins are spooled but very slender, with a domed head at each end. The auxiliary chambers are only in the cylinder body, stopping at the interface with the plug. The plug itself is entirely standard, with bores for the five pins. How then does the system work?

Figure 2.53: The M&C cylinder has auxiliary pin chambers.

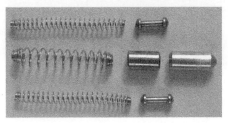

Figure 2.54: (Top) Inside of M&C cylinder showing auxiliary chambers. (Bottom) Conventional and trap pins.

When the correct key is inserted into the lock, it raises all five pins to the shear line and keeps them there as the plug turns. Regardless of which way the key is turned, the top edges of the lower pins in chambers 2–5 pass directly underneath the trap pins in the auxiliary chambers. The trap pins remain in their chambers since the key is providing the upward force necessary to balance the auxiliary driver springs.

Now consider what happens when the lock has either been picked or compromised by a bump-key. In either case we can assume that all five top pins are initially at the shear line as the plug begins to rotate. Since neither the lock-pick nor the bump-key (which has maximum depth cuts in all positions) is able to maintain the lower pins at the shear line simultaneously, one or more of these pins will be at their rest positions. As soon as the plug rotates fractionally clockwise or anticlockwise, pin chambers 2–5 in the plug will move into positions directly underneath the left or right auxiliary chambers. At this point one or more of the trap pins under the action of their driver springs will force its way into the free space above the lower pins in the plug. Further rotation of the plug is then blocked. The trap pins are under sufficient tension to prevent the plug from being turned rapidly past the auxiliary chambers by a plug-spinner.

More to the point, the plug cannot be returned to the locked (12 o'clock) position without raising the trap pins to the shear line. This is bad news for lock-picks and worse news for bump-key attacks since the bump-key cannot easily be removed from the plug once it is captured by the trap pins.

2.3 Inline with Passive Profile Pins

Winkhaus VS/VS6

(DE) 6-pin + 10 passive profile pin (3–4)

Winkhaus started out as an ironmongery business in 1854 whose main product line was padlocks. Traditionally based in Münster, Germany, the company released its first profile cylinder in 1956. We deal in this section with the Winkhaus VS ("Versatile & Secure") cylinder, produced since the 1970s.

On first inspection, the VS cylinder (Fig. 2.55) appears to be a conventional pin-tumbler lock. However, in addition to the regular pin-tumbler pairs, there are lateral profile pins on each side of the keyway at 3 and 9 o'clock (Fig. 2.56). The VS model has five pins and up to eight profile pins in two rows of four. The VS6 has six pins and up to ten profile pins in two rows of five. The first pin pair is hardened, and there are further hardened inserts flanking both the keyway and the pin chambers at the front of the cylinder. Like DOM and EVVA locks, the highly paracentric keyway broaching makes pin manipulation difficult with conventional tools.

The profile pins interact with dimples on each side of the key; these are all at the same height and staggered with respect to the pin cut centers to allow an increased bitting depth. A key that is not side-bitted will push the profile pins outward into

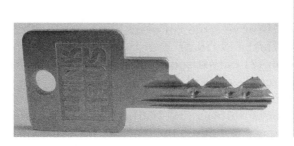

Figure 2.55: Winkhaus VS key and cylinder with highly paracentric keyway.

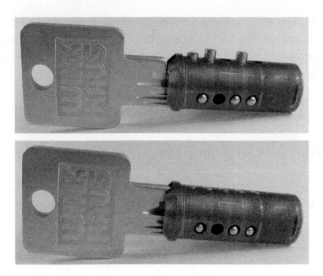

Figure 2.56: (Top) Winkhaus plug with key partially inserted. (Bottom) Key fully inserted.

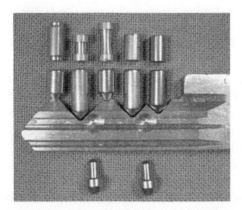

Figure 2.57: Winkhaus pins and antipicking drivers of various types. Profile pins at bottom.

the longitudinal groove on the inside of the cylinder housing, preventing rotation of the plug.

The driver pins in this lock deserve some comment (refer to Fig. 2.57 and also Fig. 2.110 in the section on Winkhaus Titan). Four types are used: (1) conventional cylindrical; (2) straight spool; (3) curved spool (hourglass); and (4) "rolling-pin" drivers with a reduced diameter at each end. Both types of spooled pin will block if incorrectly raised. Rolling-pin drivers, the central part of which is of full width, do not skew like ordinary spool or mushroom pins, but instead cannot be impressioned like ordinary pins since they have a reduced diameter at the shear line. The combination of the various spooled drivers and the very restrictive keyway broaching make the Winkhaus cylinder very difficult to manipulate.

EVVA DPS/DPX

(AT) 5-pin + profile side-bar (3)
(IT) **Mottura PX** (equivalent)

The EVVA DPX (Figs. 2.58–2.60) is a high-security 5- or 6-pin cylinder supplemented by two profile bars at 4 and 8 o'clock. It is an update on the EVVA DPS system, which has only a single profile bar at 4 o'clock. Each profile bar has multiple (five or more) ribs that must simultaneously register with elongated bittings on both sides of the key. The ribs may be in various positions along the length of the profile bar, greatly increasing the number of system combinations in the same way as passive profile pins. The advantage of such a system over conventional multiple keyway broachings is that the profile control is hidden inside the keyway. Moreover, the profile-control points can vary in their positions, giving further flexibility for differing.

Figure 2.58: EVVA DPX/Mottura PX 5-pin cylinder and key.

Figure 2.59: (Left) EVVA DPX 5-pin core and profile bars. (Right) Blocking action of plastic insert in side-bar.

Figure 2.60: Operation of the EVVA DPX.

The initial concept was described in K. Prunbauer's international patent application filed in 1991 (WO 93/09317). A U.S. patent was filed for the DPS/DPX system in 1996 (US 5,797,287). The marketing documentation suggests that there are over 32 billion possible key combinations. However, this must be understood in the context that the system provides only the usual number of 5- or 6-pin differs, multiplied by the number of possible passive profiling options.

Any passive profile-control system is subject to bypass with a skeleton key, that is, a key with profiling cuts in all possible positions. To circumvent this obvious deficiency, in the DPX system the central third portion of each profile bar contains a plastic insert that must be pushed radially outward to the full width of the key blade. Failure to achieve this results in the lower pin in chamber number 3 blocking in the profile-bar slots as the key is turned in either direction to around 90 degrees (see Fig. 2.59). This adds protection against the use of an undercut key with the correct pin-tumbler bittings.

Hardened inserts and a frontal crescent are present to protect the pins and profile bars from drilling. Since the profile bars are passive, they cannot prevent the plug from turning if the cylinder were picked open, although full rotation is not possible due to the blocking feature described earlier. Picking is rendered difficult, however, by keyway design and the use of spooled drivers.

Gege AP 3000

(DE) 6-pin + profile cog (3–4)

Gege (pronounced "geh-geh") is an Austrian lock manufacturing company, established in 1862 and now owned by Kaba AG. Among their range of pin-tumbler locks we find the AP 2000 and AP 3000 5- and 6-pin Europrofile cylinders. These locks all include highly paracentric keyways, drill protection of the front of the plug and cylinder, and doubly spooled drivers. Cylinders may also incorporate one or more hardened driver pins. The key designs are registered and duplicate keys are only supplied on proof of ownership from a Gege accredited dealer.

While the AP 2000 can be considered a "standard" high-security cylinder for operational purposes, the AP 3000 (Figs. 2.61–2.63) incorporates additional nonstandard features. As well as six inline pins, the plug incorporates a small cog wheel with four

Figure 2.61: Gege AP 3000 key and 6-pin security cylinder.

Figure 2.62: Cog on underside of AP 3000 (left) engages the key stem (right).

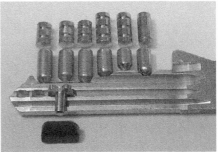

Figure 2.63: Partial insertion of key in Gege plug (left). As well as a cog, hardened and serrated drivers are used (right).

teeth, mounted in a recess at 6 o'clock midway along its length (see Fig. 2.62). There is a steel ball (also at 6 o'clock) to check the bottom edge of the blank while preventing the cog from being drilled out. The cog wheel acts like a set of passive profile pins to verify the perforations on the key blade. A key with the correct pin-tumbler bittings that does not contain the correct set of four holes in the blade cannot be inserted fully into the plug. The wheel is set at a height just below the maximum cut depth, which makes it hard to manipulate the rear two pins with conventional lock picks. Since the cog has four teeth, no matter what the orientation, there is always at least one tooth protruding into the keyway.

Vachette VIP

(FR) 5-pin + 18 passive profile pin (3–4)

Vachette, a household name in France for locks and architectural hardware, was founded in 1864. Its mainstay products include padlocks and surface-mounted door locks. Since 1997, Vachette has been part of the ASSA Abloy Group. In recent times Vachette has developed a multiple inline system called Radial, which we cover later in this chapter along with an earlier and more unorthodox lock called the 2000 SM.

The Vachette VIP (Figs. 2.64–2.66) is a one star A2P-rated 5-pin security cylinder produced in Europrofile and other formats. The system is distinguished by its capacity to accommodate up to 18 passive profile pins, which results in a huge number of potential system combinations. There are also a number of other important security features.

Both the cylinder and plug contain substantial drill proofing: two hardened pins along the length of the third pin chamber and two axial pins straddling the keyway. The keyway has a transverse side-ward that overlaps the narrow top section; this is

Figure 2.64: Vachette VIP key and profile cylinder.

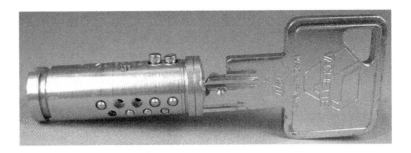

Figure 2.65: Vachette VIP plug with key partially inserted.

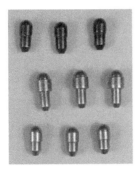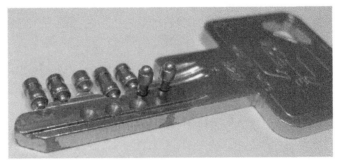

Figure 2.66: (Left) Vachette VIP profile pins. (Right) Key with antipicking lower pins and two profile pins.

teamed with a special lower pin design that makes it particularly difficult to navigate with lockpicking implements. Although the driver pins are of conventional type, the lower pins are spooled with a shallow-sloped base and a very small nipple-shaped tip. This design counters both picking and impressioning.

On each side of the key blade, near its base and midsection, there are two rows of profile dimples. Viewing the key as it is inserted with the cuts topmost, there are five midheight dimples and four lower dimples on the right and four midheight dimples and five lower dimples on the left. There are correspondingly a total of 18 horizontal bores, nine in each side of the plug, to accommodate the profile pins. The profile control is entirely passive: there is no spring biasing or secondary mechanism associated with the profile pins.

The profile pins are inserted into the bores in the plug with their thin ends toward the keyway. The enlarged portions of the profile pins impinge on channels milled into the inner surface of the cylinder. There are four such channels, two just below 3 and 9 o'clock, and two at 4 and 8 o'clock. The profile pin bores are of a smaller diameter than the chambers for the main pin-tumblers. Thus there is no need to insert the full complement of profile pins, since the driver pins cannot enter the profile bores as the plug turns. Each of the 18 profile pins can either be present or absent, yielding an effective number of profile-control options of $2^{18} = 262,144$.

The base part of the key blade is thick enough to accommodate the lower profile dimples on each side. The dimples for the midheight profile pins overlap somewhat with the ridges of the key blank.

The remainder of the mechanism is conventional. The system is able to utilize eight pin sizes from 0.135″ to 0.295″ with a depth increment of 0.020″. The MACS is 6, so, for example, a number 7 cut can be next to a number 1 cut. The theoretical number of pin-tumbler combinations, not counting MACS and other constraints, is $8^5 = 32,768$. This is multiplied by the number of profile options, resulting in around 8.5 billion keying combinations. The Vachette VIP lock is therefore well-suited to arbitrarily large MK systems. Note, however, that since the profile pins are passive, a key could be ground down or drilled to accept all 18 profile pins. Naturally, the key blanks are restricted, so as to minimize the chances of this happening.

2.4   Inline Horizontal Keyway

Alpha, Azbe, Codem

(JP) **Alpha** 6-pin + 5 profile pin (1–2)
(ES) **Azbe** 5-pin (1)
(FR) **Codem** 5-pin (1)

In conventional pin-tumbler cylinder locks, the blade of the key is inserted so that its wider face is in the plane of the pin chambers. Many modern pin-tumbler security locks have a so-called horizontal keyway where the key is inserted so that its blade is perpendicular to the plane of the pins. Instead of V-shaped cuts along the narrow edge of the key stem, the key blank is drilled along its wider face, leaving the edges of the blade intact. The operating principle is the same as in conventional pin-tumbler locks: the borings or dimples in the key blade raise the pin-tumbler pairs to the shear line, enabling the plug to be turned.

Horizontal keyway locks have the disadvantage that the amount of lift on each pin is limited by the thickness of the key blank (which is a lot less than the width of the key blank). Nonetheless, with tight manufacturing tolerances, an acceptable number of differs should result, as evidenced in high-end systems like the ABUS EC "Extra Classe" (Fig. 2.67), FTH Thirard SHG8, and DOM iX-5. From a manipulation point of view, the system offers certain advantages. Access to the pins is more limited than in conventional pin-tumbler locks, and this means that special lockpicking tools are required. The positioning of the pick may also be hampered by the broaching of the keyway, although the pins do not have to be raised very far to reach the shear line. An advantage of the construction is that it effectively precludes the use of pick guns, which need more space than is available to flick the pins. A second advantage is that keys can be symmetrically bitted for operating the lock either way around, similar

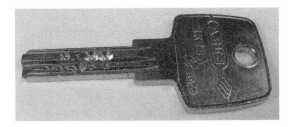

Figure 2.67: Reversible key from ABUS EC800 horizontal keyway cylinder.

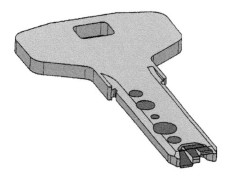

Figure 2.68: Many new key designs feature active elements that extend beyond the plane of the key (US 5,724,841 by R. Botteon).

to the convenience keys used in car locks (Chapter 7).[10] If reversible operation is required, the pin chambers are usually offset from the middle of the keyway so that the borings in each side of the key blade do not interfere with each other.

Access to the pins can be further controlled by a "mezzanine" keyway. In this idea, detailed in Silca's 1995 patent (Fig. 2.68), the beveled edges of the pins rest in a longitudinal channel that is below the bottom edge of the keyway. A key with a mobile element resembling a scoop is then required to access the tips of the pins and raise them onto the key blade as it is inserted.

An example of the horizontal keyway system is provided by the Alpha 6-pin cam lock made in Yokohama, Japan. The lock, shown in Figs. 2.69 and 2.70, is typically used for vending machines, lockers, and as a key switch for control panels. The pins are located at 12 o'clock with a set of up to five passive profile pins at 9 o'clock. The pins have a diameter of about 80 thousandths of an inch, roughly equal to the thickness of the key blade. There are five depths of cut with pin lengths varying from 0.158″ to 0.222″ in increments of 0.016″. With six pin positions and five pin sizes, the theoretical number of differs is 15,625. In Table 2.5 we have listed the number of codes for various values of MACS and for two bitting rules applied either separately

---

[10]Conventional pin-tumbler and wafer lock keys can also be symmetrically bitted if the keyway warding allows this.

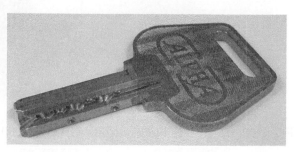

Figure 2.69: (Left) Alpha 6-pin horizontal keyway cylinder. (Right) Alpha key with profile pin dimples.

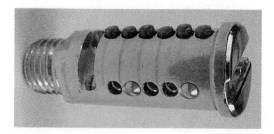

Figure 2.70: Alpha plug showing pin-tumblers and profile-control pins.

MACS	I	II	III	IV
1	707	462	454	382
2	4,569	3,880	4,130	3,740
3	10,727	9,642	10,164	9,462
4	15,625	14,300	15,000	14,100

Table 2.5: Number of codes for 6-pin locks with five depths of cut subject to various bitting rules: (I) MACS only; (II) MACS and ≤ 3 cuts the same; (III) MACS and ≥ 3 cuts different; (IV) MACS and both rules.

or together (along with the MACS constraint). The number of usable differs can be read off the table, taking into account the actual value of MACS, which is 3 for the Alpha lock. The code series runs from (0 0 0 1 1 2) to (4 4 4 3 3 2) when the bitting rules mentioned in the table are accounted for. Inclusion of profile pins makes up for the relatively small number of pin-tumbler codes.

A system with five possible profile pins has a maximum of $2^5 = 32$ profiling combinations. Taking profiling options into account increases the number of available codes for the Alpha lock to around 300,000. The profile pins may also be applied in master-keying without the need to change the actual broaching of the plug. Thus a change key may contain some, but not all, of the profile borings, whereas the master key would have the full set of profile borings. This is a clear advantage in overall

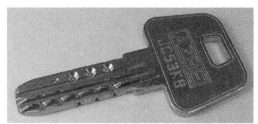

Figure 2.71: Azbe 5-pin horizontal-keyway cylinder, core, and key.

Figure 2.72: Codem 5-pin horizontal-keyway cylinder and reversible key.

system cost since only a single broaching and blank need be manufactured, with keyway control achieved by using different sets of profile pins.

In assessing the level of security provided by a horizontal keyway system, it should be remembered that the profile-control pins are passive and therefore do not hamper manipulation. Thus, it is important that the keyway possess a profile that renders the insertion of manipulation tools difficult. The inclusion of hardened inserts to guard against drilling of the cylinder is also a feature that is sometimes overlooked with this kind of design.

Further examples of horizontal keyway locks include the Azbe (HS-4) and Codem 5-pin cylinders. Both of these have reversible keys. The Azbe lock (Fig. 2.71) may also contain up to 10 passive profile pins, located in rows of 5 at 6 o'clock and 9 o'clock. The broaching of the keyway is, however, not severe enough to prevent manipulation.

The Codem cylinder (Fig. 2.72) is an early-model horizontal keyway lock in a Europrofile format with a keyway shaped like a squashed "H". The main feature is

Figure 2.73: Insertion of key into plug of Codem lock.

that its key is rotationally symmetric, with five cuts along one edge on the top and bottom surfaces. The Codem cylinder does not have profile-control pins.

Having a reversible key is advantageous; however, with rounded-top pins and no drill protection, the lock provides only minimal pick resistance and a level of security comparable with a standard 5-pin domestic cylinder (see Fig. 2.73). Codem locks are now obsolete, having been replaced by more secure and flexible dimple key systems like Kaba.

In later sections we revisit the horizontal keyway lock to see what various manufacturers have done to enhance the level of security that it can provide (see Mul-T-Lock, DOM, ISEO).

Laperche Diam

(FR) 7-pin + 12 profile pin (3)

Laperche, traditionally manufacturers of a push-wafer Bramah-type axial cylinder, also make a pin-tumbler cylinder with a horizontal keyway (see Fig. 2.74). The cylinder comes in two varieties called Diam and Diam XL, both having seven pin-tumblers. The Diam XL contains active elements in the key. The cylinder uses bottom pins that are torpedo-shaped to increase the manipulation resistance (see Fig. 2.75). Up to 13 profile-control pins are included, spaced in between the regular pin locations: seven at 3 o'clock and another six at 6 o'clock. The profile pins are passive: when present, the key must be side- or bottom-bitted in the corresponding positions so as not to block rotation of the plug. The profile pins do not add to the manipulation resistance of the cylinder.

The key for the Diam XL has a two-part active element situated at the end of the slotted stem (see right side of Fig. 2.74). Two sliding pins are mounted in opposition in a hole bored through the key tip. The stem pins are limited in their travel and may protrude either into the central slot in the key stem or past the outer edges of the key blade. While the key is being inserted into the keyway, the stem pins are flush with the outer edges of the key stem, but as the key nears the end of its travel, the central slot meets an obstruction at the end of the keyway that forces the two

Figure 2.74: Laperche Diam 7-pin cylinder and dimple key with active elements in end of key blade.

Figure 2.75: Laperche Diam 7-pin core with two passive profile pins installed. Note the spooled lower pins.

stem pins radially outward. The outward motion of the stem pins causes them to raise two recessed blocking pins to the shear line. A straight key blade without the active element cannot reach the blocking pins.

2.5 Twin Inline

DOM iX-10

(DE) 10 pins in 2 rows (3–4)

The DOM Company of Brühl, Germany, has been manufacturing innovative security locks since 1936. DOM is now part of the Black and Decker group of companies. DOM's flagship product is the DOM iX series, recognizable by its horizontal keyway. This series includes a number of single inline and twin inline high-security locks. We focus in this section on the DOM iX-5, which has a single row of five pins, and the DOM iX-10, which has two staggered rows of five pins. These locks have the particular feature that the pins are not aligned on a plane that passes through the center of the cylinder plug. Before presenting the details of these unusual locks, it is worth mentioning some relevant development history.

The idea of using multiple rows of pins in a pin-tumbler cylinder lock is an old one (see the section on Kaba locks). Krühn [66] mentions a German patent from 1927 (DE 453,824) for a lock with a U-shaped key and two rows of pin-tumblers. A number of German manufacturers have since produced twin inline cylinder locks with bilateral keys, examples of which include the BKS Multipin from 1975 [66] and the DOM D, pictured in Fig. 2.76. The DOM D was patented in 1985 in Germany (DE 85 33 406) and in 1986 in the United States by H. P. Häuser and A. Stefanescu of DOM Sicherheitstechnik (US 4,787,225). It uses the same pin-tumbler design as the DOM iX-10 but with a V-shaped key formed by joining two blanks of identical section at their base. In terms of locks where two or more rows of pins have been positioned on one face of a flat key, the patent of A. Crepinsek (US 3,393,542) in Fig. 2.77 is highly relevant. Crepinsek's design, from 1965, is closely allied with the modern DOM iX-10 lock: it had 4 to 6 rows of pins with adjacent pin pairs linked by a spline joint.

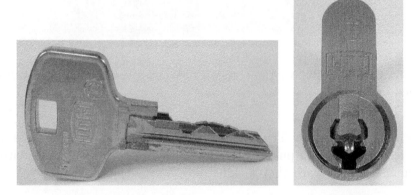

Figure 2.76: DOM D bilateral key and 10-pin cylinder.

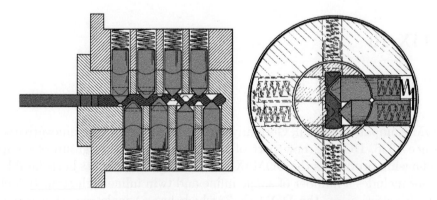

Figure 2.77: A. Crepinsek's 1965 patent (US 3,393,542).

Through the 1960s, DOM performed much of the necessary development work in realizing a high-security lock with off-axis pins. As can be appreciated from Fig. 2.77, once a row of pins is displaced so that it is not radially aligned with the plug, the shear line interface is not at right angles with the line of action of the pins, therefore the usual method of a perpendicular cut is not effective. This problem is not encountered in multiple inline designs like Kaba since the chamber axes are radially aligned. As evidenced by DOM's 1972 patent (US 3,731,507), this technical difficulty was overcome through clever design of the of the pin-tumblers. DOM pins are teardrop shaped in cross-section, so they cannot rotate in the pin chambers, and the ends that form the shear line are contoured to match the curvature of the plug at the point where they are chambered. The fabrication of these components clearly requires high-precision engineering.

The DOM iX-5 cylinder uses a single row of 5 pin-tumblers plus up to 10 vertical/lateral profile-control pins. DOM iX-10 cylinders have up to 10 profile-control pins in addition to the 10 pin-tumbler pairs; these are arranged as five vertical and five lateral control pins. Lateral control pins act on one edge of the key blade and the key must have corresponding dimple cuts to allow the pins room to retract and clear the shear line as the plug is turned. Vertical control pins act in the center of the underside of the key. The key itself, which is made of nickel silver for durability, is symmetric and three of its four sides are active simultaneously when it is inserted. In addition, there are many keyway profile variations to supplement access control. As the reader is no doubt aware of by now, the profile pins do not increase the lock's resistance to picking: rather, they are an adjunct to the fixed keyway broaching.

The DOM iX-10 KG lock, the high-security variant of the DOM iX-10, is pictured in Figs. 2.78–2.82. The plug and barrel can contain up to five hardened roller inserts and a ball to protect against drilling. Optionally, the active end of the lower pins may be cut away on both sides, leaving only a thin wedge that is operated by the indentations in the key blade. The lower pins come in five different lengths, so that the number of differs is theoretically $5^{10} = 9,765,625$ for the 10-pin version, not counting profiling

Figure 2.78: DOM iX-10 KG Europrofile and mortice cylinders.

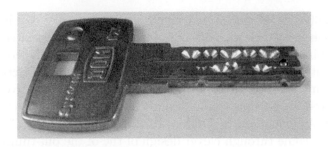

Figure 2.79: DOM iX-10 KG key with movable insert.

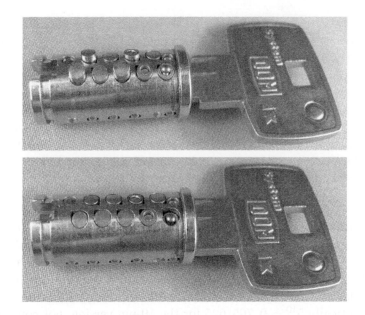

Figure 2.80: DOM iX-10 plug in operation. Pin in position 1 is construction-keyed to allow a rapid key change.

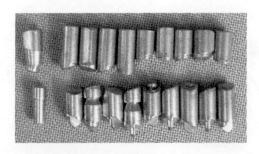

Figure 2.81: Collection of teardrop-shaped pins from DOM iX-10 lock. Blocking pin pair on left.

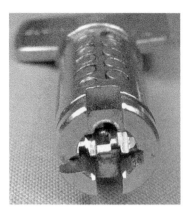

Figure 2.82: Rear view of DOM iX-10 plug showing operation of floating ball on blocking pin.

options. Driver pins come in several styles, including hardened pin cores, spooled drivers, and even special stacked-spool drivers of multidisc construction that can skew at several different levels to prevent picking.

The KG model utilizes the floating-ball system, introduced around 1980. There is a longitudinal channel in the key with a hole bored through the blade between the fourth and fifth pin positions. A steel ball is mounted in this hole and is free to move a limited distance either side of the key blade. As illustrated in Fig. 2.28, the plug contains a fixed obstruction, the ball-deflection pin, which slots into a channel in the key as it is inserted. Behind the ball-deflection pin is a movable blocking pin that must be displaced to enable plug rotation. The keyway opposite the blocking pin is enlarged slightly to form a ball-deflection chamber. The floating ball in the key blade is able to maneuver around the ball-deflection pin and into the ball-deflection chamber, finally repositioning itself in the keyway so as to displace the blocking pin. The mode of operation is similar to what happens when a feeler pick is used to raise a single pin. Clearly, a fixed-blade key cannot change its width to mimic this effect.

DOM also makes a split-bladed key for safe deposit boxes and evidence rooms, requiring both halves of the key to be inserted together to open the lock. The halves of the key would normally be carried by two different people (e.g., the bank manager and the client in the case of a safe deposit box). The design is covered in Wolter's 1974 patent (DE 2,433,918), which also describes profile-control warding for the cylinder and key.

The floating-ball system invented by DOM sparked a craze among European lock manufacturers, many of whom now offer key systems containing active or movable elements.[11] The active element may take the form of a ball, wheel, free pin, sprung pin, pair of pins, or a pivoting member embedded in the key blade. The rationale

---

[11] *bewegliches Element* in German or *élément mobile* in French.

is that the key is much harder to copy without authorization. However, in many cases the addition of active elements to the key does not add significantly to the manipulation resistance of the lock since it adds at most one further blocking pin to the mechanism.

Tover 2F30

(ES) 10 pins in 2 rows (2)

The Spanish lock manufacturer Tover, founded by Talleres Oliver in 1960, produces high-security mortice locks operated by double-bitted keys (see Chapter 5). Recently, Tover released a twin inline pin-tumbler lock cylinder with a horizontal keyway, similar to the DOM iX. The model in question, shown in Figs. 2.83–2.85, is the

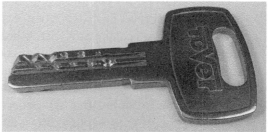

Figure 2.83: Tover 2F30 10-pin profile cylinder and reversible key.

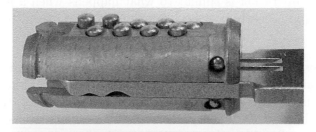
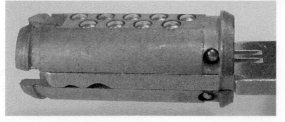

Figure 2.84: Tover 2F30 plug with key at different stages of insertion.

Figure 2.85: Pin pairs from the Tover 2F30, including spooled top and bottom pins.

Tover 2F30: a 9-pin lock with four pins in the left-hand side of the keyway and five on the right. Mushroom-shaped drivers and torpedo-shaped bottom pins may be included to provide an increased level of security. The bottom pins vary in length from 0.200″ to 0.265″, in five increments of 0.015″. With nine pins, ignoring pinning restrictions, the theoretical number of combinations is $5^9 = 1{,}953{,}125$.

The key is flat and reversible with two rows of staggered cuts occupying one half of the real estate on each side of the blade. Drill proofing is provided by balls inserted on both left and right sides above and below the keyway as well as by a hardened rod to cover the pin chambers. Unlike DOM, the pins have a circular cross-section. Since the plane of both rows of pins is slightly off-center, there is a need for a modest bevel on the faces of the pins at the shear line (see Fig. 2.85). Consequently, there is a slight loss in manipulation resistance, although this is offset by the presence of spooled pins.

Lockwood V7

(AU) 7 pins in 2 rows (3–4)

The V7 is a high-security 7-pin cylinder designed by the Master Locksmiths Association (Australia) and produced by Lockwood for commercial and public-sector applications. Pictures of the V7 lock appear in Figs. 2.86 and 2.87. It is based on a 1980 patent by Ogden Industries (AU 521,634 or US 4,320,638) introduced after the expiry of the patent on the previous Lockwood 6-pin security profile series. Since it has seven pins, it offers a large degree of flexibility for master-keying. The Lockwood product documentation specifies that the V7 is capable of 40 million usable differs, although this includes profile variation options. A 7-pin lock with 10 pin sizes and no MACS cannot have more than $10^7 = 10{,}000{,}000$ differs, and for the practical value of the MACS this would be reduced significantly (refer to Table 2.4).

As implied by its name, the seven pins are arranged in an alternating pattern on the two sides of a "V" in the keyway (see Fig 2.88). There are four pins situated at 15° to the right of 12 o'clock and three at 15° to the left. Correspondingly, the key has seven bittings that alternate on slopes of ±15° along the blade. Pin stack length is

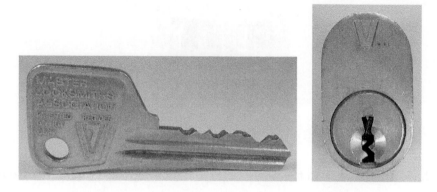

Figure 2.86: Lockwood V7 key and 7-pin cylinder.

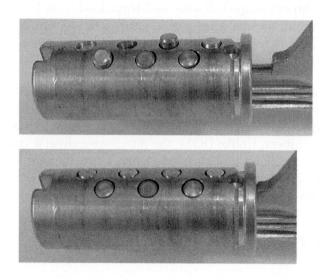

Figure 2.87: Lockwood V7 plug with key partway (top) and fully inserted (bottom).

compensated to achieve uniform spring tension across the seven pin-tumbler pairs. This feature prevents the detection of pin length by feeling the driver spring tension. The keyway broaching is designed to make it hard to insert manipulation tools. The plug also contains frontal drill-resistant ball bearings. Key blanks for the V7 are restricted: Lockwood will only supply key copies against a signed authorization from the registered owner.

Head

(CN) 9 pins in 2 rows (1–2)

As well as the usual row of five pin-tumblers at the top of the keyway, the Head cylinder in Fig. 2.89 has an additional four active profile pins projecting into the

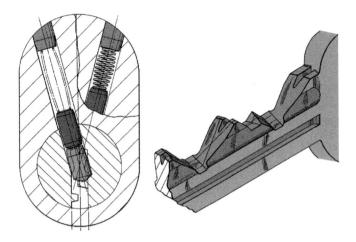

Figure 2.88: G. F. Dunphy and D. J. Newman's design of the V7 cylinder (US patent 4,320,638).

Figure 2.89: Keyway and key from a $(4 + 5)$-pin Head padlock.

keyway at 9 o'clock. The key is single-sided with bittings in the top of the blade and dimples in the left side that address the profile pins. Despite the presence of side pins, the lock is not difficult to pick due to the relatively low precision of the mechanism.

2.6 Inline with Active Profile Pins

Lockwood 7

(AU) 7-pin + blocking pin (2–3)

Lockwood produced a 7-pin inline cylinder, the "Lockwood 7," incorporating an antiraking device on the seventh pin-tumbler. The design was intended for use with a key profile hierarchy presented in Australian patent 258,614 (1963) by J. P. Hynes, T. L. Rawlings, B. Tescher, and G. F. Dunphy, assignees to Ogden Industries. The rationale for this mechanism is manipulation resistance rather than profile control, but we include a description in this section since the extra pin is an early form of active profile control.

The lock is illustrated in Figs. 2.90–2.92. The antiraking device is in the form of a spring-loaded hollow cap, resembling a bowler hat, fitted in a counter-sunk boring just past the seventh pin position at about 5 o'clock in the plug (see Fig. 2.91). The driver spring for the device is very stiff compared with a normal driver spring. As a result, when the plug is in the locked position, the device impinges on the tip of the seventh pin and overraises it. This causes the seventh pin to act as a blocking pin. Because the device is positioned right at the back of the keyway, it renders the manipulation of the lock considerably more difficult. The plug incorporates three ball bearings at the front at 5, 7, and 12 o'clock to guard against attack by drilling.

The keys for this lock are distinguished not only by their seven cuts, but also by the presence of a pick-up slope on the right-hand side at the tip of the key. As the key is inserted, its bittings contact the seven pin-tumblers in the usual manner. The pick-up slope catches the antipicking device and pushes it back into the side-boring of the plug. The seventh pin is then freed from the action of the antipicking device

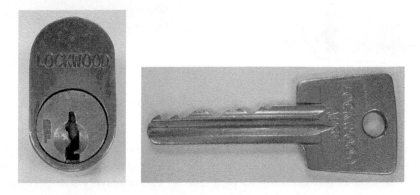

Figure 2.90: Lockwood 7 oval 570-series cylinder and key.

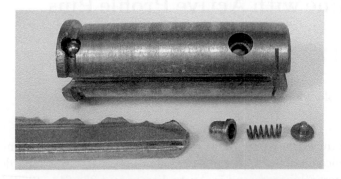

Figure 2.91: Lockwood 7 plug with antipicking device on seventh pin.

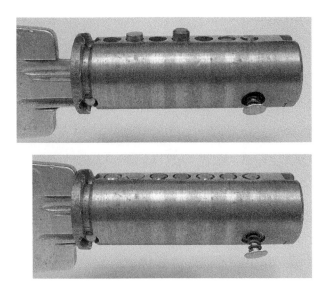

Figure 2.92: Lockwood 7 with key partially and fully inserted.

and can be brought to the shear line by the seventh bitting of the key. The device is stopped by a thumbtack-shaped pin the diameter of which is larger than that of the pin chambers; thus as the plug rotates, the device is prevented from obstructing the seventh driver.

ABUS TS 5000

(DE) 6-pin + 5 profile pin (4)
(IT) **CISA TSP** (equivalent)

The ABUS TS 5000 ("Top security") cylinder, pictured in Fig. 2.93, appears to be a conventional pin-tumbler cylinder apart from the rather oddly shaped keyway. However, the lock has several clever security features. The key profile is overhanging, that is, it actually turns back on itself at one point (see Fig. 2.19 in the chapter introduction). Further inspection reveals that there are up to five profile pins at 7 o'clock and up to five profile balls at 5 o'clock along the sides of the keyway (see Fig. 2.94). The cylinder has internal grooves at 5, 8, and 9 o'clock that accommodate the profile pins and balls in the locked position.

The profile balls are passive and will prevent rotation unless a matching profile bitting in the key is encountered. The profile pins may be either passive or active. A passive profile pin drives a captive ball that sits in a recess at 9 o'clock. In contrast, the active profile pins are sprung from inside the plug and are similar in shape to the pawn in a chess set (see Fig. 2.95 and also Fig. 2.27 in the chapter introduction). The base of the active profile pin (the bottom of the "pawn") normally blocks a hidden ball at

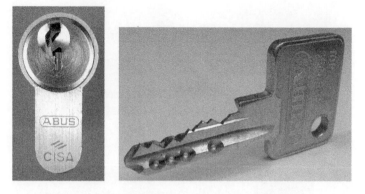

Figure 2.93: ABUS TS 5000 profile cylinder and key with overhang in profile.

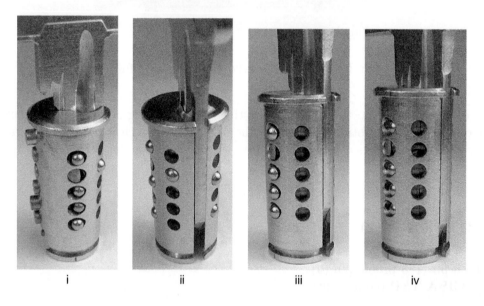

Figure 2.94: ABUS TS 5000 plug detail. (Left to right): (i) Passive profile balls at 9 o'clock; (ii) active profile ball at 7 o'clock and passive profile balls at 5 o'clock; (iii) passive profile balls at 9 o'clock in locked position; (iv) key fully inserted to retract passive profile balls and create space for active profile ball in position 2.

8 o'clock, causing it to protrude into the corresponding cylinder channel. When an active pin is depressed, which happens when there is *no* profile cut on the key, the skinny part of the pin allows room for the profile ball to retract and clear the channel in the plug at 8 o'clock. So in summary: a passive profile pin or ball requires a bitting on the key blade, whereas an active profile pin requires no cut in the key blade.

An additional security feature is the blocking pin: a hardened pin of near maximal length that requires next to no lifting to reach the shear line of the plug. That is, it is effectively a dead-lift pin. Due to the shape of the keyway, it is virtually impossible to manipulate the pins behind the blocking pin without actuating it. Spooled drivers and hardened inserts further enhance the security of the cylinder.

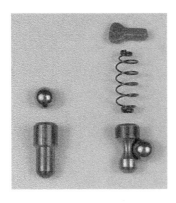 

Figure 2.95: (Left) Passive and active profile pins and balls (active on right). (Right) Abus TS 5000 key showing positioning of profile pins by key blade.

ISEO R11

(IT) 6-pin + 11 profile pin (3)

ISEO Serrature S.p.A. is one of Italy's leading security lock manufacturers. Founded in 1969, it is now part of ISEO Holding, a group of companies providing integrated security solutions including high-security locks.

The ISEO product line also includes a number of horizontal-keyway dimple key cylinders, one of which, the R11, is pictured in Figs. 2.96–2.98. This lock deserves mentioning due to its innovative profile pin design. First, we focus on the key: this is symmetric and has dimple bittings on all four sides. As with other horizontal-keyway designs, there is a ramp on each side that picks up the top pins.

The pins are arranged in three rows: top, side, and bottom. There are six top pins, and up to five bottom pins and six side pins. The top pins are conventional with a spring, driver, and lower pin. The driver pins may be spooled. The side pins are profile-control pins and a side-bitting is only required in the key blade when a profile pin is present. The cylinder has longitudinal grooves at 3 o'clock and 6 o'clock to accommodate the side and bottom pins.

The innovation is in respect of the bottom pin design. There are two bitting depths. The deeper one of these two is matched by a passive double-pointed "male" profile pin. The shallow bottom bitting accepts a hollow "female" pin that contains a small spring and a secondary driver pin (see Fig. 2.98). The driver pin is spooled and has its point facing radially outwards. If the female pin encounters a shallow profile bitting, then its driver pin will be compressed and it will act in much the same way as a male pin, not blocking rotation. If on the other hand the female pin encounters a deeper profile bitting, then the top spool of the driver is permitted to extend under spring pressure so that it blocks on the edge of the hollow female pin as the plug

Figure 2.96: ISEO R11 horizontal-keyway cylinder and symmetric dimple key.

Figure 2.97: Two views of ISEO R11 plug.

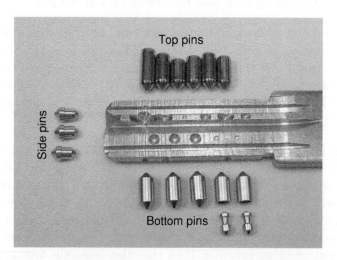

Figure 2.98: ISEO R11 key and pin set: top and bottom pins act on wide faces of key.

begins to turn. Thus in this case the driver pin cannot retract into the hollow part of the profile pin, which blocks rotation.

The plug and cylinder also contain hardened inserts to resist drilling. The R11 has been upgraded to the ISR100—an 11-pin lock with an active element in the form of a floating ball embedded in the key. This system is certified to EN 1303 and supports in excess of 600,000 differs.

Vachette 2000

(FR) 5-pin + 8 profile pin (3–4)

The Vachette 2000 lock in Figs. 2.99 and 2.100 looks deceptively like a conventional five pin-tumbler cylinder. However, in addition to the quite restrictive broaching of the keyway, there are between five and eight cleverly concealed, active profile pins: up to four pins on the lower left and four on the lower right side of the keyway.

Figure 2.99: (Left) Vachette 2000 lock cylinder. (Right) Plug contains five ordinary and five active profile pins.

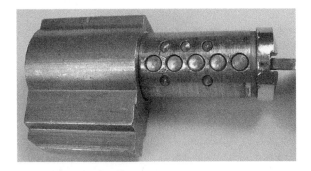

Figure 2.100: Vachette 2000 plug with key inserted to align pins.

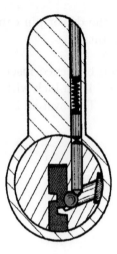

Figure 2.101: Operation of the Vachette 2000 active profile pin from 1973 Swiss patent 578,105.

The design, shown in Fig. 2.101, was patented in Great Britain in 1972 (UK 1,408,340) and in Switzerland (CH 578,105), although the inventor was not named on the patent.

The profile pins are ball driven and are of a smaller diameter than the main pin-tumblers. The bore for each profile pin is parallel to the conventional pins, but off the main axis of the plug and in line with the midpoints of the five pin-tumbler bores. Profile pins consist of a profile rod and driver ball, with the ball barely visible in the keyway due to the presence of the main pins. The rod has tapered ends and rests atop the ball, being raised when the driver ball is displaced by the profile dimples in the key. The rods must be raised by varying degrees to bring their corresponding driver pins to the shear line. The off-axis driver pins are cylindrical with conical ends.

Since the profile rods are not centrally located in the plug, only rods on one side will be active depending on the direction of opening (the other rods will slide under their drivers as the plug is turned, as long as they are not overraised). This fact is a slight aid in picking the lock, but, despite this, it is a difficult job picking the four profile pins on one side and the five top pins—there is very little room to maneuver a picking tool. There is also a hardened insert in the front of the plug to deflect drill bits.

MLA Binary Plus

(AU) 6-pin + 6 profile pin (4)

The Binary Plus system, shown in Figs. 2.102–2.104, is an enhancement of the basic 6-pin cylinder by the Master Locksmiths Association (Australia). Previous MLA systems had a registered key blank requiring authorization for key copying. However,

Figure 2.102: MLA Binary Plus 6-pin cylinder and key. Profile pins are concealed in narrow slot at 9 o'clock.

Figure 2.103: MLA Binary Plus plug: location of side-trap springs (left); antipick pins overraised and caught by side-trap springs (middle & right).

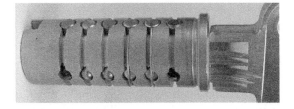

Figure 2.104: Operation of Binary Plus cylinder: key releases antipicking pins from side-trap springs.

the keyway and cylinder design offered only limited protection against manipulation and unauthorized key duplication. The Binary Plus cylinder incorporates up to six profile pins and "side-trap" springs offering much greater resistance to manipulation and illicit key duplication.

A row of profile pins is present at 9 o'clock in the core. Profile pins may be either short or long. The lower cylinder pins are either normal or torpedo-shaped, with a groove milled around their periphery. Each torpedo-shaped pin has a corresponding side-trap spring located in a circular groove milled into the plug and anchored at 7 o'clock. Each trap spring is C-shaped and of such a length that its free end just protrudes into the corresponding pin chamber. If the pins are overlifted, they become trapped by the side springs, in much the same way as in a Chubb detector lock (refer to Chapter 5).

Long profile pins are passive and function in the normal manner: a bitting is required on the side of the key blade to prevent the profile pin from engaging a longitudinal channel in the cylinder housing. Whereas long profile pins may be used in tandem with the conventional pin-tumblers, short profile pins are used with the torpedo-shaped pin-tumblers. The function of the short profile pins is to unset the side-trap springs by forcing them radially outward to the edge of the groove in the plug. This action clears the lower pin chambers and allows the top bittings on the key to raise the pins pairs to their respective shear lines. The trap springs do not cause the pins to become set at their shear line when the key is withdrawn since the profiling on the key blade, which is uncut at the end of the key, pushes the springs outward and allows the driver pins to return to their normal rest positions.

With two different sizes of profile pin, a lateral ridge of variable height is required on the key blade. The long profile pins give extra degrees of freedom for key control and master-keying, while the short profile pins add manipulation resistance. The presence of a ridge also ensures that a thinner key blank will not activate the short profile pins, rendering key duplication more difficult.

Schlage Everest

(US) 6–7 pin + profile pin (2–3)

The Schlage Everest is produced in 6-pin and 7-pin varieties for mortice and interchangeable-core cylinders. The lock is illustrated in Figs. 2.105–2.107. The keyway features an undercut groove with a single "check pin" chambered in a vertical bore to the left of the conventional pin chambers at the rear of the plug. Because the keyway contains an overhanging section, the key blade must have a corresponding raised milling in order to contact the check pin. This undercut milling is in a different plane to the standard side-millings in the key (as in the ABUS TS 5000); that is, it cannot be reproduced by a milling machine without tilting the key.

Figure 2.105: Schlage Everest mortice cylinder. Blocking pin visible on underside of cylinder (right).

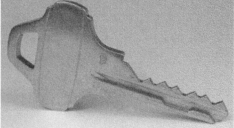

Figure 2.106: Front and rear views of Schlage Everest key: undercut groove visible on right.

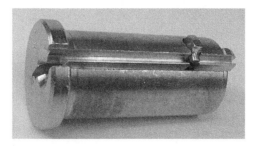

Figure 2.107: Underside of Schlage Everest plug showing location of blocking pin.

A standard side-milled Schlage key with the correct top bittings will not lift the check pin, and thus the plug will remain locked. Only an Everest key with the undercut milling in the correct place and the correct top bittings will raise the check pin and operate the lock. The system therefore has a high degree of resistance to unauthorized key duplication. The Everest design has been integrated with the Schlage Primus side-bar lock, which we cover in Chapter 4; the resulting combination is called the Everest-Primus.

Winkhaus Titan

(DE) 6-pin + 10 profile pin (4)

Despite the rather small amount of unused space in the Winkhaus VS series, the Winkhaus Titan (Figs. 2.108–2.110) is proof that some further security features can still be added. It was released in 1987 by Winkhaus Sicherheits-Systeme GmbH

Figure 2.108: Winkhaus Titan key and profile cylinder.

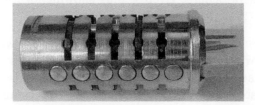

Figure 2.109: Winkhaus Titan core with key partially inserted (top & middle) showing overlifting of rocker arms and profile pins. (Bottom) Key fully inserted.

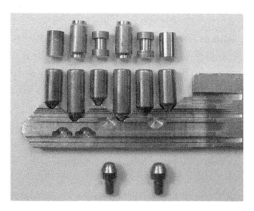

Figure 2.110: Winkhaus pins and tamper-proof/drill-resistant drivers.

(Münster). The following material is incremental to the Winkhaus VS description presented earlier in this chapter.

The Winkhaus Titan contains, in addition to the usual six pins and 10 profile-control pins of the VS series, up to 10 lateral profile-control pins (see Fig. 2.109). These independently functioning profile pins are arranged in two rows of five along the bottom sides of the keyway.

Each lower profile pin drives a spring-loaded rocker arm. The rocker arms are deployed in two rows of five and pivot on axles placed longitudinally at about 2 and 10 o'clock in the plug. The outer edges of the rocker arms are barbed. The right-hand rocker arms can snare the channels at 1 and 5 o'clock. Similarly, the left-hand rockers can snare the channels at 7 and 11 o'clock.

In their natural positions the rocker arms, which are sprung from the top, protrude into the channels at 1 and 11 o'clock. An unbitted key will overraise the rockers so that they stick in the channels at 5 and 7 o'clock. Thus the key must have the correct set of lower profile bittings to set the rocker arms, as well as the correct midprofile bittings and pin-tumbler cuts.

The Winkhaus VS, VS6, and Titan may also incorporate an electronic key-top transponder for additional security. This electronic upgrade, called "Blue Chip," was released in 1999. Further details on the Winkhaus Titan appear in [37], which also covers a number of other European high security locks.

2.7 Cruciform

Cruciform locks, from the Latin word "crux" meaning "cross," have a four-sided keyway. They are used for various applications including padlocks, utility locks, cabinet locks, and light-duty commercial door locks. A number of different examples of cruciform locks and keys are given in Figs. 2.111–2.114. Of the four channels comprising the keyway, one is usually wider than the others. This provides positive location for insertion of the key, which is generally nonsymmetric. The key is stopped either by its tip or by shouldering around the stem. In some models, there are no pins in the key locating channel. Locks made by IKON, Moreaux, and a number of other manufacturers have between three and four rows of pins with up to four pins per row, with a key bitted on three or four sides as appropriate. Some cruciform keys are made for two-sided operation and therefore have a symmetrical bitting pattern with respect to the midplane of the key. The design and operation of a cruciform lock are illustrated in Fig. 2.115 from a 1959 UK patent by Pearson.

Figure 2.111: Cruciform keys: BKS-Yale 4 × 4 pin (left); Moreaux axially symmetric 4 × 2 pin (right).

Figure 2.112: Twelve-pin padlock taking a four-sided key.

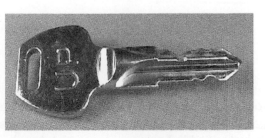

Figure 2.113: Four-sided utility lock and key.

Figure 2.114: A cruciform lock with 10 pins.

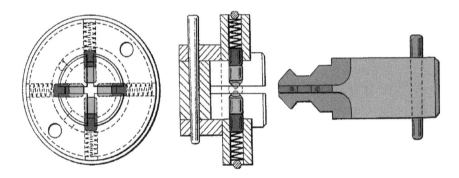

Figure 2.115: Cruciform lock and key design from a 1959 patent by T. F. Pearson (UK 940,778).

It is not uncommon for a cruciform lock to have a pin count that is not a multiple of three or four since the pins may be arranged, as illustrated in Fig. 2.114, in rows with unequal numbers of pins. A cruciform lock typically has from 8 to 18 pins, but the number of combinations is limited by the width of the key blade, which may only support two or three depths of cut. Since they are relatively expensive and have bulky keys, cruciform locks are a poor choice for large systems, which means that they would not normally be master-keyed. They are also somewhat difficult to reassemble.

Despite a somewhat formidable appearance, with ordinary driver pins installed, these locks are not especially difficult to pick open provided they can be effectively tensioned. Sometimes they are used in safes, in which case the lock is set a long way back and the key has an extended stem; this greatly increases the manipulation resistance against manual lockpicking.

Helason

(AT) 16 pins in 4 rows (3)

Helason is an established Austrian company based in Vienna whose product range includes a number of security door locks. Traditionally, the locks were built around a 16-pin cylinder. The sample that we have included in this section is from a RFZ 5116 deadlatch, pictured in Figs. 2.116–2.118. This particular model has been superseded, and Helason is now producing locks with dimple key cylinders.

Figure 2.116: Helason RFZ 5116 deadlatch with 16-pin mechanism.

Figure 2.117: Helason key and cylinder.

Figure 2.118: (Left) Helason cylinder with cover removed. (Middle & right) Core with key partially and fully inserted.

The RFZ 5116 cylinder comprises a chrome-steel cover over a die-cast zinc body that houses a plug of the same material. A steel drill-pin is mounted at the rear of the cylinder that accepts a hollow (pipe) key. The plug contains four rows of four pins arranged symmetrically in a round keyway. Each ring of four pins is at the same offset from the front of the cylinder and fills the keyway to within 30 thousandths of an inch from the drill-pin. The keyway also has a shallow channel to guide a locating fin in the key. The key, which is made from a length of round brass stock, has four rows of flat-bottomed V-cuts in the stem, similar to a Renault TS car key (which has cuts on two sides only).

Four depths of cut are used, with pin lengths ranging from 0.160″ to 0.205″. The maximum depth of cut is determined by the inner diameter of the pipe key. Due to the large spacing of cuts, there are no restrictions on the adjacent cut specification. The resulting number of key combinations is therefore huge (4^{16} or 4,294,967,296). Although the tolerances on the pin-tumblers are not exceedingly tight, the lock offers a high level of pick resistance due to the limited keyway access around the drill-pin.

2.8   Multiple Inline

Kaba, KESO

(CH) up to 26 pins in 2–8 rows + blocking pin (3–4)

The Kaba AG Company, which originated in Switzerland, is now a multinational corporation and, along with ASSA Abloy, one of the largest in the global security industry. The Kaba Group includes Silca, Ilco, Elzett, Gege, Unican, Mauer, and Mas-Hamilton. Kaba's product range includes a number of multiple inline pin-tumbler locks with up to 26 pins arranged in a radial formation having between two and eight rows. The lower pins protrude about half the width of the keyway. The distinctive feature of these locks is that the key, which is reversible, contains a multiplicity of "dimples" or small indentations. For this reason the locks are often referred to as dimple-key locks. They typically have a high manipulation resistance and require high-precision machining for key duplication. Other brands of locks in this grouping are produced by KESO (Switzerland), BKS (Germany), JPM (France), Lips (Netherlands), Lori (US), Sargent (US), and Showa (Japan). ASSA Abloy acquired Lips in 2000 and KESO in 2001.

Like many other commercially successful locks, the Kaba lock has a long design evolution that can be traced through its patent history. One of the earliest references to a lock with a dimple-type key is a 1913 French patent by C. Renaux (US 1,224,021). This described a 3-pin padlock with ball-driven pin-tumblers, operated by a flat key with holes of varying depths drilled into one of its faces.

The first recognizable Kaba-type lock with a reversible key was described in a 1934 patent (UK 421,715) from the Swiss firm A. G. Bauer, founded in Switzerland in 1862. This was an 8-pin lock where the pins were arranged in two opposing rows of four at either side of the vertical keyway (we denote this as a $(4+4)$-pin lock). A 1946 Swiss patent by O. Rüegg (CH 252,097) presented a circular-keyway lock with four rows of pins, closely resembling the Helasaon lock. Bauer's design of the $(4+4)$-pin lock was continued in its 1947 Swiss patent (CH 260,517), which described a key design with elongated millings. The familiar form of the Kaba dimple key appears in this patent.

A parallel thread in the development of multiple inline pin-tumbler locks was started in a series of patents by E. Keller of Zurich, who founded KESO in 1963. Keller's patents, from 1960 (CH 344,637) and 1963 (CH 372,947), specify $(4+4)$-pin and $(4+4+4)$-pin locks with, respectively, two and three rows of pins. The six-faceted form of the KESO keyway is immediately recognizable in these patents (see Fig. 2.119). The technology was licensed to Sargent in the United States in 1965 and soon afterward to Lips in the Netherlands. The idea of offsetting the key bittings (or borings, as they are more accurately described) to form left- and right-handed keys is depicted in these patents. This idea has important ramifications for master-keying, where both left- and right-handed lock cylinders exist in the same MK system. Bauer also contributed to the KESO development (see 1966 US patent 3,303,677), going on to develop their own three-row lock: the $(5+5+5)$-pin Gemini. It was around this time also that the idea of using nonradially aligned rows of pins with a curved shear line was mooted (as in Crepinsek's 1965 patent and the DOM iX-10 lock).

The mid-to-late 1960s also saw the development of the four-row radial pin-tumbler lock. The original form of what is now the Kaba 20 lock appeared in a 1965 patent by Bauer (DE 1,553,294, issued in 1969). A key from a Kaba 20 lock appears in Fig. 2.120. Two of the four rows of pins are set at 3 o'clock and 9 o'clock, with two

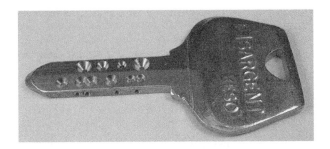

Figure 2.119: Sargent Keso dimple key.

Figure 2.120: Reversible key from a Kaba 20 cylinder lock.

additional rows at a 45-degree offset from these toward the bottom of the keyway. This design was later modified by the addition of a fifth row of top pins (at 12 o'clock) in the Kaba Star. The modern form of the Kaba Quattro lock is encapsulated in the 1980 patent by Bauer Kaba A.G. (US 4,343,166), which fully utilizes the sides and (thin) edges of the key blade. This development allowed a smaller, lower cost dimple-key lock to be produced.

Owing to the prevalence of the dimple-key lock and its significant differences from conventional pin-tumbler locks, as well as describing the operating principles, we also present some more detailed technical information, accounting for the fact that each row of pins in a Kaba lock may be offset forward or backward to create a multiplicity of different bitting geometries. We subsequently cover five types of dimple locks:

1. Kaba Gemini (illustrated by the KESO/JPM 2002).

2. KESO 1000 S.

3. KESO 2000 Omega.

4. Kaba Quattro.

5. Kaba ExperT.

This is followed by a brief section on some variants of the Kaba principle (Vario, Nova, and Elolegic).

Kaba Gemini

The Kaba Gemini and KESO cylinders contain nominally 15 miniature pins arranged in three rows: one at the top of the keyway and two opposing rows symmetrically placed on either side in the upper half of the keyway. KESO locks produced by JPM (as pictured in Figs. 2.121–2.123) are made in a Europrofile format with five pins in the top and both side rows. The driver pins and springs are mounted in detachable housings for ease of assembly. KESO side pins are at ±90 degrees to the top pins. By comparison, Gemini side pins are at a 15-degree angle down from the

Figure 2.121: KESO 2002 Gemini 3 × 5 pin Europrofile cylinder and key.

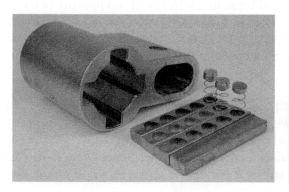

Figure 2.122: (Left) KESO 2002 cylinder body with driver inserts. (Right) Core with pins.

Figure 2.123: KESO 2002 core with key at various stages of insertion.

normal to the key face [95]. Top pins are actuated by borings on the edge of the key blade, while side pins are actuated by the borings in the flat faces of the key. When the key is inserted, only the borings in the top edge and upper part of the key faces come into contact with the locking pins in the plug. The remaining borings are a symmetric copy so that the key functions in either orientation.

Kaba Gemini locks use various bore patterns. The top row of pins can be offset to the front (odd) or rear (even) and accommodates five pins, using every second position out of the 10 possible ones. There are 11 bore positions for side pins. In a given bore pattern, there are five positions on one side and six on the other. Top pins are available in three sizes, while side pins are supplied in four different sizes with an increment of 0.35 mm. A fourth depth of top pin is used in MK applications [95]. As an example, a Gemini cylinder could be supplied with the following bore pattern: 2, 4, 6, 8, 10 (top); 1, 3, 5, 7, 9 (left side); 1, 3, 5, 7, 9, 11 (right side).

Although only every second bore position can be used in a given cylinder, keys may be bitted in all the bore locations, like the KESO 1000 S key in Fig. 2.124. A single key may therefore be used to address multiple locks with different bore patterns, which is a great asset for master-keying. With 15 pin positions, four side depths, three top depths, and no MACS or other pinning constraints, the number of system permutations is very large: $4^{10} \times 3^5 = 254,803,968$ (more than a quarter of a billion).

Master-keying is accomplished by replacing some of the lower pins by blind pins that are already at the shear line and so do not need to be lifted. Blind pins do not extend into the keyway. The use of blind pins, while convenient in MK systems, slightly compromises the security of the lock by reducing the number of active pins and also reduces the available number of combinations.

KESO 1000 S

The KESO 1000 S, shown in Fig. 2.124, is an enhancement of the KESO/Kaba lock with three rows of five pins. In addition to the standard conical-tip lower pins, a number of profile-control pins may be added. The function of the control pins is

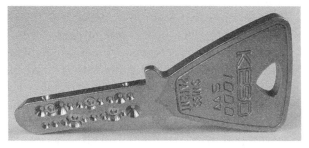

Figure 2.124: KESO 1000 S cylinder and key with both left- and right-handed bore patterns.

Figure 2.125: KESO 1000 S pins and drivers: standard on top, profile-control on bottom.

different to conventional passive profile-control pins (in a DOM iX lock, for instance). The key end of a Kaba control pin (see Fig. 2.125) has a flat outer portion and a raised inner portion. The presence of these control surfaces is twofold. First, the overall length of the control pin measured from the inner tip must be compatible with the depth of the boring on the key blade in order to place the top surface of the pin at the shear line. Second, the bore in the key must be of a larger-than-normal diameter to accommodate the flat outer portion of the control pin. These two features, taken together, validate the profile and depth of the bores on the key blade. A key with the same bore pattern and depths of cut will not operate the lock unless it also has the indented dimples required for the control pins. Blank keys can be supplied with specified control pin borings. This provides an effective means of copy protection since the control pin borings cannot be made by standard key-copying equipment. The idea is further described in a 1994 patent by Kleinhaeny of Bauer Kaba A.G. (US 5,438,857).

KESO 2000 Omega

In addition to the arrangement of 15 mini pin-tumblers in three rows of five, the KESO 2000 Omega (Figs. 2.126–2.130) incorporates a further security feature: the presence of twin active elements in the edges of the key blade. The elements are in the form of two opposing pins with an intervening spring. Normally, the pin ends are flush with the edges of the key, but they may also be pushed inward against the spring. (This arrangement is symmetric, so that the key may still be inserted either way round.) The plug contains a special conical blocking pin at 12 o'clock at the rear of the keyway with a downward protruding spindle (shown in Fig. 2.128). The blocking pin is spring-biased. If a key with a fixed blade is inserted, even with all pins correctly raised, the blocking pin will not allow the plug to rotate since its head remains wedged against a recess in the cylinder. On the other hand a key with a hole that admits the spindle of the blocking pin will turn the plug, but the blade of the key will be trapped by the spindle. Thus it is necessary for the key to have an active element that allows the spindle to enter the key stem when the plug is turned, but also pushes the blocking pin upward so that the spindle does not bind the key blade.

Figure 2.126: KESO 2000 Omega 15-pin cylinder and dimple key with twin active elements.

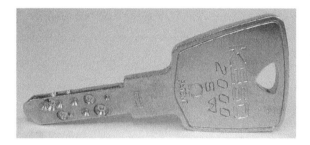

Figure 2.127: KESO 2000 S Omega key with both indented dimples and active elements.

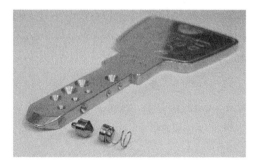

Figure 2.128: Close-up of a KESO Omega hidden pin with blocking pin pair.

The presence of an active element in the key stem renders unauthorized copying even more difficult.

The idea of utilizing profile-control pins actuated by the edge of the key blade was the subject of Bauer Kaba's 1990 patent by H. Kuster (US 5,101,648), although this patent did not consider active elements in the key. The active elements from the Omega and indented dimple cuts from the KESO S are able to be combined, as

Figure 2.129: Operation of KESO 2000 Omega plug showing action of blocking pin.

Figure 2.130: Incorrect lifting of active element results in engagement of top-hat pin in groove.

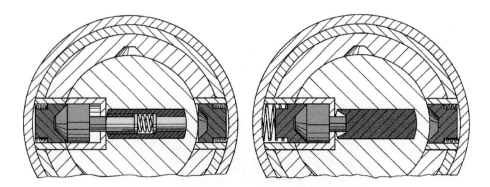

Figure 2.131: US patent 5,457,974 by E. Keller for KESO Omega lock: key without active element is trapped by top-hat pin (right).

evidenced in the KESO 2000 S Omega model (Fig. 2.127). The design of the active element key and the blocking pin is the subject of Keller's 1994 US patent 5,457,974, diagrams from which appear in Fig. 2.131.

Kaba Quattro

The Kaba Quattro lock, shown in Figs. 2.132–2.134, is a small-format cylinder that can be adapted to retrofit most existing brands of lock using rim or mortice cylinders

Figure 2.132: Kaba Quattro 22-pin cylinder and key.

Figure 2.133: (Left) Kaba Quattro plug and core with outer sleeve removed. (Right) Core with pins loaded.

Figure 2.134: Kaba Quattro tumbler pins (left) and counter-pins (right).

as well as knob-sets. A large range of adaptors and tail-pieces are made for this purpose. The Quattro is characterized by its four rows of pin-tumblers, arranged in an "X" around the top part of a rectangular keyway with shallow side wards. The lock cylinder comprises, in order of decreasing diameter: an outer sleeve, a brass core, and a plug. The core is fixed, while the plug is rotatable. The function of the sleeve is to retain the driver pins and springs and to allow easy removal of the core assembly for servicing.

The nomenclature used in this section for the Quattro cylinder is outlined in Figs. 2.135 and 2.136. The lower pins (or "tumbler pins") are flanged at the shear line end so that they do not push through into the keyway under the tension of the driver springs (see Fig. 2.134). Since the pin stacks are radially disposed and the core is of small diameter, the drivers (or "counter-pins") are made from hollow caps with an internal spring. The drivers are also compensated; that is, longer pins have shorter drivers and vice versa.

Tumbler pins are supplied in four sizes for side pins and three sizes for corner pins. As in Gemini locks, the depth increment is 0.35 mm (0.0138″). Sizes are numbered from 1 to 4, with 1 corresponding to the longest pin (unlike conventional pin-tumbler locks). The tumbler pins are made of nickel silver for durability, although hardened steel mushroom pins may also be substituted for the longer side pins. This gives a degree of drill protection to the cylinder. Pinning the lock requires a special loading jig that retains the counter-pins and driver springs until the sleeve is fitted.

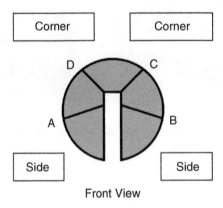

Figure 2.135: Naming convention for pin chamber rows in Kaba Quattro and ExperT locks.

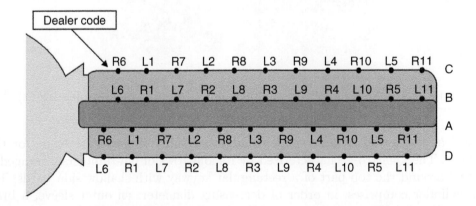

Figure 2.136: Naming convention for bore positions on Kaba Quattro and ExperT keys. R is for right-handed and L is for left-handed cylinders. Dealer code (in AUS & EU) is along row B for Kaba ExperT locks and row C for Quattro.

For each row of pins 11 bore positions are available. Cylinders may be constructed to have either left- or right-handed borings for the pin chambers in any particular row, but not both. This amounts to saying that the spacing between chambers must be two bore positions. Thus either the six odd-numbered positions $(1, 3, 5, 7, 9, 11)$ or the five even-numbered positions $(2, 4, 6, 8, 10)$ may contain active pins. On the other hand, keys can be made with both left and right borings in any row to operate both left- and right-bored cylinders. Since there is not enough space for adding master-keying pins, it is necessary to omit some tumbler pins to allow a cylinder to be operated by more than one different key. As with Kaba Gemini, blind pins are used to block the holes left by omitted tumbler pins.

Handedness applies to each of the four rows, so it follows that there should be 2^4 or 16 different possible cylinder bore geometries or bore patterns. For instance, one possibility is A—odd, B—odd, C—even, D—even. This is assigned Quattro bore pattern QR and has $6 + 6 + 5 + 5 = 22$ pins in total. The number of pins therefore varies from 20 $(= 5 + 5 + 5 + 5)$ to 24 $(= 6 + 6 + 6 + 6)$, depending on the bore pattern. In particular, row C is reserved for the dealer code or "dealer perm." Blank keys are supplied by the factory with this row already cut. The local locksmith then has the remaining three rows to complete the key combination according to a "permutation code." This method ensures that there is no accidental key interchange between local suppliers. Ignoring any pinning constraints, the Quattro system admits around $4^{11} \times 3^5 = 1,019,215,872$ key combinations per bore pattern and dealer perm, which is over one billion codes. In practice, the factory may reserve some of the other bore positions (e.g., row D) to exercise further control over the distribution of key blanks. This still leaves approximately $4^6 \times 3^5 = 995,328$ key changes per dealer and factory permutation for each bore pattern.

Both Kaba Gemini and Quattro locks can be picked using appropriately fashioned tools, albeit with considerable difficulty, but very light tension is required to prevent pins from binding at the wrong height. Once picked, care must be taken not to allow the driver pins, whose diameter is less than the width of the keyway, to spring out of their chambers as the plug is turned. Users of Kaba locks need not worry, however, as picking a lock with such tight tolerances is very time consuming and therefore not a practical option for burglars.

Kaba ExperT

The expiry of the Kaba Quattro patent in 2004 was countered by the release of the Kaba ExperT system, ensuring continued copyright protection of the highly successful Kaba Quattro. It can be appreciated from Figs. 2.137–2.141 that the Kaba ExperT is closely based on the original Quattro design. Thus while minimizing the burden of change on dealers and locksmiths, the design is sufficiently novel to acquire a new patent.

The new system, which was released on a worldwide basis, utilizes the same diameter plug as Quattro and the same bore positions. There are also a number of differences,

Figure 2.137: (Left) Kaba ExperT 22-pin oval cylinder. (Right) Dimple keys for Kaba ExperT (front) and Quattro (back).

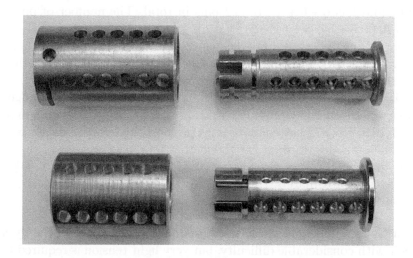

Figure 2.138: Comparison of core and plug from Kaba Quattro (top) and Kaba ExperT (bottom).

the most significant of which is the patented blocking-pin mechanism. The core is also of a slightly smaller diameter than the Quattro. As with other Kaba systems, keys are registered to the owner, and proof of registration is required to authorize the duplication of keys. Pin lengths for the ExperT are the same as the Quattro with identical corner pins, but having a different side-pin design (see Fig. 2.139). ExperT side pins have a flat section with a pointy tip at the center. Top and side tumbler pins are made of nickel silver. Although the keyway profile is rectangular with no side wards, the key blanks are compatible with the Quattro profile so that they can be made to operate the older-style cylinders. The dealer code has been moved from row C to row B (see Fig. 2.136). All 16 possible bore patterns may be used.

The function of the blocking pin is described next. This is a hard steel pin with a flat base. One or more blocking pins may be installed in side row B, adding a "block code" to the dealer code. The blocking pin has several roles. First, it requires special

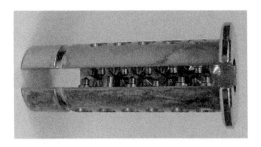

Figure 2.139: (Top) Kaba ExperT plug with pins loaded. (Bottom) Side pins with blocking pin (upper row); corner pins (lower row).

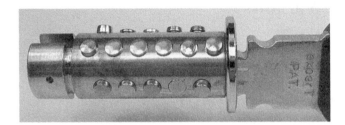

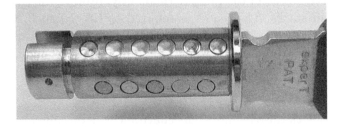

Figure 2.140: Two views of Kaba ExperT plug showing insertion of key.

milling on the key blade as well as the inclusion of a pick-up slope at the end of the blade for row B. This ensures that the keys cannot be copied by Quattro key-duplication machines. Second, it adds active profile control to the system: checking the depth and shape of the corresponding bore in the key (this is equivalent to the workings of the KESO S system, which was described previously). Lastly, the blocking pin is designed so that a key without the required pick-up slope cannot be fully inserted into the keyway (see Fig. 2.141). The presence and function of the blocking pin is the novel feature in the ExperT system on which the new patent is based.

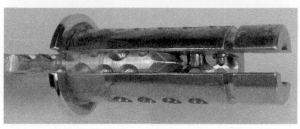

Figure 2.141: (Top) Raising of blocking pin by pick-up slope of ExperT key. (Bottom) Bevel on Quattro key cannot raise blocking pin.

Other Kaba Variants

A modification of the Kaba lock is the Kaba Vario code-change cylinder. This comes with a code-change key, which, when inserted and turned 45 degrees to the left and then withdrawn, allows a new key to be inserted to recombinate the cylinder. A total of eight recombinations are possible. The principle employed is that of the construction key (also applied in DOM iX locks): a master-keying pin or ball becomes trapped in a hole in the plug, permanently altering the composition of one or more of the pin stacks. The reader is referred to US patent 3,234,768 (1963) or to Kaba's 1993 UK patent 2,271,807 for further details. This feature is particularly useful in the construction industry when a building is to be handed over to its new owner. Equally, the cylinder code may be changed when a working key is lost or stolen.

The Kaba range of locks also includes a mechatronic version called Kaba Nova. In this format the dimple key has an extended blade, the top of which supports 14 electrical contacts (seven on each side) that connect to an integrated circuit encapsulated in the head of the key. This system provides for programmable access control in addition to the normal features of the mechanical key system. Since galvanic contacts on a key are prone to failure due to oxidization, Kaba AG has recently released a transponder-based mechatronic system called Elolegic that built around the Kaba ExperT cylinder. A chip embedded in the plastic key head is interrogated by wireless RF electronics in a control module (mortised into the door). If the code carried by the key is verified as correct, an enabling signal is sent to a miniature servo-motor that releases a blocking pin at the rear of the cylinder. This frees the lock for operation by the mechanical part of the key in the usual manner. The system supports in excess of four billion chip codes.

BKS

(DE) 20 pins in 4 rows (3)

The German company BKS ("Boge & Kasten in Solingen") was founded in 1903. Originally producing architectural hardware, the company also supplies cylinder locks for its door and window systems. Since 1983, BKS has been part of the Gretsch-Unitas Group. The current range of BKS security locks includes a number of high-grade 5- and 6-pin inline Europrofile cylinders (series 88 and 31) accredited to German standard DIN 18252 classes P2 and P3. The P3-grade cylinders contain drill and forced-extraction protection. The BKS series 33 incorporates four profile discs that register with a secondary side-bar locking mechanism. The series 50 is a 6-pin cylinder featuring a reversible key, which is unusual for vertically oriented keyway locks. The key for the series 50 is similar to the Holden Commodore two-track key, dealt with in Chapter 7.

The BKS series 45 "Janus" dimple key cylinder is pictured in Figs. 2.142 and 2.143. Janus, the Roman god of gates and doors, was portrayed with two faces allowing him to look in both directions. Janus relates to the BKS 45 cylinder because of the reversible figure "8" keyway that allows the symmetrically bitted key to be inserted either way round. The design of the lock closely resembles that of the Kaba Quattro described in the previous section, and we will only give a brief coverage here. Further details may be found in BKS's 1990 patent US 5,131,249 (see Fig. 2.144).

The cylinder consists of a hollow shell anchored to the internal "frame" of the profile cylinder. This type of construction has the advantage of offering a higher degree of resistance to breakage than the solid brass profile cylinder. The shell surrounds a brass core consisting of a fixed barrel and rotatable plug. There are four rows of pin chambers, equally distributed from 9 o'clock to 3 o'clock around the top of the figure "8" keyway. Rows are staggered longitudinally to allow a greater depth differential between adjacent borings on the key. Each row may contain up to five pin-tumblers.

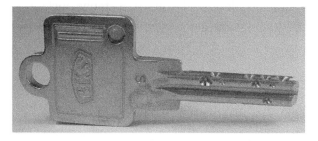

Figure 2.142: BKS Janus cylinder and dimple key.

Figure 2.143: (Top) Core from BKS Janus cylinder. (Bottom) Pins, drivers and springs.

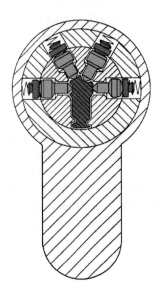

Figure 2.144: The design of the BKS Janus lock from US patent 5,131,249 by H.-D. Baden and M. Hinz.

The lower pins are T-shaped in section, which stops them from pushing through into the keyway. The drivers are very short and have a reduced-diameter end to support the driver spring. When the key is inserted, only the borings on the upper half of the figure "8" are active. Since a circular geometry has been used and the chambers are axially disposed, all borings are perpendicular to the surface of the key blade. This fact facilitates manufacture, but also makes it easier to make unauthorized copies of the key. There is insufficient space in the barrel to house mushroom drivers, which increases the chances of picking the lock.

BKS also produces an electronic version of the Janus called the ESI 58. This has a key-top transponder teamed with an electronic front end for the lock cylinder, allowing additional information such as user codes and access times to be programmed into the lock.

Bricard Chifral

(FR) 13 pins in 3 rows + blocking pin (3–4)
(IT) **CISA RS3** (equivalent)

The French company Bricard has a long history of lock making. Founded in 1782 during the reign of Louis XVI, Bricard is famous for its metallurgical and metal-working capabilities in the reproduction of medieval locks, a skill that is still in demand in a country with many châteaux to maintain. The Bricard Museum in Paris holds an impressive collection of such works. Bricard is also well known for its 7-wafer lock with a three-sided key (see Bricard SuperSûreté in Chapter 3).
A recent addition to the Bricard range of locks is the Chifral, shown in Figs. 2.145–2.147. The Chifral is also marketed as the CISA RS3, since Bricard was bought by CISA S.p.A. (CISA has since been acquired by Ingersoll-Rand.) The lock is similar in construction to a Kaba Gemini cylinder and accepts a symmetric dimple key. The cylinder houses 13 ordinary miniature pin-tumblers and a pair of special pins that constitute a blocking mechanism. The inclusion of the blocking mechanism requires that the key have a pivoting element in the blade (like a hinged version of DOM's floating ball).

The layout of the pins is as follows: five pins are located at 6 o'clock, four at slightly past 3 o'clock, and four more at slightly before 9 o'clock. At 9 o'clock and 3 o'clock,

Figure 2.145: Bricard Chifral cylinder and key with pivoting member.

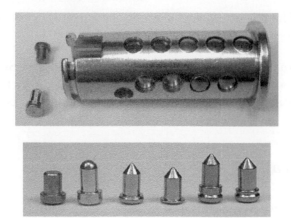

Figure 2.146: (Top) Bricard Chifral plug with blocking pins on left. (Bottom) Selection of Bricard Chifral pins.

Figure 2.147: Operation of Bricard Chifral cylinder.

marginally above the positions that the fifth pins would occupy in the left and right rows, there is a bore that traverses the plug. The left half-bore is slightly offset toward the front of the plug with respect to the right half-bore, which accommodates a long blocking pin with a rounded end. The blocking pin has a driver spring that is stronger than the other driver springs in the cylinder, and thus it acts as a deflector pin for the pivoting element. The left bore accommodates a short blocking pin that has a flat end and does not protrude into the keyway. A vertical bore at 12 o'clock intersects the axis of the left bore. The function of the vertical bore is to create a space at the forward end of the pivoting member in the key. This space also houses the flat end of the short blocking pin.

When the key is inserted, the right blocking pin, which is more strongly sprung, actuates the pivoting member to displace it toward the left and into the vertical bore. At this point, the pivoting member is outside the plane of the key blade and can make contact with the short blocking pin (a fixed key blade cannot do this). Because of the difference in spring tension, the long blocking pin is pushed radially inward, while the short blocking pin is pushed outward. The dimension of the pair of blocking pins together with the pivoting member is such that both shear lines are attained simultaneously when the member is hinged over to the left-hand side of the plug. With the other 13 pins raised to the shear line, the plug is free to turn.

The presence of the blocking pins does not hamper picking since the end of the pick is not constrained to stay in the plane of the key blade. However, the longer pins in the Bricard Chifral are spooled so that they tend to wedge against the chambers if overlifted. This considerably enhances the lock's manipulation resistance. The key is not practical to duplicate since it contains a movable part.

Vachette Radial

(FR) 10–32 pins in 4–6 rows + blocking pin (3–4)

The Vachette Radial, depicted in Figs. 2.148 and 2.149, is a multiple inline pin-tumbler lock with a horizontal keyway. It is available in a number of models including the Radial S and the Radial Si. Both models can house between 10 and 32 miniature pin-tumblers. The cylinder takes a dimple key that is symmetric and of roughly rectangular section with some profiling features. The cylinder and plug are protected from drilling by rows of hardened inserts and crescents in the front of the plug. Additional drill protection is in the form of a hardened cap at the front of the joining rod that runs along the top of the Europrofile cylinder. The use of a joining rod, rather than a solid cast brass body, serves to protect the cylinder from being snapped in half.

The stainless steel pins have a smaller diameter than usual dimple pins and are arranged in up to six rows (refer to Fig. 2.149). Pairs of rows, each forming a 30-degree V, are clustered at 3, 6, and 12 o'clock (viewing the cylinder from the front with the main pin chambers up). The top two rows at 12 o'clock can contain six pins each, whereas the other four rows can house up to five pins. Drivers in the top two rows are solid, whereas the other four rows of pins use hollow counter-pins (as in Kaba locks) due to the limited space. Vachette gives the figure 905×10^6 (close

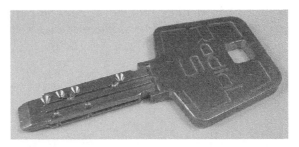

Figure 2.148: Vachette Radial S cylinder and reversible dimple key.

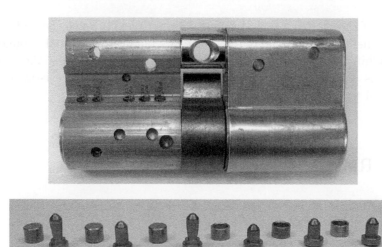

Figure 2.149: (Top) Vachette Radial cylinder with cover removed. (Bottom) Selection of pins and drivers.

to one billion) as the number of possible differs. The lock mechanism follows the Kaba principle, which has been described previously.

The Vachette Radial Si, whose patent was filed in 1985 (US 4,667,495), has an additional active element in the reduced-width end of the key blade (similar to the key in Fig. 2.150). The element is a captive pin that traverses the blade and protrudes a small amount on either side, but no more than the nominal width of the blade. A blocking pin is located at the rear of the plug, aligned with the major axis of the keyway, and there is a ramp opposite the pin. The blocking pin is recessed and is not visible in the keyway. A standard Vachette Radial S key will not contact the blocking pin. However, when a key with the movable pin is inserted, the ramp causes the pin to be deflected by an amount sufficient to allow its other end to contact and raise the blocking pin.

Vachette has upgraded the Radial cylinder from the S and Si models to the NT, ensuring protection for the new design by a patented blocking pin mechanism that interacts with a movable element at the end of the key blade. The design is described in French patent 2,619,149 (1987) by F. Debacker and J. Girard. The inclusion of a movable element in the key protects against unauthorized copying. The system, which sports a three-star A2P rating, is in other respects identical to its predecessors, so in the next section we mainly focus on the blocking mechanism.

Vachette Radial NT

The Vachette Radial NT cylinder in Figs. 2.150–2.153 has four rows of pins in two pairs centered at 12 o'clock and 3 o'clock. The key blade is roughly

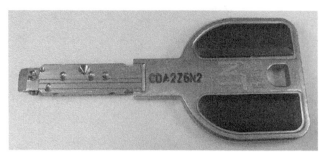

Figure 2.150: Vachette Radial NT cylinder and key.

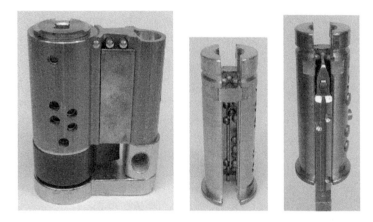

Figure 2.151: (Left) Radial NT cylinder with cover removed. (Center) View of plug with bridge for mobile. (Right) Key contacting check and blocking pins.

Figure 2.152: Vachette NT lower pins and drivers (blocking pins rightmost).

rectangular, with a reduced-width end and a milled profile on both edges. The keyway warding comprises pairs of shallow ridges on the top and bottom sides, and a V-shaped projection that matches the milled profile on the edge of the key.

Figure 2.153: Key detail for Vachette Radial NT. Points E and F refer to ends of movable element. H is the hole.

The blocking mechanism is in the form of three extra pin-tumblers located at the rear of the plug. Two of these are best described as blocking pins, while the third is a profile-control or check pin (see Fig. 2.152). The blocking pins are situated behind bore position 6 at 3 o'clock and at just left of 12 o'clock. The check pin is at 6 o'clock. The blocking pins interact with the movable element in the key blade in a rather complicated manner that we now address.

The reduced-width end of the key houses a transversely mounted, slideable rod. The travel of the rod is limited to the uncut edge of the key blank. On each face of the key, a hole is provided in the blade end, offset from the center, exposing a portion of the movable element. Note that the hole does not extend wholly through the key blade. According to Fig. 2.153, we refer to the end of the rod nearer the hole as point F and its opposite end as point E. When the rod is fully displaced toward point F, a small depression in the rod is revealed. Lastly, a bridge or ramp is fitted into the plug at 9 o'clock at the same depth as the blocking pins (see Fig. 2.151 center).

During insertion of the key, the keyway ward at 3 o'clock pushes the movable bar toward the 9 o'clock position. Since the ward stops short of the rear of the plug (prior to the blocking mechanism) at the point where the rod end E encounters the bridge, it is free to surmount this obstacle. The bridge therefore displaces the bar so that the rod end F contacts and lifts the blocking pin at 3 o'clock, bringing it to the shear line. At the same time, the rod is shifted so that its depression aligns with the hole in the key blade. The hole also registers with the second blocking pin near 12 o'clock, which enters the depression in the rod. The depth of the depression is calculated to bring the second blocking pin to the shear line. The final piece of the puzzle is the check pin at 6 o'clock. This pin acts to ensure that the key blank has full width at the point underneath the hole; thus a blank with a hole all the way through the blade will not work. The overall principle is similar to the DOM floating-ball system with some added checks.

The latest upgrade in the Vachette range is the Radial Cliq, which is an enhancement of the Radial NT carrying a key-top transponder that sends an encrypted code to the control electronics in the lock.

YBU

(JP) 10 pins in 5 rows (1–2)

The YBU lock shown in Figs. 2.154 and 2.155 is used primarily on security shutters. It is a double-entry lock taking a key that is symmetric around the midpoint on the key stem. A similar construction is found on some Club-type car steering wheel locks that have a gun-shaped tubular key with eight dimple cuts, although these can only be inserted from the front of the lock.

There are five rows of pins arranged radially around the keyway. Four rows have two pins, with the remaining row having only a single pin in the front position. The key has a solid cylindrical stem with dimples cut in five rows of three, only nine of which are active when opening the lock. A small recess in the end of the key accepts a stud protruding into the keyway that provides turning force to plug.

When inserted from the front, the bittings in the front two positions contact the pins, with the bittings in position 3 being inactive. When inserted from the rear, the last two positions are active. The bittings in positions 1 and 3 are mirror images since both must operate the pins at the front of the lock, depending on the direction the key is inserted.

Figure 2.154: YBU 10-pin radial lock and key.

Figure 2.155: YBU lock with cover removed (left) and close-up of pin-tumbler (right).

As the YBU pin-tumbler mechanism has quite loose tolerances, the only difficulty with picking it is in fashioning a suitable tensioner for the job.

2.9   Tubular

ACE/GEM

(US) 7–8 pin (2–3)

Tubular or axial pin-tumbler lock are often used as cam locks on coin-operated equipment such as telephones and vending machines. The most common locks of this type, such as ACE and GEM, have 7 or 8 pins arranged around a central plug, although models with as many as 10 pins have been made. Figs. 2.156 and 2.157 show an ACE 7-pin axial lock produced by the Chicago Lock Company, which first introduced this type of lock in 1933 [50]. In some models, the pins are offset to the left or right of the locating slot.

The lock consists of an outer shell in which a plug and barrel assembly is coaxially mounted, with the barrel secured by a retaining pin to the shell. The plug (or spindle) has a threaded end to which a locking cam is attached. A series of axial bores,

Figure 2.156: ACE 7-pin tubular cam lock and key.

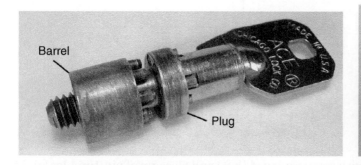

Figure 2.157: Core from ACE cam lock (left) and pin set (right).

seven or eight in number and usually equispaced, is provided to chamber the split pin-tumbler pairs and driver springs. The drivers and lower pins reside in the barrel, with the upper pins located in the plug. An annular keyway is formed by the lip of the shell and the rod-end of the plug, in which the ends of the pins are partially visible. The pins are retained by the lip at the front of the shell.

The plug usually contains a spline, with a matching recess in the lip of the shell to accept a locating fin in the key. The fin serves both to provide turning force to the plug (via the spline) and to retain the key in the shell during operation of the lock. The key stem is hollow and cylindrical, with bittings milled into its periphery to varying depths.

The pin-tumblers prevent the rotation of the plug with respect to the fixed barrel and shell until their shear planes are brought into coincidence by insertion of the correct key. There are typically seven pin sizes ranging from $0.020''$ to $0.110''$ in increments of $0.015''$. The theoretical number of key changes is therefore $7^7 = 823,543$. Some systems use eight depths of cut, in which case there are theoretically $8^7 = 2,097,152$ permutations for a 7-pin lock. Although the MACS is effectively unlimited, in practice there may be other constraints, such as the progression step, that reduce this number. Some manufacturers specify as few as 50,000 usable key combinations.

Master-keying of axial pin-tumbler locks can be accomplished in a number of ways. The most straightforward is via master-keying pins, the idea being the same as for inline pin-tumbler locks, but with a more severe constraint on the available space due to the effective pin chamber length. Another approach, suggested in a 1972 patent by the Fort Lock Corporation (see Fig. 2.158), uses a special top pin design teamed with a compound bitting on the key: the master-key actuates the pins via the reduced-diameter part of the bitting.

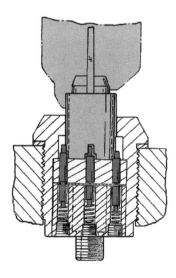

Figure 2.158: Master-keying of an axial pin-tumbler lock (US patent 3,738,136 by M. Falk).

Numerous patents have been sought for devices to manipulate the axial lock: examples are provided by US patents 2,059,376 (1935) and 2,070,342 (1936). The first of these describes a lock-pick for 7-pin axial locks and the second a decoder. US patent 3,251,206 describes a lock-pick similar to the HPC model (see Fig. 1.10). Tubular lock-picks are very effective on standard axial locks, which are in a sense easier to manipulate than conventional inline pin-tumbler locks since their pins are partially exposed and therefore more readily accessible in the keyway. The tubular lock-pick has the added advantage that, once picked, it may be used as a working key. Note that if the lock is picked with conventional flat tools, the plug will relock at multiples of one-eighth of a turn since the pins will spring back up.

Designers have made a concerted effort to enhance the level of security provided by axial pin-tumbler locks, and many special features have been suggested. For instance, US patent 3,267,706 proposed an obscured-fin axial lock. Various restricted keyway shapes have been proposed (e.g., 1976 US patent 4,069,696) as well as serrated top and bottom pins (US patent 4,099,396), both of the previous examples being due to the Chicago Lock Company. A different approach was taken in 1987 US patent 4,802,354 by the Fort Lock Corporation (see Fig. 2.159). In this design, the faces of the plug and barrel at the shear plane are machined to leave a series of ridges to hinder manipulation. Yet another modification involves the use of springs of varying tensions to thwart the tubular lock-pick, which relies to some degree on springs with the same tension and drivers of the same length.

A recent ACE/GEM model features eight pins having a flower-shaped keyway with eight "petals." In this model, the central stem of the cylinder is slotted to accept a fin on the key. The keyway design is similar to that of the Fort Apex lock (pictured in Fig. 2.160), requiring fluting on the key. The Apex lock uses an extra cylinder ring and intervening plates to create multiple shear planes. This frustrates lockpicking and decoding attempts by introducing additional setting points at incorrect levels.

Some tubular locks have a hardened ball embedded centrally in the face of the plug to guard against attack by drill or hole-saw. The barrel retaining pin should also be hardened to protect against drilling. Axial pin-tumbler locks may be combined with other types of lock to improve security or increase the number of available key combinations. The American Locker Company produces an ACE variant (pictured in Fig. 2.161) that has a conventional 5-pin flat key extending through the tubular key. ACE also makes a UL-rated tubular lock with coaxial pins at 3, 6, and 9 o'clock [128].

Van Lock

(US) 7-pin (2–3)

The Van lock (Fig. 2.162) has seven axial pins that are flush with the front of the plug and are depressed by prongs on the key. The lock is produced as a padlock and

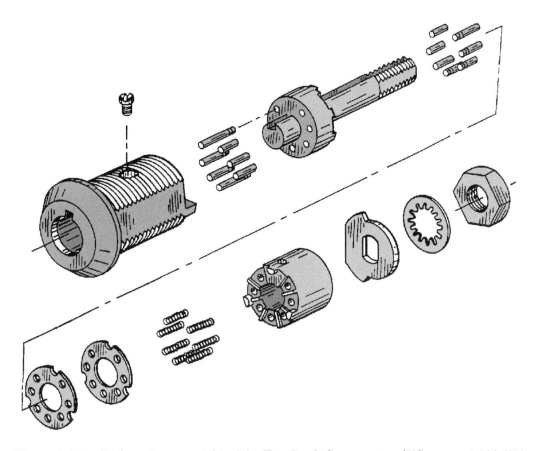

Figure 2.159: Pick-resistant axial lock by Fort Lock Corporation (US patent 4,802,354 by G. Johnson).

Figure 2.160: Fort Lock's Apex 7-pin axial lock and key.

as a cam lock for vending machines. The plug is mounted in a fixed barrel having a flanged edge and a pair of grooves that guide the locating stubs on the inner wall of the key. The turning force is provided by the key prongs themselves, with the stubs serving to orient the key correctly and retain it during operation. Since there is nothing other than the pins on the front face of the lock, tensioning the lock for

Figure 2.161:   American Locker Company lock with combined tubular/flat-bladed key.

Figure 2.162:   Van Lock 7-pin padlock with key.

picking is slightly more difficult than in a standard ACE or GEM axial pin-tumbler lock. The key is also more difficult to duplicate than standard tubular keys due to the "inside-out" construction. An Allen-keyed screw on the underside of the key allows the skirt to be detached from the handle for removal of the seven key prongs. The design is covered in US patent 2,993,361 (1961) by L. E. Van Lahr. This type of lock can be traced to Johnson's rotary 6-pin lock of 1861 [57].

A rekeyable version of the Van lock called Vanmatic is also produced that accepts up to eight different keys. The system has the feature that it can be rekeyed *in situ* without removing the lock cylinder. Two different types of keys are supplied for this purpose: operating keys and change keys. Operating keys have locating stubs, and can therefore only be inserted when the plug is in the locked position (12 o'clock), while change keys, which are numbered and marked with a black spot, do not possess stubs. Counting the pin positions in a clockwise direction from 1 o'clock in Fig. 2.162, we note that there is no boring in the plug for a top pin in position 8 (12 o'clock). There are, however, a driver spring and bottom pin at this position that come into play when the combination of the lock is changed.

The rekeying process works as follows: Suppose that the lock is initially set to operating key number 1. In this setting, pin 7 is at 11 o'clock. The number 1 change key is inserted and turned to 3 o'clock, at which point it is removed. The pins all spring back up to the top surface of the plug, including the pin at position 6, which is now aligned with the eighth pin chamber. When change key number 2 is inserted, it allows the plug to be rotated anticlockwise by an eighth of a turn and removed. This action places pin 7 at 12 o'clock and repositions all the other pins by one position in the clockwise direction. Operating key number 1 will no longer operate the lock since its stubs prevent it from being inserted to match the shifted pin positions. On the other hand, operating key number 2 is a CW circular shift by one position of key number 1, so it now operates the lock. The combination can be circularly shifted in this manner eight times until the lock returns to its original configuration. A set of eight operating and change keys is provided for recombinating the lock.

Izis Arnov

(FR) 5-pin (3)

The Izis or Izis Arnov lock (Figs. 2.163 and 2.164) is another axial lock with a pronged key like the Van lock. It is now produced by the French company Cavers. The lock

Figure 2.163: Key and cylinder from Izis Arnov 5-pin axial lock.

Figure 2.164: Rear part of Izis Arnov core with set of pins and drivers.

distinguishes itself in a number of ways from standard axial pin-tumbler locks. The cylinder consists of a flanged brass body inside a steel sleeve that features a six-sided petal-shaped keyway. The body in turn houses a core with five axial chambers for the springs and driver pins.

Whereas in an ACE or GEM lock the plug turns while the barrel remains fixed, in the Izis lock the reverse occurs. The top pins are chambered in bores at the top end of the stationary brass body (see Fig. 2.164). The bores have a reduced-diameter opening to retain the pins against the action of the driver springs. The shear plane in this lock is the interface between the inner edge of the brass body and the front face of the core.

Normally, an arrangement such as this would not work in an axial lock since the front part of the lock cannot be turned. The novel aspect is the design of the key. The stem and its locating fin are joined to the key head. As well as the obvious prongs of varying lengths that complement the lengths of the respective top pins, the brass skirt in which the prongs are mounted is, in fact, rotatable around the key stem. During operation, the skirt, which is also slideably mounted on the stem, is pressed against a strong spring in the shoulder of the key. At its point of maximum insertion, the key stem becomes free to turn with respect to the skirt. The skirt, on the other hand, is temporarily fixed to the front of the body since its prongs are pressed into the pin chambers. The fin on the key stem is at this point fully engaged in a recess in the face of the core. If all the pins have been correctly depressed to align them at the shear plane (as in Fig. 2.165), the head of the key can directly turn the core and hence the tail-piece of the lock. The key cannot be retracted while the core is being turned since its fin is pressed against the upper surface of the shear plane.

The lock has a good degree of manipulation resistance since it has quite strong driver springs and both top pins and drivers are contoured. It is also more difficult to tension since the core is recessed and the face of the lock is fixed. As might be expected, however, a lock of this type is not hugely popular due to the size and

Figure 2.165: Underside of Izis cylinder with key inserted and turned.

shape of the key, the cost of key duplication, and the nonstandard cylinder format.[12] The lock also develops wear patterns on the inner contact surfaces due to the heavy spring tension.

Central, JPM, Pollux

(FR) **Central** 7-pin (2–3)
(FR) **JPM 505** 5-pin (3)
(FR) **Pollux** 5-pin (2–3)

JPM 505 and FTH Thirard locks (Figs. 2.166–2.168) are also axial pin-tumbler locks, again from France. They have a key with a cylindrical central stem and fins spaced at five points of a regular hexagon. Bittings are milled into the end of the key so as to depress the pins to the appropriate depths. The operating principle is very similar to that of the Bramah lock covered in Chapter 3, although axial pins are used instead of sliders. The original 6-pin design was published in a 1978 French patent (FR 2,415,185) by R. Frank.

Whereas in a conventional GEM or ACE tubular lock the blocking function is provided by the action of split pin-tumblers at the shear plane between the plug and barrel, in JPM 505 locks the pins are not split. Instead, each of the five pins is provided with a peripheral slot at a certain point or points along its length. On each pin, there is a locating tab that engages a longitudinal channel around the central bore in the plug. The presence of the locating tab ensures that the slot in the pin always faces radially outward. The plug contains a circumferential channel roughly midway along its length. The channel is occupied by a two-part ring, which forms

Figure 2.166: JPM 505 5-pin profile cylinder and key.

---

[12]In fairness, similar remarks apply to many other locks featured in this book!

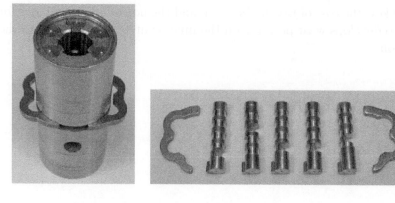

Figure 2.167: (Left) JPM 505 core. (Right) Ribbed pins and detainer ring.

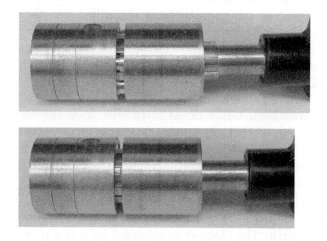

Figure 2.168: Operation of JPM 505: key aligns slots in pins with channel.

a fixed blocking plate as in the Bramah lock. The ring is basically petal-shaped, with its inner arcs skirting around the pins and its outer arcs lodging in corresponding chambers in the cylinder body. Clearly, no rotation of the plug is possible without the pins being simultaneously depressed so that their notches register with the stationary ring. The pins are provided with ribbing to confound attempts at manipulation.

The other two locks in this section closely resemble their conventional ACE/GEM counterparts except for the key and keyway design. The first of these is the Central lock, pictured in Figs. 2.169 and 2.170. The second is the Pollux 5-pin axial cylinder in Figs. 2.171 and 2.172. Central produces 5- and 7-pin axial locks. The keys have six or eight radial fins, with one fin having an enlarged base to provide turning tension and locate the key in the keyway.

The Pollux cylinder has five pins and a sixth fin on the key to provide tension to the plug as well as locating the pipe key in the keyway. As with the Central lock, all pins

Figure 2.169: Central 7-pin cylinder and 8-fin key.

Figure 2.170: Mechanism of Central lock and operation by key.

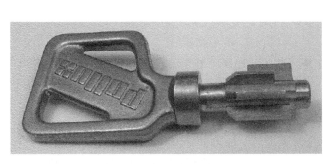
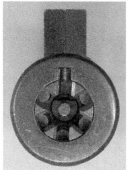

Figure 2.171: Pollux 6-fin key and 5-pin cylinder.

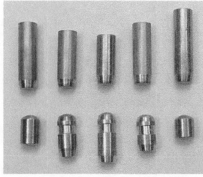

Figure 2.172: (Left) Two halves of a Pollux core. (Right) Top and bottom pins.

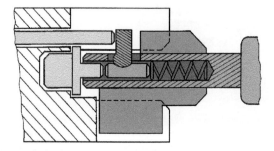

Figure 2.173: The Pollux Interactive uses a mobile fin on the key (FR 2,678,670 by P. Bonnard and J.-L. Millier).

must simultaneously be depressed to the shear plane depth to operate the lock. The design is closely related to Schlage's 1967 8-pin axial lock (US patent 3,411,331).

The keyways of Central and JPM locks are such that the fins are obscured by the fixed front-piece of the lock when the key is turned, making tensioning by external means more difficult (in a similar vein to US patent 3,267,706). So, unlike conventional tubular locks, the pins do not remain accessible as the core of the lock rotates. Picking tools called "umbrellas" (*parapluies* in French) exist for these kinds of locks, but the presence of incorrect notches in the pins makes picking considerably harder. Central locks also have an armored collar that surrounds the cylinder to prevent it from being sawn off.

Following the current trend in key copy prevention for high-security locks, many of the French lock manufacturers have devised systems with active elements in the key. Some examples of these for the axial variety include Pollux Interactive, Pacific Interactive, and Cobra Axira. The Pollux Interactive, described in French patent 2,678,670 (1991), has a key with a movable extension on one of its fins slideably mounted in the stem. A diagram from this patent is shown in Fig. 2.173. The height of the movable fin is set by contact with the drill-pin in the keyway. The action is controlled by the length of both the drill-pin and the pin that is being actuated by the movable fin. These locks, along with many other French high-security locks, are featured on the "Montmartre" Web site [7].

Zenith Cavith

(FR) (5 + 3)-pin (3)

The Zenith Cavith lock shown in Figs. 2.174 and 2.175 is an embellishment of the Central 5-pin cylinder with three radial pin-tumblers in the groove occupied by the locating fin of the key. The key has three V-shaped cuts on this fin to operate the radial pins, while the five other fins are end-bitted to depress the axial pins to the correct depths. The axial pin-tumbler pairs are of the same construction as the pins in the Izis lock (Fig. 2.164).

Figure 2.174: Zenith Cavith 5-pin cylinder and key.

Figure 2.175: Zenith Cavith core (left) and operation by key (right).

The additional radial pins increase the number of overall codes in proportion to the number of differs they provide. For example, assuming three depths of cut for the radial pins yields 3^3 or 27 times more key codes. The extra bittings in the key add a degree of copy protection while at the same time making it infeasible to open the lock with standard tubular lock-picks.

The original Zenith design is from a 1967 French patent by Établissements Cavers (FR 1,533,953). An improved design with a fixed front plate, as in Fig. 2.174, and antipick axial pin-tumblers is presented in a 1994 French patent by Lucas and Édouard of Cavers (FR 2,716,484).

ISEO R6

(IT) 6-pin (3–4)

The ISEO R6 lock (Figs. 2.176 and 2.177) has a 6-pin cylinder that takes an end-bitted key with a profile resembling the letter "E." As set out in French patent specification 2,491,531 (1980), the pins are arranged in two clusters of three at the

Figure 2.176: ISEO R6 cylinder and end-bitted key.

Figure 2.177: ISEO R6 core comprises three sections.

top and bottom ends of the E-shaped keyway. There are six cuts on the key bit, with the central portion reserved for depressing a spring-biased tail-piece that links the plug to the locking cam. The locking principle is similar to that of the ACE lock, but the construction is considerably more robust and the pins more difficult to manipulate.

The core of the cylinder comprises three sections, as shown in Fig. 2.177. The front-most of these is a hardened ring carrying the keyway cut-out. The middle section contains six top pins and has a cut-out for the tail-piece. Top pins are shouldered in the middle section of the core with a reduced-diameter shank visible in the keyway. The top side of the middle section is socketed with the front section so that these two parts turn together.

The lower section houses the six driver pins and springs as well as the tail-piece, which is sprung from the rear of the cylinder. The tail-piece is normally disengaged from the locking cam, requiring the key to be inserted to displace it. Some of the driver pins are spooled to render manipulation more difficult. There is also a seventh driver pin (not visible in Fig. 2.177) anchored in the lower section at the same radial distance as the two outermost pins.

The front-plate of the cylinder is fixed, so that the key must be fully inserted before it can be turned. This fact makes it difficult to apply tension to the core in the case of a manipulation attempt (similar remarks apply to Bricard SuperSûreté, Fichet-Bauche 787, Chubb AVA, Mottura, and many axial pin-tumbler locks).

Tover 27A

(ES) 6-pin (3–4)

The Tover 27A is a heavy-duty lock cylinder designed for multipoint locking systems of the type produced by the Spanish company Tover. The lock, shown in Figs. 2.178–2.180, is of nonstandard dimensions, comprising a cylindrical steel sleeve and cover fastened with a grub screw. Removal of the cylinder cover reveals a two-part plug of round section with a shear plane at the halfway point. The plug, which is made of cast zinc alloy, contains a rectangular cavity that houses six brass axial pin-tumblers.

There is a central hole in the plug through which a linkage rod passes. The rod has a rectangular section at the front end and a flat portion on one face at the other end. The rod is spring-loaded and must be depressed by a key in order to provide coupling to the locking cam at the rear of the plug.

What is special about this lock is the construction and operation of the pin-tumblers (see Fig. 2.179). These are made of flat, stamped brass and are arranged in two rows of three in the cavity of the plug. Since they are flat, we will refer to them as bars rather than pins. The bar-tumblers are in two parts: front and rear, according to which part of the plug they usually inhabit. The front bars are rectangular and

Figure 2.178: Tover 27A large-format profile cylinder and key.

Figure 2.179: Two halves of Tover 27A core.

Figure 2.180: (Left) Key pushes bar-tumblers to correct depths. (Right) Not all key bittings are active.

have different lengths. The rear bars or drivers are mounted on thin stems and are spring-biased toward the front of the plug. There are two sorts of driver shapes, differing in their end profiles at the shear plane. The first of these is shaped like an inverted L. The second resembles the letter "t" and has an antipicking function.

The locking principle is that of an axial pin-tumbler lock except that the geometry is linear rather than circular (as in the majority of cases covered so far). Each bar-tumbler pair must be depressed by the correct amount in order to align their interfaces with the shear plane of the front and rear plug halves. The key that achieves this is end-bitted with a rectangular blade. The blade is spot-welded onto the key stem, as in the NS Fichet key, with the bit set slightly off-center so that it may only be inserted one way. The key appears to have nine bitting positions; however, the middle three of these are cosmetic since their only function is to actuate the linkage rod. When the correct key is inserted and turned, the blade is retained by the front cover of the lock until it is returned to the locked position. Note that the key cannot be withdrawn at 180 degrees since the blade is offset from the center of the key stem.

The lock enjoys a surprisingly high level of manipulation resistance, despite appearances to the contrary. The plug has a high tendency to skew and block during picking due to incorrect setting of the bars with t-shaped drivers. This mimics the effect of mushroom or spooled pins but in a shear plane rather than at the rim of a plug. The system is not a prime candidate for master-keying since the bars are adjacent to each other and would interfere if more shear lines were introduced. Although there is no drill protection on the cylinder, in a high-security installation, a cylinder-guard could be added to compensate for this.

2.10   Concentric Pin

For the pin-tumbler locks we have so far encountered, the main differences in operating principle can be put down to the arrangement of a set of rod-shaped pin-tumblers. The shape and function of the pin-tumblers has remained largely unchanged, although they can be arranged in the cylinder to create one or more shear lines.

The class of tubular or axial pin-tumbler locks operates via a shear plane defined by the circle of the pins. The ISEO R6 and Tover 27A also operate via a shear plane that cuts the cylinder in the transverse direction. In locks having concentric pin-tumblers, the shear plane is defined by the interfaces of two or more coaxially located pins. There are two fundamentally different ways to implement a concentric pin-tumbler, illustrated in Fig. 2.181. The first, known as a tube lock, uses a set of coaxial sleeves that are open at both ends; the second uses sleeves that are capped at the end further from the keyway.

In a concentric pin-tumbler lock with open-ended sleeves, or tube lock, springs with the same tension may be used on all components since they function independently. Now consider a system of one pin and one capped coaxial sleeve, as in Fig. 2.181B. If the distance through which the central pin moves is x_1 against a spring with force constant k_1, the force applied to the surrounding sleeve is $F_1 = k_1 x_1$. The spring constant k_2 for the sleeve must be such that it will be displaced by the force F_1 an amount that is less than the tolerance of the lock. In a hypothetical lock with

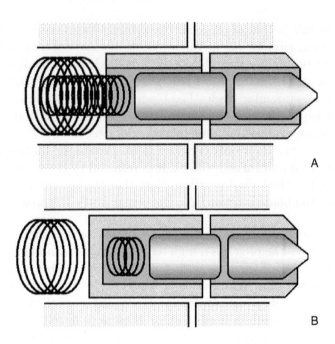

Figure 2.181: Two different types of concentric pin-tumbler mechanisms.

a pin and two concentric sleeves, the sum of forces due to both the pin (F_1) and the first sleeve (F_2) is applied to the outer sleeve. Thus the spring constant of the outer spring must be large enough so that it is only compressed a "small amount" by the combined force $F_1 + F_2$. Because of the design constraints imposed by the interconnected system of springs, capped concentric pin-tumbler locks are limited to a small number of sleeves in each pin chamber. In the Mul-T-Lock only two-part pins (inner and outer) are used. Additional security is obtained by increasing the number of concentric pin chambers to five.

We next present a tube lock that has three-part coaxial pins (inner, middle, and outer). This is followed by a discussion of the Mul-T-Lock, which includes multiple, closed-end concentric pins.

AGE

(JP) 3-pin (1)

The concentric pin-tumbler mechanism of the Japanese AGE padlock in Figs. 2.182 and 2.183 consists of two of coaxial sleeves or tubes around a solid central pin. The design is based on a U.S. padlock called Wiselock, patented in 1920 by S. Wise (see Fig. 2.184). Although in no way a high-security lock, it serves to demonstrate the tube lock principle. The padlock body is in two parts connected by a hinge. The pin and sleeves are cut transversely in one or more places to provide a set of

Figure 2.182: AGE concentric 3-pin padlock: key (left); locked position (middle); open position (right).

Figure 2.183: Two views of the concentric tube mechanism.

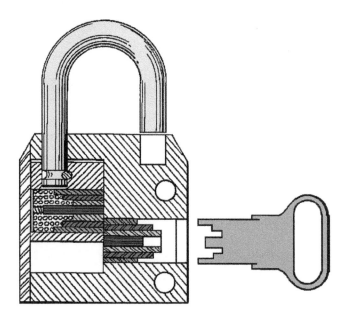

Figure 2.184: S. Wise's 4-tube lock with push-key from a 1920 patent (US 1,390,222).

shear planes. When all of the shear planes are coincident, the lock body may be pivoted, opening the shackle.

The central pin and sleeves are brought under tension by an arrangement of concentric springs. All the components act independently; the springs are isolated from each other by the thin walls of the driver pins. The key for a single-chamber concentric lock such as this is end-bitted with symmetric cuts across the blade (as in the ABA Pagoda lock covered in Chapter 3). The bittings depress the central pin and sleeves to the correct depths so that a single shear plane is created.

Mul-T-Lock

(IL) 10-pin (3–4)

The Mul-T-Lock, depicted in Figs. 2.185–2.187, is a horizontal keyway lock with five concentric or telescoping pin-tumblers. It was originally patented in Israel in 1976 (IL 50,984) and is also described in US patent 4,142,389. The lock design experienced

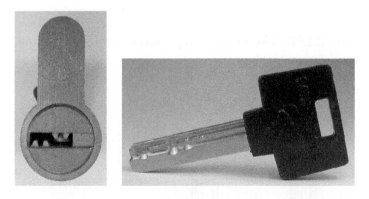

Figure 2.185: Mul-T-Lock 10-pin cylinder and reversible key.

Figure 2.186: Mul-T-Lock plug with key partially inserted.

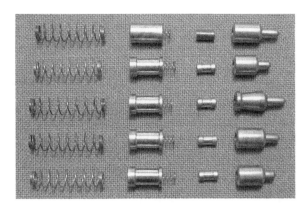

Figure 2.187: Mul-T-Lock inner and outer pins, drivers, and springs.

rapid commercial success. Mul-T-Lock, founded in 1973, is now a global player in the security industry and part of the ASSA Abloy Group. As well as rim, mortice, and Europrofile cylinders, Mul-T-Lock produces a large range of padlocks for various applications.

The pins are centrally located in the cylinder. Each pin pair consists of an inner pin, with its own driver and spring, operating inside a hollow outer pin. The outer driver is capped at the top to house the spring for the inner pin and may also be spooled, as in Fig. 2.187. The keyway broaching has wards flanking the pins on each side, making them harder to manipulate, although in some locks the warding may not run the whole length of the keyway. The nickel silver key is symmetric, with the key bittings offset to one side of the blade so that it can operate the lock either side up.

As we explained before, the use of capped coaxial pin-tumblers results in mechanical coupling between the concentric pins. In order to ensure that the operation of the inner and outer pins is effectively independent, the driver springs must have different tensions, with the outer pin having a strong spring and the inner pin a very light spring. The balance of spring tension is such that the inner pin can be depressed fully with negligible impact on the outer pin.

Five sizes of inner pins in increments of 0.5 mm and four sizes of outer pins can be used to combinate the plug. The inner pins are either of stainless steel or nickel silver. Because of the horizontal keyway construction, there is no MACS restriction. Master-keying is effected by including master pins in one or more inner pins and master rings (master pins with a central hole) in one or more outer pins. The plug and key are supplied in both left- and right-hand models, depending on whether the top cuts are on the left- or right-hand side of the key blade.

As is the norm in high-security locks, the front of the plug and cylinder contain hardened steel rollers to thwart drilling. The key is hard to duplicate without specialized machinery. This is because some of the inner pins typically have to be raised more than their respective outer pins, requiring a peak or inverted cut for the inner pin.

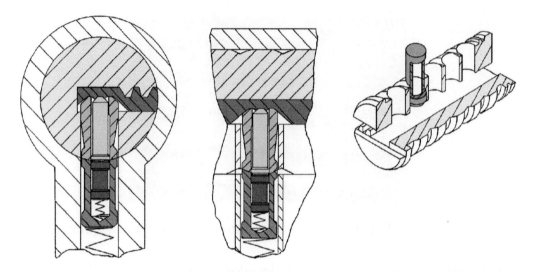

Figure 2.188: N. Eizen's design for an enhanced Mul-T-Lock (US patent 4,856,309).

As mentioned in the 1987 Mul-T-Lock patent (see Fig. 2.188), the lock is susceptible to picking with suitably shaped tools, although it is a difficult task especially if spooled drivers are present. It is also susceptible to shimming from the front if the lip of the plug is ground down. These two weaknesses were addressed in the same 1987 patent by redesigning the plug to have a circular groove on either side of the outer pin bores. This creates a nonlinear shear plane between the plug and the barrel that prevents shimming. The contact surfaces of the outer pins are also matched to the curvature of the plug, ensuring a more precise fit than can be achieved with flat-bottomed pins.

The original Mul-T-Lock patent has now expired, and a new one has been taken out on a version of the lock that incorporates a movable element in the key blade.

Mul-T-Lock Interactive

(IL) 10-pin (3–4)

In line with the recent trend of adding movable or floating elements to keys, Mul-T-Lock has introduced the "Interactive" system, illustrated in Figs. 2.189–2.192. The relevant patent in this case (US 5,839,308) was filed in 1997 by Eizen and Markbreit. The lock is also marketed as the Mul-T-Lock Gamma and Picardie Interactive. The movable part, which we will call a floating pin, replaces one of the key bittings (e.g., the first) and takes the form of a pin constrained to slide vertically in the key blade. There are two such elements so that the reversibility of the key is preserved. Instead of the usual pair of concentric pins in the first bore in the plug, there is an undersized

Figure 2.189: Mul-T-Lock interactive cylinder and key.

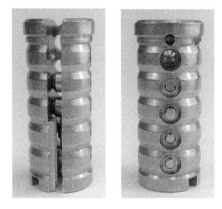

Figure 2.190: Mul-T-Lock Interactive plug with inverted driver pin in position 1.

Figure 2.191: Key partially inserted: movable insert not yet in contact with pin 1.

pair of pins of equal length. This pin pair is actually so short that a blank key will not raise it sufficiently to attain the shear line.

In the bottom part of the keyway, directly underneath the first pin position, is a spring-biased pin with its conical end pointing upward, as in Fig. 2.190. The floating pin in the key blade is limited in its downward travel; however, it can be displaced

Figure 2.192: Mul-T-Lock interactive key with movable insert in position 2.

upward. This is what happens when the floating pin contacts the inverted driver pin, with the result that the short upper pin pair is raised to the shear line. The presence of the floating pin does not significantly affect the lock's pickability, but it does make key duplication more difficult since standard Mul-T-Lock blanks cannot be used and the Interactive blanks are more tightly controlled. There is also a decrease in the number of available system codes since the lengths of the inner and outer pins driven by the floating pin must be the same.

2.11 Rotating Pin

Emhart

(US) 6-pin (3–4)

Invented in 1975 by L. Raskevicius, the Emhart is an ingenious pin-tumbler lock using six specially constructed, rotating-interlocking upper and lower pins. Pictures of the lock appear in Figs. 2.193–2.195. The original design called for magnetized driver pins, although this idea did not eventuate in the production model, a diagram of which is shown in Fig. 2.196. The Emhart cylinder was produced by Corbin-Russwin, now part of the ASSA Abloy Group. Production of the Emhart is being wound down.

As can be seen from Fig. 2.194, the driver pins are cut to form a T at the bottom, which mates snugly with a T-shaped gap in the top of the lower pin. It follows that the pins can only be disengaged by shear (transverse) motion. The active end of the lower pins is V-shaped, with the axis of the V aligned at one of number of possible angles with respect to the T cut at the top. The base of the T-shaped cut in the lower pins must be simultaneously raised to the shear line and rotated to the correct angle ($\pm 20°$) by the angled bittings in the key so that it can disengage from the driver pins (see Figs. 2.195 and 2.197). The system does not support number 1 key cuts since these are too shallow to reliably rotate the pins. Further bitting constraints

Figure 2.193: Emhart 6-pin cylinder and key with angled cuts.

Figure 2.194: (Left) Emhart plug. (Right) Rotating interlocking pins and drivers.

Figure 2.195: (Top) Emhart plug with key partially inserted. (Bottom) Key fully inserted to align pins with grooves in plug.

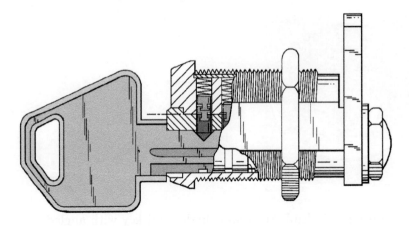

Figure 2.196: Emhart cam lock from US patent 4,208,894 (1978) by W. E. Surko Jr.

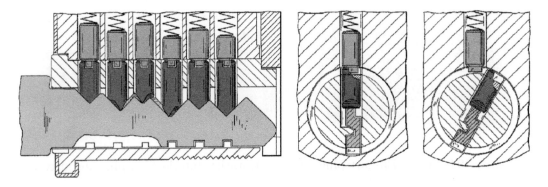

Figure 2.197: Operating principle of the Emhart lock (US patent 4,098,103 by L. Raskevicius).

include a variable MACS due to the two possible cut angles. If adjacent cuts are at the same angle, a MACS of 4 is possible, whereas adjacent cuts at different angles reduce the MACS to three depths of cut.

Since the top of the T-section protrudes past the shear line, grooves are milled in the circumference of the plug to allow clearance for the pins. In addition, the bottom edge of the key must have crenellations so that it will not be obstructed by the bottom ends of the driver pins as it turns through 180 degrees. Master-keying pins may be added (see Fig. 2.196) that change both the depth and angle of cut. This gives a second degree of freedom in developing a master-keying scheme (the same is true of Medeco locks, which are covered in Chapter 4). The plug also contains hardened rods and a drill-resistant crescent to deflect drill bits (see also [106]).

A competing design was proposed in 1987 by J. M. Genakis (US patents 4,932,229 and 4,998,426). Like the Emhart, Genakis's lock utilized mating top and bottom pins. The bottom pins also included a slot intended to register with the prongs of a side-bar, as in the Medeco lock.

2.12 Pin Matrix

Vingcard

(NO) 32-pin (3–4)

The Vingcard series of mechanical locks (models 1040, 1050, and 1060), manufactured by the Norwegian company Trioving, is designed primarily for high-traffic applications requiring frequent rekeying. They are thus ideally suited to hotels where a given key may be used only a few times before another key must be issued to an incoming guest using the same room. The Vingcard 1050 lock described in this section is illustrated in the series of Figs. 2.198–2.204. A forerunner of the lock, patented in

Figure 2.198: Front and rear of Vingcard 1050 lock with auxiliary keyed access.

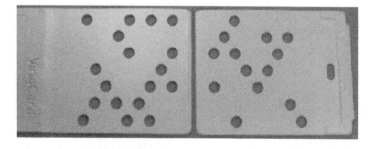

Figure 2.199: Vingcard pass and control cards are complementary.

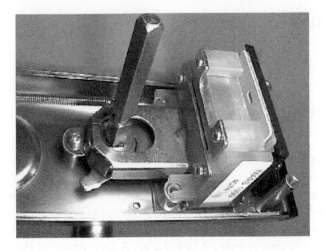

Figure 2.200: Handle mechanism and control box with control card installed. Handle is decoupled from spindle in locked position.

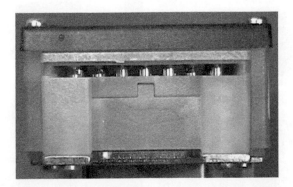

Figure 2.201: Rear of control box with pass card inserted from front showing slot for control card.

 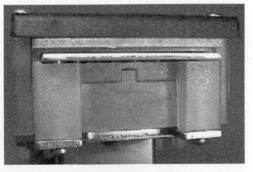

Figure 2.202: (Left) Both cards correctly inserted and carriage in open position. (Right) Pass card inserted upside down, the control card is pushed down too far.

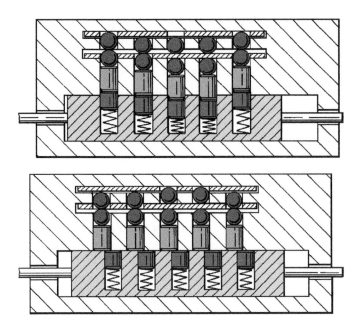

Figure 2.203: Vingcard lock principle from 1979 US patent 4,149,394 by T. Sornes.

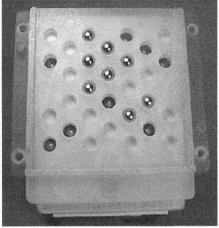

Figure 2.204: (Left) Control box with cover removed to show pin matrix. (Right) Separator and eight ball bearings installed showing masking of some pins by control card.

1977, featured a 5×5 matrix of pins. Diagrams from the U.S. version of the patent are given in Fig. 2.203.

The Vingcard lock (Fig. 2.198) incorporates a conventional 6-pin security cylinder that gives keyed access to the room for cleaning, maintenance, or in an emergency. When locked, the exterior handle moves freely, preventing the lock from being opened by brute force applied to the handle. The lock is equipped with an antithrust bolt so that the primary bolt cannot be shimmed or opened by "loiding" with a credit-card.

Two punched plastic cards (see Fig. 2.199) are required to operate the Vingcard lock. The first card is a "key" or "pass" card, which is issued to the guest and inserted right-way-up in the slot at the front of the lock. The second card is a "code" or "control" card; this is inserted into the rear of the lock and is normally concealed behind an access flap fastened by a hex bolt. The control card is used to "program" the lock, that is, to set its combination. Because the cards are made from perforated plastic, they are inexpensive and easy to replace, thus providing a practical means of recombinating the lock. A given pass card will only operate the lock while the control card for which it is designed is inserted. Once the control card is changed or removed, the previous pass card will no longer work.

We now turn to the internal operation of the lock. The front and rear handles are connected by a spindle with a square shank (see Fig. 2.200). The spindle is in three parts that are free to turn relative to each other. The central part of the spindle actuates the bolt. A slideable coupling ring is located on the front portion of the spindle; normally, this ring does not overlap the middle part of the spindle, and thus the front handle is decoupled from the bolt. For the front handle to operate the lock, the coupling ring must be displaced so that it couples both the front and middle parts of the spindle. The displacement of the coupling ring is governed by a polycarbonate control box constructed like a drawer. The remainder of this section focuses on the functioning of the pass card, control card, and control box.

The control box (Fig. 2.201) consists of a fixed bottom half and slideable top half, or carriage, to which a flange is attached. The flange is needed to drive the coupling ring. The control box contains 30 conventional sprung pin-tumbler pairs, arranged in a nonrectangular array with their tips facing upwards. There are also two larger-diameter pins that are activated only by the control card (see Fig. 2.204). This means that the lock cannot be operated by a blank pass card when the control card is not installed.

When both the control card and pass card are correctly inserted, and all 30 pins are depressed to their respective shear lines, the carriage is free to slide as the pass card is pushed to the back of the slot. The motion of the pass card is transferred to the coupling ring, which engages the front handle of the lock. Similarly, withdrawing the pass card, which is held in position during opening, moves the carriage of the control box back to its original position.

How is it that both the pass and control cards are needed to operate the lock? Interestingly, if either card is inserted upside down, the lock will not operate (see Fig. 2.202). To delve further, note that the carriage has a separator that divides the space under the cover of the control box into two slots or chambers. The control card is inserted from the rear into the lower chamber, while the pass card is inserted from the front into the upper chamber. Thus there is a slight vertical offset between the two cards equal to the thickness of the separator.

Each of the 30 pins has its conical end protruding into the lower chamber. The borings for the pins extend through the separator, creating a space for a ball bearing that sits atop the pin and is limited in its vertical travel by the ceiling of the upper chamber.

There are two ways to set a pin-tumbler so that its shear line is at the correct depth. First, a pin may be depressed by one card's width from the top of the lower chamber. Equivalently, the ball bearing may be depressed by one card's width from the top of the upper chamber. The first action is achieved by inserting the uncut portion of the control card into the lower slot at the rear of the lock. The second action is achieved by inserting an uncut portion of a pass card in the upper slot at the front of the lock (see Fig. 2.203).

The dimensions of the pin plus ball stack within the control box are such that any downward displacement of the pass or control cards will cause the shear line of the pin to be below the interface between the fixed and sliding parts of the control box, preventing opening.

It should be clear that a completely blank control card (i.e., one with no holes) on its own could move all 30 pins into their correct positions to allow opening. However, the lock always operates with a control card installed that has at least some holes in it. The pins that encounter holes remain fully raised and must be actuated by the pass card in the upper slot. It is important, however, that the pass card not contain any holes in the same positions as the control card. If this were to happen, then the pass card would press the ball bearing down against the top of the control card, misaligning the corresponding pin-tumbler. It is therefore necessary for the pass card to contain holes in precisely those positions where the control card is uncut. In other words the pass card and control card must be *complementary* across all of the 30-pin matrix. (Remember that the two larger pins in the lower chamber can only be operated by the control card.)

When a correct pass card is inserted, the remaining pins are brought to the shear line without upsetting the alignment of those pins actuated by the control card. With all 30 pins correctly depressed, the pass card pushes against the backstop of the upper chamber, causing the carriage to slide and engage the lock mechanism. During opening, the ball bearings engage holes in the pass card, capturing it. The pass card remains captive until the carriage is returned to the locked position, at which point the ball bearings slip down into their borings as the card is withdrawn.

With 30 essentially binary pins, it is not hard to see that there are $2^{30} = 1,073,741,824$, or well over one billion different combinations of control and pass card holes (although some trivial combinations have to be excluded). Naturally, the two cards are not required to have the same number of holes.

In terms of the vulnerability of the lock, we make the observation that the pin matrix is susceptible to decoding. This is the case since there is a difference in tension between a ball that rests on an undepressed pin and one that rests on an uncut portion of the control card. Nonetheless it would take a considerable amount of time to decode the control card accurately by hand and hence fabricate a pass card. The decoding idea is explored further in a 1994 patent by M. W. Tobias (US 5,355,701) where a matrix of pressure-sensitive resistive elements is proposed.

Another popular system produced by Trioving is an electronic version of the hotel lock that utilizes a plastic card with a magnetic strip. The card is inserted into a magnetic card reader at the front of the lock. This system, which is gaining popularity, has the advantage that cards, once used, may be reprogrammed instead of being disposed of. A further system, the Ilco Marlock, works using infrared light transmission through three perforated strips on a key. The perforations are opaque in the visible part of the spectrum but transparent at infrared wavelengths [127].

2.13   Key-Changeable

Code Lock

(UK) 6-pin (3–4)

Around the time of World War II, a requirement emerged for a compact and effectively pick-proof lock that could also be easily recombinated. The U.K. Ministry of Defence adopted the so-called Code lock, shown in Figs. 2.205–2.208, to fulfill this requirement. It had theoretically one million user-settable combinations on a miniature key. The design patent for the Code lock was filed in 1938 by B. Sterner in Belgium (US patent 2,424,514), the final version of the patent appearing after the war in July 1947. A more up-to-date specification is provided in French patent 974,712, which was published in 1951 under the company name of Code Designs.

Sterner made improvements to an earlier design of lock and key by O. D. Von Mehren, described in US patents 1,819,853 (1928) and 1,899,739 (1929). The improvements

Figure 2.205: The mechanically reprogrammable 6-pin Code padlock and key.

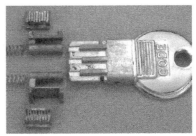

Figure 2.206: (Left) Code cylinder lock. (Right) Key, saddle pins, and wedges.

Figure 2.207: (Left) Armature of Code lock. (Right) Key inserted with one pin aligned with wedge.

Figure 2.208: Rear views of Code cylinder during operation.

rendered the lock more difficult to decode by concealing the tumblers. The Code padlock was subsequently used to secure military and ministerial despatches. The lock was also produced as a cylinder lock for doors. Although the lock is now obsolete, having become somewhat of a collector's item, we include it here because of its historical significance and innovative design.

Code padlocks and cylinder locks (see Figs. 2.205 and 2.206) have a rectangular keyway with three channels milled on each long side. Axial pin-tumblers are arranged in the channels and depressed by the corresponding ribs on the key. The appearance of the keyway belies the complexity of the lock's internal construction. From the outside, one might be tempted to believe that the operating principle is the same as that of an ACE or GEM tubular lock or perhaps a side-bar lock such as Tubar. However, this is not so. Much of the lock's complexity stems from the fact that it is recombinatable or key-changeable.

The lock, which is made exclusively from die-cast components, comprises a stator and an armature,[13] as illustrated in Figs. 2.207 and 2.208. Both the stator and armature are molded with six longitudinal channels. The channels in the armature straddle the keyway as well as its outer rim leaving six edges that are serrated, each forming a rack with 18 teeth. This rack plays a crucial role in setting the combination of the lock.

The six pins or "U-shaped pushers" as Sterner referred to them, ride in the channels of the armature and are spring-biased from the rear of the cylinder. Each pin is coupled to a saddle in the outer rim of the armature that follows the axial motion of the pin. The saddles have ends that extend into an annular gap between the armature and stator. The armature would be free to turn if not for the presence of a further component, which we shall refer to as a wedge (see Fig. 2.206). There are six of these also, each one constrained to move in a channel of the stator. Wedges are T-shaped in cross-section with the top of the T fitting neatly into the profile of the stator channel. The inner face of each wedge is serrated, having 10 teeth, the spacing of which matches the racks around the armature. The overall length of a wedge allows it just to pass between the raised ends of the pin saddles.

The operation of the lock is best understood by first fixing the positions of the wedges in each channel and subsequently allowing these positions to vary. Suppose that all six wedges are held in position in the stator channels by the meshing of their toothed edges with the racks on the armature. The range of movement of the pins is such that there is always an overlap in the relative position of the wedge and saddle. With no key inserted, all the pins are fully forward at the front of the keyway. In this position the cylinder is locked since one end of each saddle is in overlap with its neighboring wedges. As the key is inserted, the pins drive the saddles to various depths, depending on the setting of the ribs in the key blade.

Assuming clockwise opening, there is a unique depth for each pin at which the ends of its saddle bracket the wedge immediately to the right. The correct key simultaneously brings all saddles into alignment with the corresponding wedges, at which point the armature is no longer obstructed and may be turned. As the armature is turned, the wedges are kept in place by their meshing with the racks and because they are bracketed by the saddles. Once the key is turned to 30°, the racks fall out of register with the wedges and each wedge is then held only by its saddle. At this point, the key stem is captured because its shoulders have encountered the notched ring around the

---

[13]Use of this unconventional terminology seems justified given the resemblance of the lock's internal mechanism to an electric motor.

keyway. The lock may either be relocked with the same key or it may be recombinated with a new key, which we consider next.

The code-changing operation involves a change in position of the wedges. The key that opened the lock can be removed in the 30° position if the code-change ring is turned relative to the body of the lock. With the key removed, the springs return the pins to the front of the lock, except that now the saddles have also drawn the wedges along with them in the stator channels. When a new key is presented, it depresses the pins to the depths at which its ribs are set. The new key is then turned counterclockwise and in so doing, brings the wedges back into mesh with the racks of the armature. The saddles disengage from the wedges, which are now set in new positions in their channels. The new key is withdrawn and, voilà, the combination has been changed. Now only the new key will open the lock.

With the metal sleeve removed from the neck of the key, the six ribs can each be set in one of 10 positions, yielding $10^6 = 1,000,000$ combinations. However, the key may be inserted either way round so that, for instance, a key with code (4 3 7 2 6 5) could open both a lock with code (4 3 7 2 6 5) and one with code (2 6 5 4 3 7). The effective number of combinations is therefore halved to 500,000 in terms of the number of different keys [88]. The same argument applies to keys with codes that are invariant to being flipped, e.g., (4 3 7 4 3 7). There are 1,000 such keys that could be inserted either way round.

The Code system, though very clever in its intricacy, has some rather serious shortcomings. The key cannot be turned more than one-sixth of a turn since at this point the saddles encounter wedges that are set at the wrong height. Owing to its complexity, the components of the lock have to be die-cast, which reduces its mechanical strength and results in a less durable product. With frequent use, the rack and wedge mechanism wears out and ceases to function reliably. The lock is also not immune to manipulation, although it is rumored that no one managed to pick it open during World War II.

Chapter 3

Wafer Locks

. . . a specification of a lock, constructed on a new and infallible principle,
. . . *Joseph Bramah, c. 1784*

3.1   Introduction

The disc-tumbler, or wafer lock as it is more commonly called, originated in the United States during the late 19th century. The earliest wafer lock patents were awarded to P. S. Felter in 1868 and to H. S. Shepardson in 1870 [50]. Felter, the founder of the American Lock Company, invented a double-sided disc-tumbler cylinder lock (US 76,066), which we cover under the heading of plate-wafer locks. Shepardson, who had worked with Linus Yale Junior in the Yale Lock Manufacturing Company, developed what is recognizably the first conventional wafer lock and subsequently formed his own company. According to Arnall [1], the first disc-tumbler padlocks also appeared around this time.

While the wafer lock design is simple in principle, the lock is not suited to production on traditional workshop equipment such as lathes and milling machines due to its slotted construction. For this reason, the body and plug of the wafer lock are generally made by die casting of zinc alloys, a process that dates from 1869. The wafers are made from stamped brass or steel. With the advent of reliable die-casting processes, like small-slot casting in the 1920s, high-volume production of wafer locks became economical due to the relatively small number of components required. However, early wafer locks suffered from cracking due to corrosion of the cast components.

Compared with pin-tumbler locks, manufacturing tolerances for wafer locks are generally not as tight, and as such they are often seen as a low-cost, lower-security alternative. This fact has led to their widespread use in the automotive industry as well as in office furniture and cabinets where they typically function as a cam lock.

In the electronics industry, wafer locks are usually incorporated in key switches for machinery control.

In its basic form the wafer lock is not a high-security lock, since it is die-cast (and hence easily destroyed) and it provides only a small number of differs. Furthermore, it has limited potential for master-keying. On the positive side, it is inexpensive, modular, and compact—not requiring an elongated housing for the driver pins and springs. Although wafer locks generally provide only a low level of manipulation resistance, various refinements to the design have been made that greatly increase their security. We examine these matters further on when we come to discuss the different types of wafer lock. The operating principle, which is easy to understand, is covered in the next section.

Construction and Operating Principles

The two most commonly encountered types of wafer lock are single-sided and double-sided, also known as single-throw and double-throw. The single-sided wafer lock, illustrated in Figs. 3.1–3.3 by a Lowe and Fletcher 5-wafer lock, consists of a plug

Figure 3.1: Lowe and Fletcher 5-wafer cam lock and key.

Figure 3.2: (Left) Wafer lock housing with two channels. (Right) Side view of cam lock.

Figure 3.3: (Left) Plug with three wafers of different sizes. (Middle) Plug with key partially inserted. (Right) Plug with key fully inserted.

fitting inside a cylindrical shell or housing. The plug is slotted in a number of places (usually five) along its length to accept flat, spring-biased wafers of an overall shape that is close to rectangular. The plug may also contain a further slot at the rear that accommodates a retaining wafer or clip. Finally, depending on the intended use of the lock, a cam may be fitted to the rear of the plug that provides the locking function. In automotive locks, the cam is connected to a linkage rod that actuates the door lock. For a desk lock, an eccentric stump on the back face of the plug drives a slotted bolt that moves in a pair of guide holes in the lock housing.

Moving to a higher level of detail, we notice that the inside of the shell is not regularly shaped, as in pin-tumbler locks, but contains two or more longitudinal channels (see Fig. 3.2). These grooves function as the chambers for the wafer-tumblers and are sufficiently deep to accept the full displacement of the wafers under the action of the key. The front of the shell has a circular ledge that fits around the enlarged diameter end of the plug to prevent shimming of the wafers.

Most wafer locks are designed to provide a fraction of a full turn, (e.g., 90 or 180 degrees), and in such cases the shell is equipped with shouldering either at the front or rear to accomplish this. For front-shouldered wafer cylinders (like the one in the illustrations), the plug is provided with a cooperating shoulder that moves in a secondary recess in the shell. Angular movement is limited to between the two points where the shoulders of the plug and shell come into contact. In the case of rear-shouldered cylinders, the rotation is stopped by a specially shaped washer mounted alongside the cam on the square shaft of the plug. In wafer locks where it is desirable to withdraw the key at angles other than 12 o'clock, a secondary set of channels is required in the shell. Thus a cylinder that allows keyed operation at 0° and 90° would have four channels on the inner surface of the shell.

The most intricate part of the mechanism is the plug (Figs. 3.1 and 3.3). A rectangular keyway, extending almost to the rear of the plug, is flanked by longitudinal wards on one or both sides that must be matched by the bullets on the key blank. The keyway has a recessed face or pair of faces against which the shoulders of the key abut when it is inserted. Combined with the warding in the keyway, these also serve to locate the key in the vertical plane.

Each wafer slot in the plug has an adjacent hole for a driver spring. Due to the proximity of the wafers, they are often arranged in an alternating sequence along the plug to allow room for the spring chambers. Wafers have straight sides, rounded ends, and a rectangular cut-out through which the key passes. The wafers are supported on one side by an arm or shoulder that rests on the driver spring (refer to Fig. 3.4).

Now although all five wafers have the same overall dimensions, the vertical positioning of the rectangular cut-out may vary from one wafer to the next. In this way the wafers are made to differ. Typically, there are five wafer sizes; however, in some systems as few as three or four may be available. Car locks may use as many as nine or ten different sizes to ensure a large number of codes (see Chapter 7).

If we view the cylinder such that the key is inserted with the cuts facing "up" (at 12 o'clock) as in Fig. 3.1, then the wafers point "down" in the sense that their driver springs cause the edge nearest the shoulder (the "bottom") to be proud of the rim of the plug. Thus prior to inserting a key, the bottom edge of each wafer protrudes into the channel in the shell at 6 o'clock. The wafers are prevented from hitting the bottom of the channel under spring tension by the presence of the shoulder (or sometimes a small retaining lug opposite it), which limits its motion in the slot. The construction is such that the wafers must usually be pulled or pushed out of

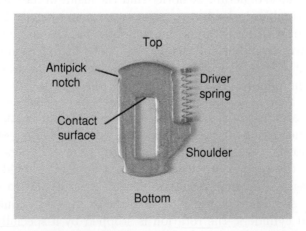

Figure 3.4: Naming conventions for standard disc- or wafer-tumblers.

their slots with a considerable amount of force (when servicing the lock). Therefore, in the locked position the bottom edges of the wafers fully engage the lower channel, preventing the plug from being turned. The top edge of the wafer will be some distance below the rim of the plug at 12 o'clock. The length of each wafer is exactly equal to the diameter of the plug, so slightly overlifting any wafer will cause its top edge to protrude into the top channel in the shell.

Clearly, it is necessary to raise each wafer by the exact amount required to make its ends flush with the rim of the plug. At this point a shear line is created between the ends of the wafers and the shouldering of the shell, allowing the plug to rotate. The function of the cut-outs in the wafers can now be appreciated: with the wafers at their lowest positions, it is the edge of the cut-out closer to the top edge of the wafer that protrudes into the keyway. As the key is inserted, its bittings contact the cut-outs of the wafers, displacing each one upward against its driver spring. With the key fully inserted, the top of each cut-out rests in the "V" of each cut in the key, and the code of the key is thus presented to the lock.

A weakness of the wafer lock mechanism is that the edges of wafers of different sizes protrude to different extents into the keyway (see Fig. 3.1). It is therefore possible, with practice, to decode the lock combination through a visual inspection of the keyway.

Codes, Permutations, and MACS

Assuming five possible wafer sizes, and therefore five different depths of cut for each position on the key, a 5-wafer lock has a maximum number of codes equal to 5^5 or 3,125. Most wafer locks have a depth increment of between 20 and 30 thousandths of an inch. In practice, the number of codes in a real key series will be much less than this due to constraints such as the MACS, undesirable combinations, and the requirement that key codes should differ by at least two sizes to minimize the possibility of key interchange (see Chapter 2). The bitting rules are typically of the form:

1. The MACS is three or four.

2. Only two adjacent cuts may be the same.

3. At most three cuts may be the same.

4. At least three cuts must be different.

5. All codes must differ by two or more depths of cut in at least one position.

The number of codes for 5-wafer locks with five sizes of wafer is tabulated in Table 3.1. The table shows the number of codes as a function of the MACS when

MACS	With MACS Applied	With Constraints Applied	Required to Differ by Two
3	2,309	1,890	274
4	3,125	2,640	399

Table 3.1: Number of permutations for 5-wafer locks with five sizes subject to bitting rules.

MACS	With MACS Applied	With Constraints Applied	Required to Differ by Two
3	2,309	2,034	296
4	3,125	2,820	589

Table 3.2: Number of permutations for 5-wafer locks with five sizes subject to relaxed bitting rules.

(1) only the MACS is applied; (2) the MACS and rules 2–4 are applied; and (3) all five bitting rules are applied. Table 3.2 shows the resulting number of codes when rule 2 is relaxed to "up to three adjacent cuts may be the same".

It can be seen that in practice only several hundred different codes may satisfy these constraints, a number substantially inferior to that offered by a 5-pin cylinder lock. With so few effective differs, there is a much greater chance of nonuniqueness of keys.

Furthermore, the less stringent manufacturing tolerances typical of wafer locks result in the possibility of what are known as *try-out keys*. This is a set of keys, usually around 50 in number, that are cut to half sizes expressly to exploit the lower tolerances of the lock. Each key in such a set approximates several original keys since it is at most half a cut away from the original in each bitting position. For instance, a try-out key with code $(1\frac{1}{2}\ 3\frac{1}{2}\ 4\frac{1}{2}\ 2\frac{1}{2}\ 1\frac{1}{2})$ can approximate $2^5 = 32$ possible key codes in the series (1 3 4 2 1), (1 3 4 2 2), ..., (2 4 5 3 2). Alternatively, a set of "computer keys" of flat section and wavy cuts (similar in shape to some lock-picks) can be inserted one at a time and tried out. It is to be expected that at least one key in a set can be made to operate a lock when it is jiggled in the keyway (hence the alternative name of "jiggler key"). Naturally, this technique works better with wafer locks that are somewhat worn through frequent use.

Master-Keying

Whereas pin-tumbler locks can have several pins in each pin chamber, wafer locks only have one wafer per chamber. This construction is very restrictive on the master-keying possibilities for the lock. The practical embodiment of master-keying for a wafer lock is to replace one or more wafers with dual-profile wafers called

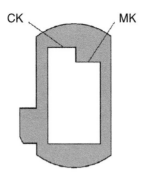

Figure 3.5: Two-profile wafer-tumbler for master-keying.

master discs. This method of wafer master-keying was patented in the United Kingdom in 1909 [21].

A master-keyed wafer has two steps in the top part of the cut-out, corresponding to the contact points for the change and master keys (as in Fig. 3.5). Each step occupies half the width of the cut-out. The second ingredient is a pair of key blanks of differing sections. One key blank, having the top part of its blade on the left, actuates the left step on the dual-profile wafers. Similarly, the other key blank, which has its blade offset to the right, operates the right step on the wafers. In this manner two different keys (with different sections and cuts) can be made to operate the same wafer lock.

The Lowe and Fletcher standard system, which has three wafer sizes numbered 1, 2, and 3, provides for three master discs with steps sizes 1&2, 1&3, and 2&3. These may be inserted in either orientation. The same principle extends to allow one key (the master key) to operate a series of differently coded locks. Multilevel master-keying of conventional wafer locks is, however, not generally possible.

Double-Sided Wafer Locks

Up to this point the discussion has centered on single-throw wafer-tumbler locks, that is, locks operated by a key with cuts on only one side of the blade. With little extra effort on the manufacturer's side, the wafer lock can be made to require a double-sided key, as in Fig. 3.6. Only the plug and key need be different, but not the barrel. Of course a double-sided, symmetrically cut (or reversible) key can be used in a single-sided wafer lock as long as the keyway is appropriately fashioned to support the key in either orientation. The matter at hand, however, is the double-throw wafer lock, for which the key cuts on the upper and lower edges of the key blade, are in general not identical. To implement this idea, it suffices to change the arrangement of wafer chambers in the plug to allow the insertion of wafers from both the top and bottom.

Figure 3.6: Lowe & Fletcher double-sided 10-wafer cylinder and key.

Figure 3.7: Plug from 10-sided wafer lock with key fully inserted.

We have already described the single-throw case in which the wafers are inserted and sprung from the slots located along the bottom edge of the plug. To create a two-sided wafer-tumbler lock, some of the slots must be reversed so that wafers can also be loaded along the top of the plug. Naturally, the spring chambers for these wafers are also on the top edge. The top-loaded wafers are then actuated by the bottom edge of the key and vice versa for the bottom-loaded wafers. Given a plug with five slots, we could have two top wafers and three bottom wafers or any other arrangement adding up to five.

In practice, the wafers are often placed in opposing pairs along the plug. Such is the case for the Lowe & Fletcher double-sided 10-wafer cylinder shown in Figs. 3.7 and 3.8. The slots, which are twice as wide as in a single-throw lock, have a spring chamber at the top and a spring chamber in a diagonally opposite position on the bottom edge. In each pair the front wafer, for instance, is upwardly sprung ($\uparrow$) and slides against the back wafer, which is sprung downward ($\downarrow$). This "doubling-up" is an economical arrangement resulting in a plug with 10 wafers in total, being only slightly longer than a 5-wafer plug. Many other loading sequences are realizable: for example, instead of $\uparrow\downarrow\uparrow\downarrow\uparrow\downarrow\uparrow\downarrow$, we could have $\uparrow\downarrow\downarrow\uparrow\uparrow\downarrow\downarrow\uparrow\uparrow\downarrow$ by appropriate molding of the spring chambers in the plug. Such systems for two-sided wafer locks are in use by numerous car manufacturers, and examples of these kinds of locks may be found in Chapter 7.

A number of advantages accrue from the double-sided design, the most notable being that the manipulation resistance of the lock is greatly increased; the number of keying combinations is greatly increased; and the cost of production is not

Figure 3.8: Side view of insertion of double-sided key as it brings wafers into alignment.

significantly affected. As an added bonus, it is still possible to have a reversible key as long as the MACS constraint is accounted for. A reversible key for a double-sided 10-wafer lock would have 10 cuts on each side of the key, since the odd-numbered bittings (say) operate the top wafers while the even-numbered ones operate the bottom wafers. Double-sided wafer locks with reversible keys are widely used in the automotive sector.

Antipicking Features

Many other design modifications for wafer locks have been introduced, mostly by the car industry, and we briefly mention a few of these here. While the overall shape of wafers for single and double-throw locks is the same, both types can benefit from antipicking features (see Fig. 3.9).

The first type of pick-resistant wafer (C in Fig. 3.9) has a number of steps cut into the side opposite the spring arm. The wafer chamber is provided with an undercut edge opposite the spring chamber. In normal operation, when the key is inserted, the wafer is lifted such that its sides remain parallel with those of the chamber. If, however, an attempt is made to pick the lock, as the top of the wafer contacts the channel in the shell, the sideways pressure causes the wafer to skew and become snagged during its upward motion.

A second type of wafer (B and D in Fig. 3.9) has the usual rectangular profile except for the bottom edge, which is chiselled on the sides such that it is slightly wider at the end, resembling a trapezium. The channels in the barrel have an inverse bevel

Figure 3.9: An assortment of wafer-tumblers. B–F contain antipicking features.

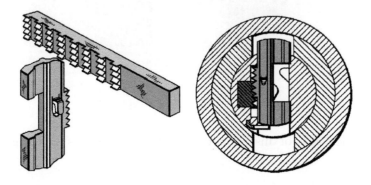

Figure 3.10: J. W. Fitz Gerald's 1933 serrated-rack wafer lock (US patent 1,965,889).

that matches the shape of the end of the wafer. The width of the channel at its narrowest point is still larger than the width of the wafer. The construction is such that keyed operation is unaffected, but underlifting of the wafer while turning force is applied will result in the end of the wafer binding in the channel.

Yet another design, which has overtaken the previous two in popularity, is to include wafers that are serrated at the top on one or both sides (E and F in Fig. 3.9). The germ of such an idea was described in Shepardson's 1870 patent (US 99,013), one of the earliest wafer lock patents, but it was way ahead of its time. According to the specification, the sides of the channels in the body of the lock are equipped with matching longitudinal serrations. This device, widely used in car locks, causes the wafer to become ensnared at various stages of overlifting, and is a highly effective deterrent to manipulation. In the case of two-sided wafer locks, the serrations may be applied to both the top and bottom of each wafer. Thus the wafer will stick in both the underlifted and overlifted states (see Chapter 7). A more elaborate version of the serrated wafer-tumbler is shown in Fig. 3.10. In this design, tensioning the plug causes the teeth of the side rack to engage the serrations in the wafers.

Further Examples

It remains for us to discuss a number of wafer lock types that differ significantly from the one- and two-sided varieties treated above. While some of these predate the modern wafer lock, we have placed them under the same organizational heading due to the similarities in design. The oldest and most noteworthy of these is the Bramah lock. Invented by Joseph Bramah in 1784 (UK patent 1,430), the Bramah lock contains a circular arrangement of wafers or *sliders*, with notches or gates in one or more places on their outer edges. The wafers are pushed axially into the lock by the key so as to align the gates and clear fixed radial obstructions in the lock. The reader is referred to Appendix E for an account of Bramah's contributions to English society and some of the colorful episodes surrounding Bramah's 18-slider lock, a picking challenge that was finally won by A. C. Hobbs around the time of the Great Exhibition. A detailed description of a modern Bramah lock is given later in this chapter.

A second major variant on the wafer lock principle is the so-called Bell lock: a lock containing a series of bar-wafers of square section [105]. The design of a Bell-type lock is succinctly described in a 4-page patent from 1918 by E. O. Bennett (US 1,328,074). In such a lock, the bar-wafers are isolated from the keyway and communicate with it only through a short stump or lug mounted perpendicularly on the midportion of each bar. The key for such a lock is not cut on either edge but instead has one or more tracks milled along one or both sides.

The Bell lock is also referred to as a Dudley lock. Although the early Dudley patents were for wafer locks, they later included Bell lock features like side-track keys, as depicted in Fig. 3.11. Subsequent patents were taken out in 1966 (US 3,263,461) and 1970 (US 3,509,749) for a recombinatable Bell lock with a plug comprising a number of rotatable sections. We later present more modern versions of the Bell lock principle including locks by SEA and EVVA.

Another type of wafer lock, which we refer to as a contoured or plate-wafer lock, possesses a multiplicity of wafers that slide alongside each other—there being no separation between them. Plate-wafer locks can be made with a variety of differently shaped wafers actuated by different facets on the key blade (as in some types of Dudley lock). The idea is illustrated in Fig. 3.12, taken from a 1918 patent by

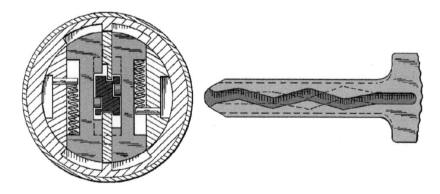

Figure 3.11: J. F. Svoboda's 1934 patent for a multiple-action Bell-type wafer lock (US 2,039,126).

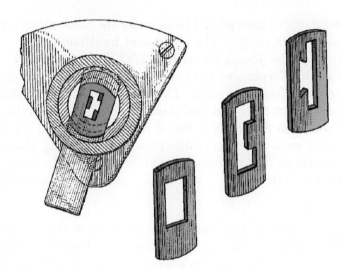

Figure 3.12: Design of a multiple-action plate-wafer lock (US patent 1,287,882 by F. A. Christoph).

F. A. Christoph. The Bricard 14-wafer lock in this chapter has a bilateral key that operates in this manner.

A further variation on the wafer-tumbler theme, popular for late-model automobile locks, is the split wafer. This system, described in greater detail in Chapter 7 on automotive locks, utilizes a series of wafers that are split vertically along their central axis, similar to Fig. 3.11. Both halves of the wafer function independently as they are picked up by ramps on opposite sides of the key stem. This doubles the number of active elements in the lock without requiring a longer cylinder plug.

Since it does not use drivers as a pin-tumbler lock does, the wafer mechanism is well suited to key-change operations. A number of interesting designs for key-changeable wafer locks have been proposed, such as the mechanically reprogrammable lock in Fig. 3.13. Other designs include the Rielda and Winfield locks, which we describe in due course. The basic operating principle is that of a two-part wafer whose halves mesh along a serrated edge called a rack. When the plug is turned to a predetermined angle and the key is withdrawn, the wafer halves may be realigned by a new key. Once the new key returns the plug to its locked position, the rack is reengaged and the combination thereby changed. As early as 1931, Sargent and Greenleaf (US patent 1,917,302) designed a recombinatable wafer lock along these lines.

Wafer Lock Classification

As the reader ventures further into this chapter, wafer locks of various geometries will be encountered. The following scheme should help the reader to understand how we have chosen to organize this material.

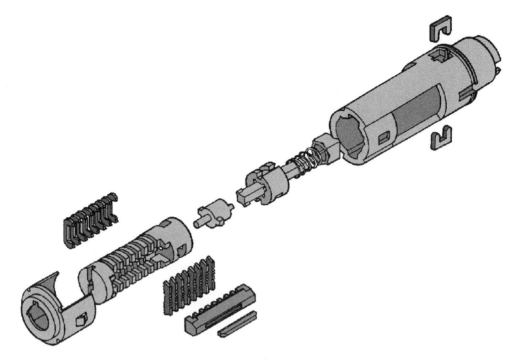

Figure 3.13: Mechanically reprogrammable 8-wafer lock (1988 US patent 4,966,021 by N. Boag).

1. Conventional: locks with inline wafer-tumblers operated by single or double-sided key. May include antipicking features. Examples: Miwa, Bricard.

2. Contoured or plate-wafer: locks whose tumblers are contiguous and therefore not in separate chambers along the plug. Examples: American, DUO.

3. Three-sided: wafer locks with tumblers that operate in three different directions with a key of T-shaped section. May include an active element in the key blade. Examples: Bricard SuperSûreté.

4. Inline push type: wafer locks whose tumblers are pushed to successive depths to clear fixed obstructions in the shell. Key is end-bitted. Example: ABA.

5. Bell: wafer locks of Bell or Dudley type having a series of bar-wafers flanking the keyway on one or both sides and requiring a side-milled key. Examples: Bell, SEA, Lori, EVVA 3KS [Chapter 4].

6. Axial: tubular wafer locks of Bramah type. Wafers arranged around a circular keyway are pushed to various depths so that their gates clear fixed obstructions. Key is end bitted. Examples: Bramah, Laperche, Vigie Picard, Fontaine.

7. Key-changeable: wafer locks that, when turned to a certain position, accept a different key to the one that opened them. Examples: Rielda, Winfield.

The following two types are presented in Chapter 7 on car locks. Both are similar in principle to the Bell and Dudley locks.

1. Reverse-cut: locks with nonrectangular cut-out in wafers that accept a key having a milled ridge along one or both sides of the blade. Examples: Holden Commodore, Mercedes two-track, Volvo.

2. Split-wafer: wafers are cut in half and have either a protruding peg that is displaced by a side track in the key blade or an active edge that contacts a ridge on the key. Key may have side tracks or ridges on both sides. Examples: Mitsubishi, Mercedes four-track, Porsche.

3.2   Conventional

Miwa

(JP) 10-wafer (2)

The Miwa Lock Co. Ltd. (Miwa Lock K.K.), established after World War II, is Japan's leading manufacturer of mechanical locks and electronic security systems. The Miwa wafer lock was introduced in 1955 and until recently was extensively used in Japan, with reports of as many as 70 million units installed in domestic and commercial premises. Miwa also produces a range of electronic locks based on magnetic swipe cards, smart cards, keypads, and fingerprint recognition.

The Miwa wafer lock (Figs. 3.14–3.16) is a precisely machined double-sided wafer lock housing up to 10 wafer-tumblers. The design is mentioned in patent JP 11-315654 (1999) and is available in a multitude of different formats for entrance

Figure 3.14: Key and core from a Miwa 10-wafer entrance set.

Figure 3.15: (Left) Miwa core and mounting rod with front cover removed. (Right) Set of wafers and springs.

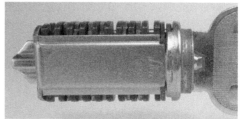

Figure 3.16: Operation of Miwa wafer lock by key.

locks. The construction of the lock is well suited to low-cost high-volume production since it does not have a solid core. A series of steel laminations are stacked inside a holder with straight edges, folded at the rear, and clipped at the front. The core is capped by an alloy front plate with a broaching for the keyway.

The wafers are slotted into the gaps between the laminations in the core. The loading sequence of the wafers can be varied to provide extra differing. For instance, the wafers in Fig. 3.15 are loaded in the order: $\uparrow - \uparrow\downarrow - \downarrow\uparrow\downarrow\uparrow$, where '$-$' denotes an absent wafer. Two steel rods, inserted from the front of the core at the top and bottom (on the far side of the keyway in Fig. 3.14), serve to retain the driver springs and also to limit the travel of the wafers. In addition to a flat contact surface on the cut-out of each wafer, the opposite edge of the cut-out is shaped to leave clearance for the other side of the key during insertion. Some wafers have a reduced-width end to hamper picking.

Key blanks are made from stamped steel with a ridge along the center of the blade to ensure that the key can only be inserted one way. The system provides a high degree of flexibility in terms of the overall number of key codes due to the large number of wafers and the different loading sequences that are supported. Master-keying is implemented by using cuts on both sides of the key together with different loading sequences, or by omission of some of the wafers. For instance, the lock shown in Fig. 3.15 has seven wafers installed, with chambers 2, 5, and 10 empty.

A number of locks using Miwa wafer and other relatively low-security cylinders have proved to be quite susceptible to picking and bypass techniques, particularly locks with thumb-turn knobs. This has been a cause of considerable concern in Japan due to the widespread use of locks of this type. Around the year 2000 there was a dramatic increase in the rate of burglaries in Osaka, Tokyo, and other major cities [67, 81]. Capitalizing on the increased demand for security, the Miwa Lock Company released a new "foreigner-proof" lock design [23, 90], insinuating that foreign gangs were to blame for the crime wave.

Many suppliers are now offering retrofitting kits for upgrading the easily compromised wafer locks. Higher security cylinders like Kaba and Mul-T-Lock Interactive are also being offered. The upgrade of the Miwa wafer lock, called the U9, bears more than a passing resemblance to the Ingersoll lock. Both the Ingersoll and the Miwa U9 are described in Chapter 4.

Bricard Bloctout

(FR) 14-wafer (2)

The Bricard Bloctout wafer-tumbler lock in Figs. 3.17–3.19 has 14 closely spaced, free-sliding wafers. The design is similar to the bilateral pin-tumbler locks proposed

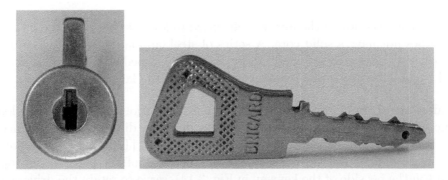

Figure 3.17: Bricard Bloctout 14-wafer cylinder and bilateral key.

Figure 3.18: Conventional wafer-tumbler and two Bricard wafer-tumblers.

Figure 3.19: Operation of the Bricard 14-wafer core.

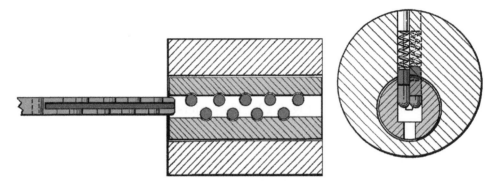

Figure 3.20: Bilateral key 9-pin cylinder design from 1981 by N. Litvin and A. I. Scherz (US patent 4,429,554).

by G. Sieg in 1975 (UK 1,543,940) and by N. Litvin and A. I. Scherz (1981) with the principal difference being the use of wafers rather than pins (refer to Fig. 3.20). The key consists of two blades that are separately cut and then either spot-welded or riveted together. The bittings are staggered and have constant width from the top to the bottom edge of the key blade.

The wafers (Fig. 3.18) are arranged in seven pairs and have a constant-height cut on one inside face and a larger cut on the other that bypasses one blade of the key. In each pair, one wafer is straddled by the left and one by the right side of the key blade. The last few wafers are sprung from the side so that they stay put when the key is removed, the other wafers being loose in their chambers.

The front part of the cylinder around the keyway is fixed, and this acts to retain the key when the lock is being operated. Picking the Bricard lock is more difficult than a conventional wafer lock because there are many wafers and less tactile feedback due to the lack of spring-biasing. However, the lock does not offer a high level of security since its core is made of plastic and the cylinder housing is made of thin steel with no hardened parts. Furthermore, the Bricard wafer lock is known to be susceptible to impressioning with a soft blank ("clé molle" in French).

3.3   Contoured

American Lock Company

(US) 12-wafer (2)

The American Lock Company was formed in the 1860s by P. S. Felter to produce a double-sided disc-tumbler lock of his own design [50]. Unlike the conventional wafer lock, which requires multiple slots in the plug to accommodate the wafers, Felter's so-called plate-wafer lock is of a much simpler construction. A modern version of the lock appears in Fig. 3.21. The keyway has a central ward, and the key is continuously milled on both top and bottom edges such that the bitting width is constant along the length of the blade.

There are typically 10 to 15 wafers mounted in a brass holder inserted through a hole in the side of the plug (see Fig. 3.22). The wafers are adjacent, unlike ordinary wafer locks where they are housed singly or in pairs, and all have the same sized cut-out, which may vary in offset. Although the wafers do not require a spring since they are guided into position by the key, they are usually spring-biased on one side by an S-shaped wire to ensure positive locking.

In a conventional wafer lock, the wafers are chosen from a set of sizes, and a key is cut to the corresponding code. Plate wafers, on the other hand, are all identical when they are first arranged in the holder. Wafers are less than 30 thousandths of an inch thick and since there is no separation between them, this sets the spacing for the cuts on the key.

Figure 3.21: American Lock Company series H10 plate wafer padlock and double-sided key.

Figure 3.22: (Left) Plug from plate-wafer cam lock. (Right) Key inserted into wafer pack.

Although the key is continuously milled rather than bitted at distinct points, for the purposes of combinating the lock the cuts are specified at the wafer spacing with a depth increment of 12 thousandths of an inch. There are 10 cut depths, and the MACS is usually taken as one depth of cut, although some codes in a series may exhibit a MACS of 2. The sum of the cut depths between the top and bottom bittings is always 9. Thus, for example, in an 11-cut system, a key may have top code (7 8 9 8 7 6 6 7 8 8 7) and bottom code (2 1 0 1 2 3 3 2 1 1 2). The key blank is cut to a preset code and then inserted, displacing the ends of the wafers either up or down relative to the edge of the holder. The ends of the wafers are then trimmed to match the diameter of the plug. In this way the set of plate wafers inherits the exact code of the key. The lock can be quite pick-resistant when the motion of the plug is heavily damped, as in a padlock with a push-to-lock shackle.

DUO

(US) 14-wafer (4)

The DUO wafer lock, mentioned in [113], is made by the Illinois Lock Company. It is typically used in applications such as vending machines and laundromats that require coins to be stored in an unattended location. The model D6416 DUO cam lock shown in Figs. 3.23–3.25 is a UL-rated lock containing a pack of 18 adjacent wafers, of which four are fixed and 14 are sliding.

The plug is mounted in a die-cast threaded cylinder equipped with longitudinal channels at 6 and 12 o'clock to accommodate the throw of the wafers. Instead of a complicated system of slots, the plug contains a single large chamber, similar in construction to the American Lock Company's plate-wafer lock in the preceding section. This simplifies the manufacture of the plug while at the same time allowing a more durable material than die-cast zinc to be used. Enhanced drill resistance is provided by the relatively thick front section of the plug, backed up by a slot

Figure 3.23: DUO 14-wafer lock and double-sided key. Key is inserted with locating notch in head at 12 o'clock and side-milling on bottom left.

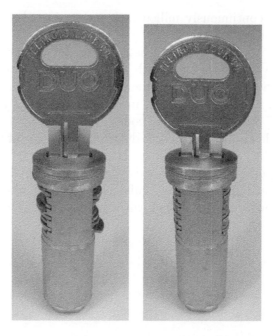

Figure 3.24: DUO plug with key partially and fully inserted.

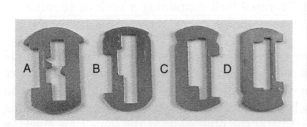

Figure 3.25: (Left) DUO wafer detail: A—profile wafer; B—side wafer; C & D—top and bottom wafers. (Middle and right) Wafers are arranged in pairs (A+B), (C+D) with spring between internal shoulders.

containing a steel wafer. Both sides of the chamber in the plug are ribbed, dividing the chamber into five sections. The four fixed wafers, which will be referred to as profile wafers, are mounted between the ribs in the chamber. The top edge of each profile wafer is larger than the width of the chamber and matches the curvature of the cylinder bore.

The key is flat and made of nickel silver. Holding the key with the locating notch facing down, one can discern a secondary set of bittings, or side-milling, on the top right edge. The key is normally inserted with the locating notch facing up when the lock is oriented as shown in Fig. 3.23. The key is stopped when its shoulder contacts the first wafer in the pack. Since the key is double-sided, there are in fact three bitting surfaces, which we refer to as top, bottom, and side. Although it is not obvious from inspection of the key, there are five bitting points on the top and bottom edges of the key blade and four on the side. The top and bottom bitting surfaces are arranged to provide a constant distance from top to bottom at the 10 corresponding bitting positions. There are two depths for the five top cuts, the five bottom cuts and the four side cuts. The resulting theoretical number of differs is therefore $2^5 \times 2^5 \times 2^4$ or 16,384. The quoted number of usable differs is 14,000.

The 14 sliding wafers are mounted in the sections of the plug chamber in between the fixed profile wafers (P). The first four sections contain three wafers each, while the rear chamber contains only two wafers. The four fixed wafers are matched to the profile of the key in the same way as the broaching of a conventional keyway. In addition, there are three types of sliding wafer, as shown in Fig. 3.25: top (T), bottom (B), and side (S). Counting from the front of the lock, the wafers are loaded in the following sequence: (T B S) P (T B S) P (B T S) P (T B S) P (B T), where wafers in parentheses are in the same section of the plug chamber. The side wafers contain a cut-out with a shoulder on the left that contacts the side-bitting of the key. Top and bottom wafers also have an irregularly shaped cut-out, only the middle portion of which contacts the key. This part of the cut-out has a constant distance from top to bottom, with differing provided by varying the vertical offset of the two contact surfaces.

Unlike a conventional wafer-tumbler lock, where springs housed in small holes in the plug act on the shoulders of the wafers, the wafers in a DUO lock act in opposition against an internally shouldered spring that tends to push them radially outward. Side wafers are paired with profile wafers and are sprung on the left. Top wafers are paired with bottom wafers and are sprung on the right. In their rest positions, the side wafers are spring-biased upward against the lower shoulder of the fixed profile wafers, protruding into the channel at 12 o'clock. The side-bittings of the key push the side wafers downward to align their edges at the shear line of the plug. In each of the top and bottom pairs, the top wafer is normally spring-biased downward, while the bottom wafer is spring-biased upward. This arrangement requires the top bitting of the key to move the top wafer in a pair up while the adjacent bottom bitting moves the bottom wafer down.

From a manipulation perspective, the lock is full of challenges. Not only are there 14 wafers to align, but all are equipped with antipick notches at one end. Furthermore, the wafers in each triple are adjacent, making it difficult to manipulate them independently. While the contact surfaces for the top and bottom wafers are relatively easy to reach, the contact surface for the side wafers is obstructed by the central warding on the profile wafers. In addition, since the top and bottom wafers in each pair are coupled by a spring, raising the top wafer to the shear line increases the tension on the bottom wafer, and vice versa. Lastly, the springs are quite stiff, requiring considerable tension to set the wafers. This last point can also make it somewhat difficult to insert the key.

3.4   Three-sided

Bricard SuperSûreté

(FR) 7–8 disc (3)

The SuperSûreté ("super security") lock, pictured in Figs. 3.26–3.30, was Bricard's flagship high-security cylinder lock since the late 1920s [7]. The key has a highly distinctive T-shaped profile with 7 to 8 bittings along each of the three sides. This gives the impression that there are a large number of tumblers in the lock.

The plug consists of a core and shell (Fig. 3.28). The core comprises seven preformed circular plastic inserts that are pinned together in a stack and capped with a steel front plate. The assembly of shell and core is mounted inside the cylinder body, which is chrome-plated to resist drilling. Each insert houses a spring-biased disc wafer that is ball-driven by the key. The cylindrical shell is made of brass and slotted at 3, 6, 9, and 12 o'clock to accept the wafers. The wafers have a cut-out with a 45-degree slope with respect to their sliding axis (see Fig. 3.29). The slope deflects the motion of the ball by 90 degrees into a translation of the wafer along its axis to align its ends at the shear line of the plug. Each wafer can be mounted in any one of three possible orientations, being operated by one of the three blades on the key. There are four different depths of cut.

Figure 3.26: Bricard SuperSûreté three-sided key.

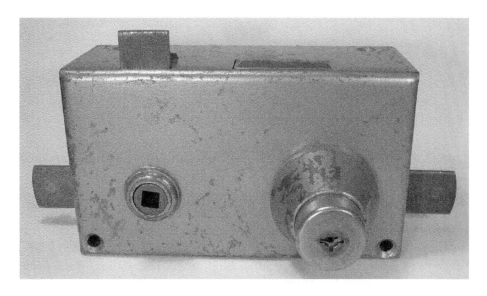

Figure 3.27: Bricard deadlock-latch with linkages for header and threshold bolts.

Figure 3.28: Bricard SuperSûreté 8-disc, 3-sided wafer cylinder (left); core and shell (right).

Figure 3.29: (Left) Bricard core with cap removed. (Right) Plastic insert with disc wafer.

Figure 3.30: Bricard SuperSûreté key with active profile ball in position 7.

Master-keying is accomplished by using extra key cuts, with seven being the minimum and 21 the maximum number in total. In non-MK systems, any extra bittings on the key blade beyond the requisite seven or eight are ornamental. This provides a level of copy protection since, without knowing which are the required seven, all 21 bittings must be reproduced to obtain a functioning key without guesswork. The four depths combine with the three orientations to provide a very large theoretical number of key changes: $12^7 = 35,831,808$ for a 7-wafer lock.

The neck of the key has a smaller diameter than the blades. The keyway is such that the key blades are obscured when the key is turned, which makes tensioning the plug difficult. The wafers may have V notches, similar to the antipicking notches in conventional wafer-tumbler locks, which catch the beveled edge of the shell if incorrectly aligned, thwarting a picking attempt. Unfortunately, the use of plastic components, no doubt facilitating construction but a poor substitute for brass or steel, potentially allows the core to be bypassed by melting.

A floating ball version of the Bricard 8-wafer lock is available called the SuperSûreté à bille, or super security with ball. In this embodiment, shown in Fig. 3.30, a floating ball is embedded in the key blade at the seventh bitting position (the bittings at this position are redundant). On insertion, the ball rides onto a driver and is deflected up and out of the plane of the key blade, actuating the seventh wafer. A key without the movable element cannot therefore operate the lock. The principle is based on the DOM floating-ball system covered in Chapter 2.

3.5   Inline Push Type

ABA Pagoda

(CN) 4-disc (3)

The ABA Pagoda 4-disc lock is one that, to some extent, defies classification. The lock is manufactured in Taiwan and comes in various packages including cam lock, key switch, and rim cylinder. An ABA cam lock is pictured in Figs. 3.31–3.34. Outwardly the lock looks like a conventional wafer lock, but on inspection some unusual structure is revealed.

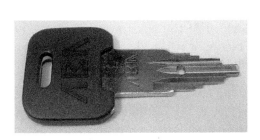

Figure 3.31: ABA Pagoda 4-disc cam lock and key.

Figure 3.32: (Left) ABA plug with discs in rest positions. (Right) Internal millings in barrel.

Figure 3.33: ABA plug with key fully inserted.

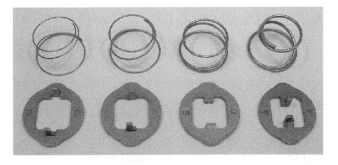

Figure 3.34: ABA Pagoda discs and springs.

A single row of eight push-type elements spans the keyway (similar to Mottura and Tover locks). The feeling that this is not a normal wafer lock is further reinforced on seeing the key, which has four stepped bittings on each side of the blade. The bittings are symmetric and decrease the width of the key toward its tip. There is also a centrally located hole through the blade, near the shoulder of the key, which receives a retaining pin as the plug is turned. Other holes may be present in the key blade, but these are ornamental.

Turning to the lock's internal structure (Fig. 3.32) we find a set of four-flanged, stamped steel discs or wafers. The wafers are circular at their periphery except for two opposing round bumps. A square cut-out in each disc allows them to be stacked along the square section of the plug. The elements that are visible in the keyway are actually the ends of prongs or posts attached to the wafers. The posts, of which there are two per wafer, are perpendicular to the face of the wafer (Fig. 3.34). The distance between the posts varies, becoming progressively smaller toward the end of the plug. The steps in this progression coincide with the steps in the width of the key blade. Thus the thinnest part of the key (at its end) contacts the posts of the fourth wafer, the first step from the end contacts the posts of the third wafer, and so on. The lengths of the steps on the key, together with the sizes of the posts on the wafers, determine to what depth the wafers are pushed when the key is fully inserted and its shoulders abut the front of the plug.

Now if the discs were circular, they would offer no resistance to the plug being turned; instead, the two crescent-shaped bumps on the discs engage longitudinal channels in the housing (see Fig. 3.32). Furthermore, the axial section of the housing is not uniform, possessing crenellations or grooves in four places along its length. These grooves, which are regularly spaced and circumscribe the inside of the housing, accept the bumps on the wafers when they are depressed to the correct depths, as shown in Fig. 3.33. An incorrectly bitted key will fail to align one or more of the discs with its corresponding groove, and thus it will not operate the lock. One can also appreciate that this structure requires both the posts on an individual disc to be pushed in by the same amount in order for the bumps to clear the channels at 6 & 12 o'clock. This helps to increase the picking resistance of the lock since a skewed disc tends to bind.

With four discs and four sizes, there are theoretically $4^4 = 256$ different combinations for the coding of keys, which is on a par with the number of effective differs for 5-wafer locks. A drawback of the design is that it is not readily amenable to master-keying. If two different keys were required to operate the same ABA cylinder, the width of the grooves inside the housing would need to be increased. This would be undesirable from a manufacturing point of view and would also lessen the level of security offered by the lock.

A final point to note concerns the tensioning springs utilized in the lock. The discs are driven in a series arrangement with the fourth disc being sprung from a rear stop washer, the third disc sprung from the fourth disc, and so on. If all four discs used springs providing the same tension, it would not be possible to position them reliably with the key, since the force applied to any particular disc would depend to a large extent on how far other discs were depressed. Consequently, in order to decouple the motion of each tumbler from the others, springs of differing tension are used. The fourth disc has the strongest tensioning spring; the third disc has the next strongest, and so on. The first disc has the lightest spring. In this way the position of each disc is effectively set by the bitting length of the appropriate key step. A similar idea applies to concentric pin-tumbler locks like Mul-T-Lock.

The lock is surprisingly manipulation resistant due to the difficulty of maintaining the discs square-on to the plug and at the right depth. The keyway broaching makes it hard to manipulate both posts of a single disc with a single-pronged tool. The plug has a provision for hardened inserts to counter drilling.

ABA also manufactures a high-security axial 7 pin-tumbler lock [122]. The special feature of the design is that the keyway cover is offset with respect to the lock spindle. The key has a correspondingly offset or kinked stem. The result is that the key must first be inserted and then displaced in order to reach the key channel, rendering standard tubular lock-picks ineffective.

3.6   Bell Locks

Dudley, SEA

(US) **Dudley** 6-wafer (3)
(CH) **SEA** 10-wafer (3–4)

Examples of the Bell or Dudley lock principle are presented in this section (see Figs. 3.35–3.37). All are of the cam lock variety and are intended for security applications such as coin boxes for lockers and parking meters. The operating principle may be traced to Bennett's 1918 patent and other patents referenced in the chapter introduction. The overall design of the lock has not changed greatly since that time, except through the addition of side-bars or mechanisms for recombination (see Fig. 3.38). Bell locks were produced in the United States by the Eagle Lock Company (Connecticut).

At the core of these locks is a system of bar-wafers or sliders, quite unlike normal wafer-tumblers. Some versions of the Bell lock use driver springs while others do not. The bar-wafers, which may be of rectangular, square, or round cross-section, have a peg or stump that protrudes about one-third of the width of the keyway. The stump is the only part of the wafer visible from the keyway. The length of the bar-wafers is equal to the diameter of the plug. The plug chambers, in which the bar-wafers move freely, are open at both ends to allow the tumblers to enter a set of holes or channels in the housing. In the locked position the bar-wafers (in the springless case) normally sit at

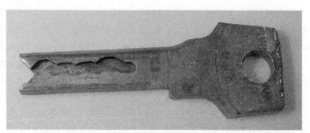

Figure 3.35: A Bell-type cam lock and two-track key.

Figure 3.36: (Left) Bell lock plug with two rows of three bar-wafers. (Right) Bar-wafers guided by key track.

Figure 3.37: Key insertion: side tracks pick up bar-wafer stumps.

their lowest position so that their bottom ends block rotation of the plug. Evidently, all the bar-wafers must be adjusted so that their ends are flush with the edge of the plug in order for it to turn and drive the cam. How the key achieves this will be examined next.

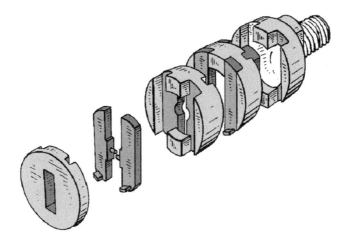

Figure 3.38: Recombinatable Bell lock from J. A. Tartaglia's 1963 patent (US 3,263,461).

Consider for the moment a Bell- or Dudley-type lock with a single row of bar-wafers. The key for such a lock will be of rectangular cross-section with a V-shaped notch at the end. The function of the V-notch is to form a pick-up ramp for the bar-wafer stumps. Depending on reversibility, one or both sides of the key will have a groove or track milled into them. This particular kind of key is variously known as a side-track, wave, or side-winder key. In some cases, as we will see shortly, the key may possess two tracks on either side, making it a four-track key.

As the key is inserted, the ramp contacts the stumps one at a time and guides them up its inclined surface into the track. The width of the track is fractionally larger than the diameter of the stump. Once the key is inserted, the vertical positioning of each bar-wafer is fixed by the height of the track at the points where it contacts the bar-wafer stumps. Insertion of the correct key results in all the bar-wafers being held such that their ends are at the interface between the plug and the shell. Because this system does not require any driver springs, it is known as direct-drive. Notice also that, by withdrawing the key, the positions of the bar-wafers are scrambled.

The coding of the lock is determined by the offset of the stump on each bar-wafer. A bar-wafer with a low-set stump will require a low bitting on the key blade and vice versa. The sequence of bitting points along the key blade determines the layout of the track that must be milled. In a given system there may be constraints on the maximum height step from one point on the track to the next (similar to a MACS constraint), which reduces the available number of system codes. An advantage of the construction is that it is very difficult to make unauthorized copies of the key since it is internally cut. A further advantage is that most Bell-type locks have a high degree of resistance to manipulation.

It is not necessary for the track to be centered in the middle part of the key blade. Locks such as SEA, which we present next, take advantage of this fact to create a reversible key. First, the stumps are set lower on the bar-wafers such that the track and ramp can be correspondingly lower down. With all the bar-wafers in their rest positions, the stumps are then in the lower half on the keyway. Since the amount of travel of the bar-wafers is limited to less than half the key blade height, two tracks can be accommodated on the same side of the blade. In either orientation, it is the lower pair of tracks that picks up and positions the bar-wafers.

Bell locks often have two rows of four bar-wafers sprung in an alternating sequence. It can be appreciated that locks with two rows of bar-wafers effectively square the number of codes when compared with locks having only a single row of wafers. The SEA-Normal lock and its subsequent variants, pictured in Figs. 3.39–3.44, is

Figure 3.39: SEA-Normal cam lock and symmetric four-track key.

Figure 3.40: SEA-2 cam lock and key.

Figure 3.41: SEA-3 cam lock and key.

Figure 3.42: Core from a SEA-3 profile cylinder.

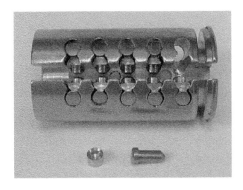

Figure 3.43: (Left) Underside of plug from SEA-3 lock with miniature pin-tumbler pair in foreground. (Right) SEA-3 bar-wafers.

Figure 3.44: Operation of SEA-3 lock: key brings bar-wafers to shear line.

of this type.[1] The lock is designed and manufactured by SEA Schliess-Systeme AG in Switzerland. It has two rows of five springless bar-wafers with a reversible four-track key, as shown in the diagram from F. Gysin's 1964 patent (US 3,264,852) in Fig. 3.45. Because there are no springs on the sliders, there is very little friction

---

[1]Note that SEA cylinders are normally mounted with the base of the keyway at the top—the opposite orientation to that shown in the figures—to ensure smooth operation.

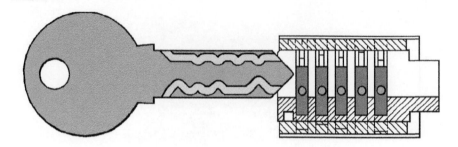

Figure 3.45: SEA four-track bar-wafer lock from F. Gysin's 1964 patent (US 3,264,852).

on the key or tumblers. The system allows for seven different bar-wafer sizes with an increment of 0.3 mm, differing in the position of the stump relative to the ends of the bar-wafer. Since there are 10 bar-wafers, the theoretical number of system codes is enormous: $7^{10} = 282,475,249$. The spacing between the bar-wafers allows for an unrestricted MACS.

The production version of the SEA lock also allows for up to four miniature pin-tumblers in a single line at 12 o'clock, spaced between positions 1 to 5 of the bar-wafers (see Figs. 3.42 and 3.43). The pin-tumblers may be either present or absent, and three pin sizes are possible, requiring different profile dimples on the edge of the key. The addition of pin-tumblers provides a multiplier for the overall number of system codes. In a similar manner to a conventional pin-tumbler lock, the profile-control pins can be used for master-keying, which would otherwise require the omission of one or more bar-wafers since these cannot be master-keyed. At the same time, the provision of active profile pins increases the resistance of the lock to attack by rapping or vibration.

The original SEA design from 1949 was upgraded in 1964 to the SEA-2 model (Fig. 3.40) and again in 1979 with the release of the SEA-3 (Fig. 3.41). Both the SEA-2 and SEA-3 are identical in operating principle to the original SEA lock, with all models employing 10 springless bar-wafers and up to four miniature pin-tumblers. However the key blanks for the three different models are not compatible: there are very slight differences in the height of the keyway and/or of the diameter of the bar-wafer stumps from one model to the next. The SEA-2 and SEA-3 both have a taller blade than the SEA-Normal, with the SEA-3 also having wider tracks than the SEA-Normal.

Some types of Bell locks have additional side-bars that slot into notches in the outer edges of the bar-wafers. An example is furnished by Vonlanthen's 1995 patent for a dual-action version of the SEA lock (US 5,956,986). This particular design adds two rows of miniature pin-tumblers at 3 and 9 o'clock that slot into dimples in the outer edges of the bar-wafers. Another example of this type is the Lori side-bar lock covered in Chapter 4.

3.7   Axial

Bramah

(UK) 7-wafer (3–4)

Axial wafer locks accept a tubular or solid key of round section with bittings at the tip. There are a number of differences with respect to conventional wafer-tumbler locks, the most significant of these being that the wafers are pushed axially into the lock rather than being lifted by sloped cuts on the key blade. Axial wafer locks, like their pin-tumbler counterparts, will relock at fractions of a turn unless the wafers are maintained at the correct depths. We present a number of such locks in this section, starting with the English Bramah lock and then proceeding with descriptions of French and German modifications to the Bramah principle.

The Bramah lock, briefly mentioned in the chapter introduction, was the first axial lock and has served as a model for many other axial push-key locks. The Bramah company, now established in the United States, produces locks for commercial and residential applications such as the MD27 mortice deadlock. The Bramah lock is also used for safes, with stainless steel blank keys supplied in stem lengths up to 6″. The construction and functioning of a model C17 Bramah lock, based on a design that has remained largely unchanged for over 200 years, is now considered. Further historical details concerning Joseph Bramah and Company may be found in Appendix E.

The lock, shown in Figs. 3.46–3.49, consists of a cylinder housing into which is set a rotatable core. The core is held in the locked position by a radially disposed system of usually seven wafers or sliders, although variants of the lock have been made with between four and 18 sliders. These are arranged around a central drill-pin onto which a strong spring and cap are mounted.

Figure 3.46: Bramah C17 7-slider mortice cylinder and key.

Figure 3.47: Two views of a Bramah lock core.

Figure 3.48: Bramah sliders.

Figure 3.49: Key depresses sliders to align gates with channel. Slotted ring in foreground.

The core (Fig. 3.47) is pierced longitudinally by seven equally spaced slots that house the sliders. In addition, the core contains a circumferential channel into which a two-part slotted ring is inserted. With the two halves of the ring in place around the core, a set of recesses is formed—one at the midpoint of each of the slider chambers. The ring is equipped with four tabs that anchor it inside the housing. Alternatively, the ring may be fastened to a fixed plate inside the lock.

With regard to the sliders (Fig. 3.48), each one is flat and rectangular except for a shoulder at the top end, closest to the front of the lock. The sliders are made of stamped steel folded into equal halves containing one or more gates along one edge. The sliders are seated in the chambers such that their shoulders are supported by the central cap and their gated edge is facing radially outward. The tension of the main spring forces all seven sliders to the front of the core. One of the slider chambers is widened toward the front of the core to accept the locating bit of the key. The remainder of the keyway is formed by the circular cavity around the drill-pin. When the key is inserted and turned, the bit is retained by the stationary front part of the keyway.

The bittings around the end of the key stem cause the sliders to be depressed as the key is inserted. By a clever economy of design, the main spring supplies tension to all seven sliders without the need for each to be independently sprung. (The original Bramah design employed coil springs on each of the sliders). Sliders whose shoulder no longer contacts the spring cap are held in position by friction due to their folded construction. Thus it can be seen that the correct key simultaneously brings all seven sliders to the precise depths at which the gates in their edges register with the recesses in the stator ring. The passage of the core to the unlocked position is then assured.

The rear of the core may have affixed to it a stump or cam to accomplish the unlocking function and communicate with the boltwork. The geometry of the Bramah lock, together with the restricted access of the keyway created by the drill-pin and front-piece, already make for a challenging lockpicking task. Add to this the presence of false-depth (antipicking) notches in the sliders, and the difficulty of the job is increased by an order of magnitude. Nonetheless, tools have been devised to defeat the lock. These are similar in construction to tubular lock-picks for ACE/GEM locks and are called *parapluies*, or umbrellas, by the French.

The Bramah system admits a very large number of differs. In each of the sliders a gate can be cut at one of eight depths with an increment of 0.020″. Since the key is end-bitted, there is no requirement to connect the cut centers by sloped edges, as in conventional pin-tumbler and wafer-tumbler locks, with the result that the MACS is effectively unrestricted. Thus, a 7-slider lock could theoretically have $8^7 = 2,097,152$ combinations. In the existing C17 system, the cut directly underneath the bit of the key is limited to two possible depths. This results in a reduced number of theoretical differs: $2 \times 8^6 = 524,288$. It may be undesirable to include patterns with many deep cuts because they tend to weaken the key. Master-keying can be accomplished by the addition of extra gates in one or more of the sliders, although originally Bramah simply omitted some of the sliders and corresponding key cuts from servant-keyed locks.

Further enhancements of the Bramah lock include the Italian Vago lock, whose production was acquired by Lips in the Netherlands [30], and German axial locks, such as those produced by S. J. Arnheim in Berlin. The Vago lock was produced with up to 30 sliders, having a mechanism reminiscent of the commutators of a universal motor. In such a lock, the contact points of the sliders are arranged on two staggered concentric circles. Sliders on the inner circle are actuated by cuts in the end of the key. Sliders on the outer circle are addressed by a set of fins around the key stem. This idea for packing more sliders into a Bramah lock is also a feature of the Doppel Bramah-Chubb lock described later in this chapter.

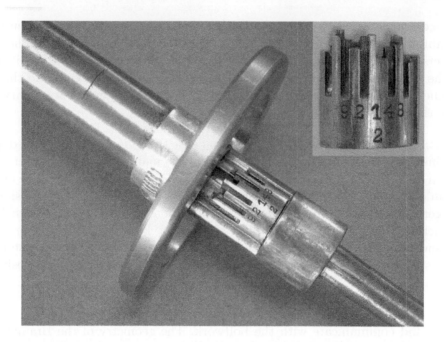

Figure 3.50: Plunger mechanism and escutcheon plate from Arnheim 14-slider axial safe lock with end-bitted stemless key loaded into the breech.

In the Arnheim axial lock, shown in Fig. 3.50 and partially described in German patent DE 287,890 (1914), the lock was set back a foot or more from the front of the safe door with a spindle and plunger mechanism, or *lafette*, to transport the key to the deeply recessed lock. Pulling out the plunger of the lock revealed a breech for the small Bramah-type key, which consisted of a curved bit with no stem. The key bittings were arranged on two concentric arcs that operated a set of 14 contiguous sliders [42]. The sliders were not gated as in a conventional Bramah lock, but split as in an axial pin-tumbler lock. Further plunger mechanisms for lever locks were also produced by a number of other German safe manufacturers, including Theodor Kromer, Carl Kästner, and Bode-Panzer (refer to Chapter 5).

Supra

(US) 6-wafer (2–3)

Before proceeding with the French equivalents of the Bramah lock, we mention another axial wafer lock called the Supra Title, which is closely based on the Bramah principle. The lock was designed by D. A. Williams in 1964 specifically for portable key safes and key boxes. It is installed on Supra models C, SA, and S5, among others. The lock is popular in the automotive and real estate sectors, where it enables access to many different vehicles or residences without the need to carry many different keys. Supra also produces a keyless push-button model. Drawings from Williams's patent (US 3,237,436) appear in Fig. 3.51, while the lock and its key are pictured in Fig. 3.52.

The lock comprises a core (Fig. 3.53) with six wafers or sliders arranged around a circular keyway with center post shaped like a six-pointed star. The sliders are U-shaped, as in the Code lock described in Chapter 2, and are spring-biased from the rear. The inner edge of each slider rests in one of the narrow grooves in the center post while its outer edge, which cannot be accessed from the keyway, normally enagages a slot in the front of the lock body. The key is tubular with six internal ribs and a locating fin along one side that registers with the corresponding slot in the keyway at 12 o'clock. The key can only be inserted and withdrawn at this position. Internal shouldering in the front of the lock limits the maximum rotation of the key to 90 degrees. It can be appreciated from Fig. 3.52 that the key is difficult to reproduce without access to the correct key blanks, which are restricted, and the appropriate machine for cutting the internal bittings.

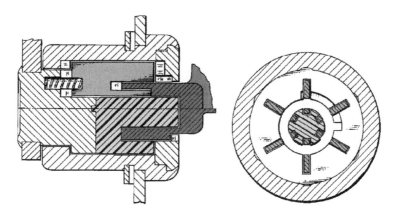

Figure 3.51: Design of the Supra Title lock from a 1964 patent by D. A. Williams (US 3,237,436).

Figure 3.52: Supra Title 6-slider key safe and two views of the tubular key.

Figure 3.53: Supra 6-slider core and rear cap.

When the key is inserted, its bittings act on the inner edges of the sliders, pushing them into the lock until the tip of the key contacts the bottom of the keyway. If a slider is depressed too far, its end protrudes into a slot at the rear of the lock body. Conversely, if a slider if not sufficiently depressed, its outer edge will not be clear of the slot at the front. Clearly, all six sliders must be depressed simultaneously to the correct depths to enable the core to turn, actuating the locking cam.

The Supra lock is made from a cast zinc alloy, which provides an adequate level of protection given that its primary use is as a key safe for attachment to an external fixture or vehicle. Differing is achieved by varying the length of the inner edges of the sliders. There are five depths of cut with an increment of 30 thousandths of an inch, providing a maximum of $5^6 = 15,625$ key combinations. Unlike the Bramah lock, the inner edges of the sliders, whose lengths determine the required key cuts, are visible in the keyway and could, in principle, be gauged to decode the lock.

An interesting attribute of the lock is that the chambers in the core can be aligned in any one of six possible orientations. This allows a key-change operation to be effected in a similar way to the Van lock in Chapter 2: a change key having no locating fin can be used to rotate the core by a multiple of one-sixth of a turn. This sets the lock up to accept an operating key whose cuts are a circular shift of the cuts on the original key.

Fontaine, Laperche, Vigie Picard

(FR) **Laperche** 5-slider (2–3)
(FR) **Fontaine** 5-slider (2–3)
(FR) **Vigie Picard** 5-slider (3)

The Laperche, Fontaine, and Vigie Picard axial locks, all manufactured in France, are closely based on the Bramah lock. A double-sided 7-slider profile cylinder from Fontaine is pictured in Figs. 3.54–3.56. This type of lock has been produced under the name "Progrès" since World War II [7]. The five push-wafer Laperche cylinder

Figure 3.54: Fontaine 7-slider profile cylinder and key.

Figure 3.55: Fontaine barrel and core.

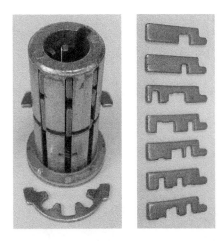

Figure 3.56: (Left) Fontaine core and slotted ring. (Right) Set of 7 sliders with multiple gates for master-keying.

appears in Figs. 3.57–3.59. Both the Fontaine and Laperche locks have removable cylinders with nonstandard dimensions. The original Vigie Picard lock, shown in Figs. 3.60–3.63, had an integral cylinder; more recent models, however, use standard Europrofile cylinders. The basic mechanism for the Vigie Picard was described in a 1922 patent by Bézard and Bézard (FR 552,963), whose translated title is "security cylinder with turning circular bolts."

Figure 3.57: Laperche 5 push-wafer cylinder and key.

Figure 3.58: (Left) Laperche core. (Right) With cover removed.

Figure 3.59: Laperche core with key inserted and pins at interface of stator ring.

Figure 3.60: Vigie Picard security door lock.

Figure 3.61: Front and rear keyway of a Vigie Picard lock.

The keys for all three locks have a round section with bittings milled into the end of the stem. Each cut has a circular footprint, and although adjacent cuts overlap somewhat, at least some part of the stem is left to actuate the sliders. Turning tension is provided by a bit or fin on the key stem. The sliders or push-wafers are constrained to a longitudinal channel and must be depressed to the appropriate depths to allow the plug to rotate.

As in other axial and push-key locks, the keyway is such that the bit is obscured by the cylinder cap when the key is turned, making tensioning by external means

Figure 3.62: Vigie Picard core and key.

Figure 3.63: Vigie Picard core: key aligns slider gates with lower channel.

more difficult. Vigie Picard locks have additional protection in the form of two drill-resistant balls that jut into the keyway from either side (see Fig. 3.61). These also retain the key and keep the sliders at the correct depths as the lock is operated.

Bramah-Chubb

(DE) 8-lever + 9-slider (5)

The Bramah-Chubb lock is a high-security lock produced in Germany in the late 1800s for safes and vaults. As the name suggests, the lock combines both the Bramah and Chubb principles in its construction. The pipe key (Fig. 3.64) is equipped with a bit for the lever cuts and is also end-bitted for the sliders. A patent by Carl Kästner from 1881 describes a Bramah-Chubb lock with five levers and nine sliders (DE 20,417).

The lever part of the lock, which is positioned further forward than the Bramah part, is a conventional Chubb-type lever lock with typically between six and eight levers. The levers have a large circular cut-out to allow space for a core containing the Bramah part of the lock, which operates axially and can have as many as nine sliders.

An enhanced version of the lock (Fig. 3.65) was patented by Max Zahn in 1892 (DE 71,766) in which the number of sliders was doubled from 8 to 16 by arranging them in two concentric circles [42]. In this case the lock is known as a Doppel Bramah-Chubb.[2] The sliders are arranged in alternating fashion with two different shoulder lengths. Long-shoulder sliders are operated from the regular end-bittings in the pipe key. Short-shoulder sliders are depressed by millings in the key stem in between the usual end-bittings and set higher up on the blade so as not to interfere with them.

As can be appreciated from Fig. 3.66, the lever and slider sections are integrated in both the Bramah-Chubb and Doppel Bramah-Chubb locks. For this reason the lock cannot be operated unless all the sliders are at their correct depths and the lever gates are in registration. The end bittings of the key first ensure that all the slider gates are clear of a blocking ring. A peripheral stump mounted on a slideable plate is

Figure 3.64: Keys for 6-lever, 8-slider Bramah-Chubb locks. (Courtesy O. Diederichsen).

---

[2] "Doppel" means "double" in German.

Figure 3.65: Doppel Bramah-Chubb 8-lever/13-slider safe lock and key with inner and outer cuts.

Figure 3.66: Internal views of Doppel Bramah-Chubb lock. (Courtesy O. Diederichsen).

then drawn toward the lever pack as the core containing the sliders begins to rotate. The stump cannot enter the lever gates unless they are aligned by the side bittings of the key. (Readers unfamiliar with these locking principles should consult Chapter 5 and the section on Bramah locks in this chapter).

A number of potential drawbacks of the Bramah-Chubb system were noted by Theodor Kromer, a competitor [116]. First, the key is not especially difficult to duplicate. Second, the complexity of the lock results in a higher cost, less robust product. Furthermore, there is insufficient space in the lock housing to employ a Chubb detector mechanism. With so many interdependent active elements, however, it is doubtful that the lock can be opened without the correct key, although it may be possible to impression the lever part of the key once the Bramah part of the lock is decoded—clearly not a job for the faint of heart!

3.8   Key-Changeable

Rielda

(IT) 7-wafer (3–4)

The Rielda lock is one of only a handful of small-format key-changeable cylinder locks, and provides ample proof of the flexibility of the wafer-tumbler mechanism. Fitting inside a standard Europrofile cylinder, it is not at all obvious at first glance how it could be made to work. The user simply inserts the current "programming key," turns it to 6 o'clock, and removes it. A new programming key may then be inserted, and on returning it to the locked position, voilà, the combination has been changed!

The design was mooted in a 1986 patent (US 4,712,399) by M. Mattossovich of Rielda Serrature S.R.L. in Italy, although the production version more closely resembles the mechanically reprogrammable wafer lock described in N. Boag's 1988 patent (US 4,966,021). The resettable lock assembly suggested in both of these patents is traceable to Raymond and Millett's 1980 patent (US 4,376,382), which is also relevant to the Winfield lock covered in the next section.

Early versions of the Rielda key and lock mechanism were quite delicate—a factor that weighed against their reliability. Although the original lock utilized two side-bars, they played no part in the locking operation, so from a manipulation perspective the cylinder was a standard seven wafer lock. Since then, Rielda has redeveloped the product in a simpler and significantly more robust format than the original cast zinc alloy construction. The lock is available in various styles with key kits comprising one gold-colored programming key and three silver-colored operating keys. The programming key, which is slightly smaller than the operating keys and can be inserted and withdrawn at 6 o'clock, is used to set the cylinder to the assigned combination.

The Rielda was introduced in the United States in 2005 by Hampton Products International in a high-security format called the Lynx lock. The new design, which uses the Rielda resettable wafers to control the shear line of a set of pin-tumblers, is summarized in a 1998 patent by A. Loreti (US 6,119,495). The Lynx lock incorporates seven wafers and an additional two pin-tumblers to increase the number of available differs. Key blanks for both the Rielda and the Lynx lock are restricted. The remainder of this section focuses on the detailed operation of a modern Rielda lock, illustrated by the model 1400S Europrofile cylinder shown in Figs. 3.67–3.71. The description assumes that the cylinder and plug are viewed end-on with the bottom of the keyway at 6 o'clock as in Fig. 3.68.

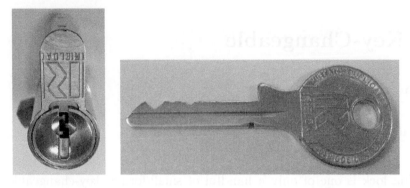

Figure 3.67: Rielda cylinder in key-change position with programming key.

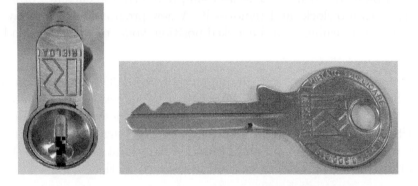

Figure 3.68: Rielda cylinder in locked position with operating key.

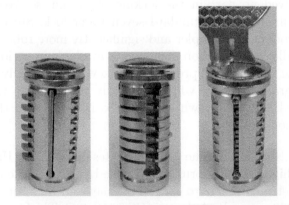

Figure 3.69: Views of Rielda plug: rod side-bar in foreground (left); comb side-bar removed (middle); key aligns wafers with rod side-bar channel (right).

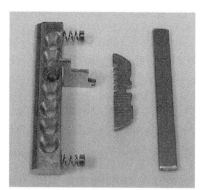

Figure 3.70: (Left) Rielda lock housing with side-bar channel at 3 o'clock. (Right) Comb and rod side-bars and two-part wafer.

Figure 3.71: Variable-height wafer (left to right): disengaged; engaged.

The Rielda lock is comprised of a plug and housing or barrel. There are a number of grooves and recesses on the inside surface of the barrel (refer to Fig. 3.70). First, there are two opposed sets of seven recesses that span the top and bottom surfaces: these are the wafer chambers. Two bores at 12 o'clock, one at the front and one at the rear of the barrel, are matched by bores on the top edge of the plug that accommodate two steel pin-tumbler pairs that double as drill-proofing pins. Finally, there is a channel that runs from the front to the back of the cylinder at 3 o'clock: this channel is central to both the locking and key-change operations.

The plug, which is pictured in Fig. 3.69, is made of brass and contains a number of intricately milled slots and bores. In addition to slots for the seven wafer-tumblers, there are two longitudinal slots that house the diametrically opposed side-bars, both of which are spring-biased at their ends. The first side-bar, which sits in a narrow slot at 3 o'clock, is a flat steel bar. In keeping with the original design, in which this component was a rod, we refer to this as the rod side-bar. The second side-bar, located at 9 o'clock, has a round apex and flat bottom and is pierced widthways by seven beveled holes. Again, in keeping with tradition, we refer to this as the comb side-bar.

The wafers themselves, appearing in Figs. 3.70 and 3.71, are of two-part construction. The right half-wafer is longer with rounded ends matching the curvature of the plug; the top of its inner edge is serrated. The left half-wafer has a rectangular portion, which is serrated at the top, leading into a U-bend with a thin arm. The two halves together form a variable-height wafer when their serrated edges are meshed. The U-bend of the left half-wafer matches the hole in the comb side-bar, which normally constrains it to slide up and down while staying in contact with the comb. The right half-wafer, which we will call the rack, sits in a bore in the plug. The bore allows the rack to slide vertically, but it does not communicate with the keyway: only the flat edge of the left half-wafer is visible in the keyway, looking indistinguishable from a standard wafer-tumbler.

The rack does, however, communicate with the rod side-bar slot, and its outer edge is endowed with a gate wide enough for the rod side-bar. Unlike the original Rielda design, the rod side-bar is a real side-bar: it provides a locking function when it impinges on the channel at 3 o'clock in the barrel. There are also false-depth notches on either side of the gate to inhibit picking.

The lock can be in any of three distinct states: (1) locked; (2) unlocked; (3) key change. Let us consider what occurs in each of these cases with reference to the diagrams in Fig. 3.72.

First, in the locked position the rod side-bar is fully deployed, engaging the channel at 3 o'clock. On the other hand, the comb side-bar is fully retracted into its slot in the plug, its flat bottom forcing the two halves of each wafer into mesh. The driver springs press down on the left half-wafers such that the elbows of their U-shaped arms are stopped against the inside surface of the barrel. While the lower ends of some of the wafer racks may enter the lower chambers in the barrel, the outer edge of the wafer racks inhibit the retraction of the rod side-bar, locking the plug. As the correct key is inserted, its cuts raise the seven wafers, aligning their gates with the slot at 3 o'clock in the plug. The rod side-bar may then move into the slot as the plug turns to the unlocked position.

In the unlocked position, the rod side-bar, which has slipped out of the channel, is now fully recessed in the plug, its edges engaging the gates in the wafer racks. The wafers cannot be displaced and the key remains captive.

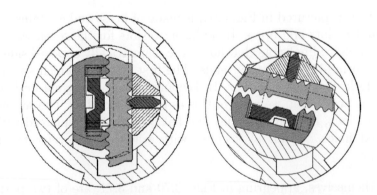

Figure 3.72: Variable wafer principle from N. Boag's 1988 US patent 4,966,021.

The key-change position is attained once the key has been turned to 6 o'clock. The rod side-bar remains fully recessed, maintaining the racks of the wafers in a fixed position. On the other hand, the apex of the comb side-bar is now resting in the channel at 3 o'clock in the barrel and moves radially outward. The left half-wafers pivot about their lower edge as their U-shaped arms slip out of the bevels of the comb side-bar (see Fig. 3.71). This action disengages the two halves of the wafers, freeing up the left halves. An operating key with a full-height blade is still bound because the notch in its stem cannot clear the drill-proofing plate at 12 o'clock (see Fig. 3.67). The programming key, being slightly smaller, can still be withdrawn. As a new programming key is inserted and turned, the left half-wafers, being under tension, adjust their heights to the new profile while their teeth reengage the racks. When the programming key is turned back to 12 o'clock, the rod side-bar returns to its fully deployed position, releasing the wafer racks. The programming key can then be withdrawn, completing the key-change process.

Winfield

(US) 7-wafer (2)

The Winfield lock (Figs. 3.73 and 3.74) was formerly used as a hotel lock due to its ability to be recombinated with a code-change key. The design is described in a 1976 U.S. patent by J. W. Raymond and J. A. Millett. The construction of the lock is particularly unusual in employing two adjacent plugs—it is therefore a bicentric lock. Each lock cylinder contains seven wafers that are set either high or low with a constant-sized cut-out. The key, which is of stamped metal construction, correspondingly has only two depths of cut in each of the seven bitting positions. This means that each cylinder can admit up to 2^7 or 128 keys.

Several different types of keys exist, each with a distinct function. A change key or guest key, with a profile allowing it to be inserted in the left keyway, can operate the lock only if the lock is mechanically programmed to accept the key. A master or maid's key, inserted in the right keyhole, is required to set the operating combination for the guest key.

Figure 3.73: Winfield keys for guest (left) and maid (right).

Figure 3.74: Winfield bicentric cylinder (left). Operation via right keyway (right).

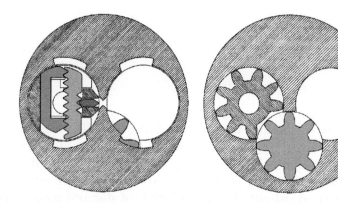

Figure 3.75: Winfield resettable wafer lock principle (US patent 4,069,694 by J. W. Raymond and J. A. Millett).

The master and guest plugs communicate at the rear of the cylinder via a system of gears as shown in Fig. 3.75. With both plugs in the locked position, the gears for each are disengaged from the bottom gear, on which the locking cam is mounted. This arrangement permits an "either-or" logic for the operation of the lock; that is to say either the guest or the maid's key can operate the lock. The key-change mechanism functions with a system of variable racks that are disengaged by the maid's key and then reengaged to accept the new guest key combination. The principle is similar to that of the Rielda lock in the previous section. In addition, an emergency key allows the lock to be operated via the right keyway.

Chapter 4

Side-Bar Locks

There is a constant effort by lock designers to design a lock which is pick-proof. However, this constant effort [...] seems to lag the increase in skill and determination of lock pickers. *C. A. Bauer*[1]

4.1 Introduction

As its name implies, a side-bar is a bar that lies alongside a set of active elements or tumblers in a lock. Until such time as the tumblers are aligned by the proper key, the side-bar prevents the mechanism from being unlocked. The side-bar plays a similar role to the shear line in a pin-tumbler lock or the bolt stump in a lever lock. By its nature, a side-bar tends to be more secure against manipulation than a pin-tumbler, wafer, or lever lock. In each of the preceding cases, the active elements have individual contact points with the locking interface: pins against the pin chambers, wafers against the cylinder body, and levers against the bolt stump. By contrast, in a side-bar lock this interface is once-removed from the active elements. Since the contact between the side-bar and the lock cylinder provides the locking function, it is harder to correlate the movement of the plug with the manipulation of a particular active element.

The concept of a side-bar lock is older than one might imagine and quite difficult to trace in English-language literature. As early as 1800 [1] padlocks like the one shown in Fig. 4.1 were in use in Scandinavia that had a rotating disc mechanism with two or more discs. An article from the *Nordisk Familjebok* encyclopedia [70, 133] attributes the invention of the Scandinavian padlock to Christopher Polhem, a Swedish scientist and inventor, born in 1661. Polhem built up an industrial plant in Stjärnsund, powered by running water, where clocks and locks were produced during the first

---

[1]Charles A. Bauer, Sargent & Greenleaf, US patent 3,181,320, filed March 14, 1963.

Figure 4.1: Scandinavian 3-disc padlock taking a pipe key.

half of the 18th century. By the 19th century, Scandinavian padlocks were being manufactured by numerous companies in the United States including the Star Lock Works, which produced these locks from 1836 to 1926 [103].

Also known as jail locks, Scandinavian padlocks were made from cast components, having a malleable iron body, shackle, and key. Later versions were made with brass discs and steel keys [1]. The shackle of the padlock served as a side-bar, being retained by two or more rotating disc-tumblers. The discs had a central rectangular hole for the key and a peripheral notch on one or both sides to correspond with slots in the inner edge of the shackle. The key itself was generally double-sided with a regularly spaced comb matching the internal warding of the lock. Cuts at various angles (usually 0, 60, and 80 degrees) were made to the bittings on both sides of the key. When all the discs were rotated to the correct angle so that their notches lined up with the shackle slots, the shackle could be withdrawn from the body of the padlock. This locking principle is not dissimilar to that of the letter combination padlock, the forerunner of today's cheap combination lock used on bicycle chains, padlocks, and suitcases. Already widespread in the 17th century, letter combination padlocks were made with as many as eight or nine wheels [25, 29].

During the late 19th and early 20th centuries, a number of lever locks utilizing a side-bar type mechanism were produced. Most of these locks were for high-security applications such as bank safes and vaults. A famous example is the NS Fichet "pompe" (meaning "pump") mechanism, which had a long barrel with a pack of parallel linkage rods operating a system of pivoting levers at the rear of the barrel. In using end-gated levers, the lock resembled the 6-lever padlocks made by the Miller Lock Company in the 1870s as well as the much later Butter's system by Chubb. We provide a description of all three of these locks in the next chapter. Another lever lock of similar principle, also having an end-bitted key, was J. H. Brennan's 10-lever lock of 1884 [57].

The side-bar principle was proposed as early as 1875 in the context of disc-tumbler locks or, more precisely, plate-wafer locks. A patent from that year by P. S. Felter of the American Lock Company (US 167,088) described a locking latch based around a plate wafer cylinder lock in which the wafers were notched at various heights on one side to accept the elongated arm of a U-shaped side-bar. A longitudinal channel in the housing of the lock cylinder prevented the plug from being rotated until the side-bar had been retracted. Felter referred to the side-bar as a "fence-bar" in this patent. The lock required a double-bitted key similar to the one used in the American Lock Company's plate-wafer lock described in Chapter 3.

One of the earliest U.S. patents for a Scandinavian padlock was by J. McWilliams in 1871 (US 116,977). A modified version of the Scandinavian padlock was patented in 1874 by Ahrend, with improvements for ease of manufacture brought by Romer & Company in 1879. Similar padlocks were also produced by the J. H. W. Climax Company and others [50]. The Romer lock, shown in Fig. 4.2, had a flat steel, one-sided comb key with teeth of differing heights. The lock comprised a stack of disc-tumblers interleaved with fixed keyway plates. The plates formed a set of fixed wards and also limited the rotation of the key within a half-circle. The discs included an extra notch that engaged a longitudinal "spring bar." The function of the bar was merely to prevent accidental turning of the discs. The frontal disc, or "trip disc" [103], turned with the key to free the other discs from the spring bar.

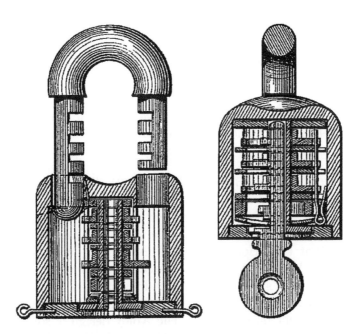

Figure 4.2: C. W. A. Romer's 1879 patent describing a key tube for a Scandinavian padlock with one-sided key (US 213,300).

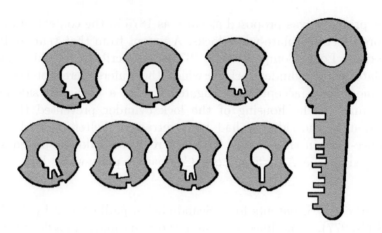

Figure 4.3: Tumbler design for 7-disc Scandinavian padlock with one-sided key from H. Ahrend's 1874 patent (US 156,113). Trip disc on lower right.

The construction differed in an important respect from Scandinavian padlocks with double-sided keys: the discs had an irregular cut-out with inner steps of differing radiuses (see Fig. 4.3). A longer tooth on the key would contact a step with larger radius, imparting a greater angle of rotation than a short tooth. The net angles of rotation of the discs were thus determined by the length of the teeth on the key. An incorrectly bitted key would either underrotate or overrotate some of the discs.

Another pre-1900 embodiment of the side-bar lock is the German "Sherlock" lock, shown in Fig. 4.4. This had a linear arrangement of 13 discs with a cut-out at either 12 o'clock or 3 o'clock. The side-bar was positioned above the discs and actuated the locking mechanism. The lock was operated by a flat key made of stamped steel, resembling a comb with up to 13 teeth. The correct key, when inserted and given a quarter turn, would leave all the 12 o'clock discs alone while rotating all the 3 o'clock discs by 90 degrees. In so doing, a channel was formed by the cut-outs in the discs into which the side-bar could move under the action of a spring. The system was in essence binary since a given tooth on the key could be either present or absent. As such there were up to 2^{13} or 8,192 different possible key patterns.

The *Encyclopedia of Locks and Builders Hardware* [21, 47] mentions a British patent from 1919 for a lock with sliding tumblers with V-shaped notches. When all the tumbler notches were brought into alignment by the correct key, the edge of a spring-loaded bolt could be retracted into the channel formed by the V's. It was in the same year that Emil Henriksson of Finland took out a patent for the now famous 10-disc Abloy lock (US 1,514,318). Instead of translational or sliding motion, the Abloy lock employs a system of rotating discs stacked inside a drum or shell. A side-bar prevents the shell from being turned until the key brings all discs to the correct angular alignment (see Fig. 4.5). The Abloy design is somewhat similar to an earlier disc-tumbler lock invented by W. G. Denn in 1901, illustrated in Fig. 4.6. This variant is interesting in that it demonstrates a connection between lever locks and disc side-bar locks: a curved lever-type key operates on the periphery of the discs to align their gates.

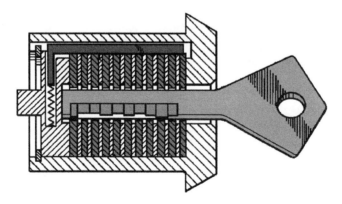

Figure 4.4: Comb key and lock from a Sherlock 13-disc side-bar lock. (Courtesy R. Loschiavo)

Figure 4.5: Abloy 10-disc mechanism from K. Martikainen's 1979 patent (US 4,267,717).

A clearer link in operating principle exists between Carl Kästner's safe lock and the Abloy lock. Kästner lodged a patent in Germany for an 8-disc lock in 1918 that was granted in 1920 (DE 323,580), so there is little likelihood that Henriksson and Kästner knew about each other's patents. The Kästner lock, pictured in Fig. 4.7, is a 12-disc side-bar lock operated by a pipe key with a detachable bit—a constraint imposed by the length of the key stem. From a design perspective, the Abloy lock

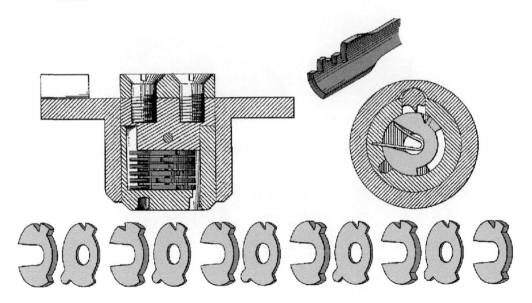

Figure 4.6: An early form of disc-tumbler side-bar lock from 1901 (US patent 688,070 by W. G. Denn).

Figure 4.7: (Left) Carl Kästner 12-disc side-bar lock in open position. (Right) Detachable-bit pipe key.

is a miniaturized version of the Kästner lock employing a solid key. The Abloy lock and its more recent variants are described in more detail later in this chapter.

Prior to World War II, a number of other notable side-bar lock designs appeared. Among them we find Jacobi's 1938 design for a tension-resistant wafer lock (US 2,182,588) and Liss's 1933 patent for a 5-pin dual side-bar lock with a conventional flat key (US 2,070,233). Liss's patent, assigned to the Briggs and Stratton Corporation, is closely linked to the more recent ASSA Desmo driverless dual side-bar lock. The ASSA Twin 6000 side-bar lock (Fig. 4.8) also utilizes this kind of side pin design. In the ASSA series of locks, the side pins are addressed by a track running along one or both edges of the key blade. The side pins must be raised to the required heights to permit retraction of the side-bar(s). Since the ASSA Twin has both conventional pin-tumblers and side pins, we refer to it as a dual-action side-bar lock. More recent versions of the ASSA lock, such as the Schlage Primus, include side pins that must be lifted and twisted to engage the side-bar.

The decision of the General Motors Corporation in 1935 to include the Briggs & Stratton side-bar lock in its motor vehicles firmly established the side-bar cylinder locking mechanism as a going concern in the car industry. The GM lock is a 6-wafer side-bar lock with the distinguishing feature that the side-bar is spring-biased radially inward instead of outward (as in Abloy and ASSA locks). The channel in which the side-bar moves is of rectangular rather than V-shaped section. This imparts a high degree of manipulation resistance to the mechanism since tensioning the plug does not assist in picking the lock. The General Motors lock is described in more detail in Chapter 7. Since World War II, side-bar locks have increased in popularity and are now widely used throughout the Western world.

Dual-action side-bar locks like the ASSA Twin provide a vastly increased number of differs compared with ordinary pin-tumbler locks. As well as being much harder to manipulate due to the presence of two independent locking mechanisms, the extra combinations are a distinct advantage in large master-keyed systems. A further benefit is protection against illegal key duplication since the side-bar bittings or tracks cannot be reproduced by standard key-cutting machines. The side-bar profile on the key also results in a much greater level of "key blank" control since keys with different side-bar bittings act as different key blank profiles. The factory exercises control by assigning the side-bar bittings or dealer permutations, with the remaining top bittings usually being cut by the locksmith.

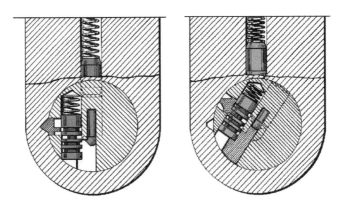

Figure 4.8: B. G. Widen's 1980 design of the ASSA Twin cylinder in locked and unlocked positions (US patent 4,356,713).

In high-security applications, particularly in the United States, the Medeco side-bar lock (Fig. 4.9), patented in 1968 by Roy Spain of the Mechanical Development Company, has become quite ubiquitous. Spain was previously employed by the Yale and Towne Manufacturing Company where he developed a disc side-bar lock similar to the Abloy lock that had spring biasing on the discs (US patent 2,578,211).

The Medeco lock is based on a twisting and lifting pin-tumbler principle. Although there is only a single row of pins, the pins have a specially shaped tip to allow them to be rotated by the angled bittings on the key (as in Fig. 4.10). The key cuts provide the required amount of lift to bring the pins to the shear line. In addition, the angled sides of the cuts twist the pins so that a longitudinal slot in each pin faces the prongs on the side-bar or fence. The side-bar is pushed radially inward as the plug begins to turn, causing the prongs on the side-bar to enter the slots in the pins.

Medeco locks proved so difficult to pick that, somewhat perversely, some locksmiths initially discouraged their use. The situation was described in a 1974 Medeco lock decoder patent by Iaccino and Idoni:

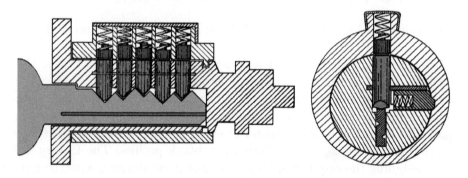

Figure 4.9: Medeco pin-tumbler plus side-bar mechanism (US patent 3,499,302).

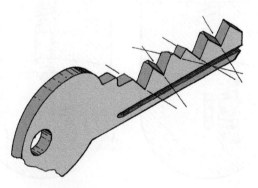

Figure 4.10: Angled cuts on Medeco key from R. C. Spain, R. N. Oliver, and P. A. Powell's 1967 patent (US 3,499,302).

In a lockout [...] the locksmith is usually unable to employ the normal picking techniques used on many other types of cylinder locks in order to [...] gain entry. Thus locksmiths often find themselves in a position where they must destroy the lock or the door [...] in order to gain entry for the occupant. Such crude techniques are repugnant to locksmiths. This has tended to discourage their recommending Medeco locks for use by their customers.

Side-bar locks with lever tumblers have been widespread in continental Europe for some time. One of the earliest examples is the Liega twin side-bar safe lock invented by Émile Fraigneux of Belgium in 1916 (DE 295,060 and UK 178,284). It utilized a system of alternating sliding levers or "frames" (like a sash window), each having a gate in two opposing edges to accommodate the side-bars. The Fraigneux lock can be thought of as a streamlined version of the 1870 Kromer Protector lock covered in Chapter 5. From Figs. 4.11 and 4.12 it can be seen that the lock was operated by an asymmetric double-bitted key that displaced the frames in order to align the gates with the side-bars. This allowed two ball bearings to be retracted from cavities in the side-wall of the lock cylinder, thereby freeing the core to rotate.

The high-security lock and safe manufacturer Fichet-Bauche, now owned by Gunnebo AB of Sweden, has produced a number of such locks including the Fichet-Bauche 484 and 666. The Fichet-Bauche 484 utilizes a pivoting 10-lever or rocker mechanism with two side-bars. The levers are arranged in two rows of five that alternate in direction. An H-profile key with four bitting surfaces is required to operate the lock. In contrast, the Fichet-Bauche 666 is a 7-wafer lock with a single side-bar. A similar concept, from T. F. Hennessy of Lori Corporation utilizing sliding wafers and pins, is illustrated in Fig. 4.13.

The Ingersoll high-security lock, produced in the United Kingdom, is another example of a lever side-bar lock. Like the Fichet-Bauche 484, it has a 10-lever mechanism and like the Fichet-Bauche 666, it uses a double-sided key. The levers are mounted

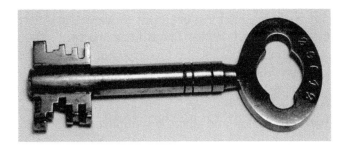

Figure 4.11: Double-bitted key from Fraigneux twin side-bar safe lock.

Figure 4.12: (Left) Fraigneux safe lock with cover removed. (Right) Key turned to align lever gates with side-bars.

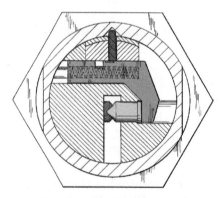

Figure 4.13: Dimple key pin-tumbler lock design with six sliders and side-bar (1981 US patent 4,404,824 by T. F. Hennessy).

on a common axis and are flanked by a single side-bar. The Ingersoll operating principle is substantially similar to Johnstone's 1966 patent shown in Fig. 4.14.

Many other side-bar lock designs have been put forward. An Australian example is the BiLock, invented in the early 1980s. The BiLock is a twin side-bar lock with two rows of driverless pin-tumblers. As in the Medeco cam lock, the side-bars are fitted with prongs that engage holes in the sides of the pins when they are raised to the correct heights. The bilateral key has a U-shaped profile, formed by folding a flat blank. The system provides a very large number of combinations and substantial protection against manipulation and illicit key duplication.

The largest lock manufacturing company in the world, ASSA Abloy AB of Stockholm, was formed toward the end of 1994 based on two companies whose flagship products

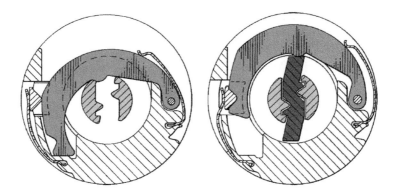

Figure 4.14: Lever side-bar principle from T. H. Johnstone's 1966 patent (US 3,367,156).

are side-bar locks. Since that time, ASSA Abloy has acquired many other major lock companies including Yale, Union, and Chubb in the United Kingdom, Medeco in the United States, and Lockwood in Australia. Before going into further details on the ASSA, Abloy, Medeco, Fichet-Bauche, Ingersoll, BiLock, and other side-bar locks covered in this chapter, we present the classification scheme that has been used to organize the material.

Side-bar Lock Classification

We have identified five fundamentally different types of side-bar locks: disc, lever, driverless pin, wafer, and dual-action. Although this categorization is certainly not the only one that could be applied, it suffices to point out the major differences in operating principles of the more than 20 cylinder locks presented in this chapter. The classifications together with the locks that fit them are listed below. Note that we have classed the Medeco Biaxial lock as a dual-action side-bar lock since it contains both conventional pin-tumblers and a side-bar mechanism.

1. Disc side-bar: cylinder locks with rotating discs and one or more side-bars. Examples: Abloy, Abloy Disklock Pro, Abloy Protec and Exec, ABUS Plus, Chubb SMI, DOM Diamant.

2. Lever side-bar: cylinder locks with pivoting or sliding levers and one or more side-bars. Examples: Fichet-Bauche 484, Mottura, Ingersoll, Miwa U9.

3. Driverless-pin side-bar: cylinder locks with pins that may be spring-biased but have no driver pins, together with one or more side-bars. Examples: Medeco cam lock, BiLock, New Generation BiLock, ASSA Desmo, Genakis, Tubar.

4. Wafer side-bar: cylinder locks with wafers or bars and one or more side-bars. Examples: General Motors or Briggs & Stratton (see Chapter 7), American Locker Co. (Lori), Fichet-Bauche 666, EVVA 3KS.

5. Dual-action side-bar: pin-tumbler cylinder locks with additional elements or degrees of freedom such as profile pins or angled cuts operating a side-bar. Examples: ASSA Twin 6000, Medeco Biaxial, Lockwood Twin, Schlage Primus, ASSA Twin Combi, Banham, Yale 5000, Scorpion CX-5.

4.2 Disc Side-bar

Abloy

(FI) 11–14 disc + side-bar (4)

Two versions of the basic Abloy lock are discernible: the Classic and the Profile (see Figs. 4.15 and 4.16). The Abloy Classic evolved from the original 1907 invention by Emil Henriksson, which was patented in 1919. The Abloy Profile was released in 1977 to provide added security against unauthorized key copying. Since the operating principles are identical, the following discussion applies equally to both the Classic and the Profile.

The Abloy lock consists of an outer brass cylinder and inner shell containing a stack of rotatable discs (Fig. 4.17). The shell is longitudinally slotted to accommodate a side-bar. There is also a cut-out section in the shell that limits the rotation of the discs to one quarter of a turn. The side-bar is L-shaped, with the shorter end of the L resting in a hole at the rear of the shell. The side-bar is spring-biased in an outward radial direction. The cylinder contains a longitudinal groove in which the side-bar rests when in the locked position. The retraction of the side-bar is controlled by the angular position of the discs. The Abloy Classic has up to 11 discs with separators. The Profile system may have up to 14 discs as well as profile-control discs (used in MK systems).

Figure 4.15: (Left & middle) Front and rear views of Abloy Classic cylinder with spring clip removed to reveal side-bar. (Right) Abloy Profile cylinder.

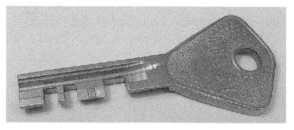

Figure 4.16: Keys for Abloy Classic (top) and Abloy Profile (bottom).

Figure 4.17: Abloy 11-disc shell (left); with first six discs removed to show side-bar (right).

The key for the Abloy Classic is half-round in cross-section with angled bittings along its length. The Abloy Profile exists in a number of half-pipe profiles with restricted blanks (see Fig. 4.16). Six bitting angles are possible from 0 (no cut) to 90 degrees in 18 degree increments, with a 90 degree cut leaving a quarter-circle of key profile.

Although an indirect or blind coding system is used for Abloy keys, for simplicity we refer to the disc codes as 1 to 6. We further assume that positions in the disc stack are numbered from the front of the cylinder. A number 6 disc requires a 0-degree rotation (or a 90-degree cut on the key), and a number 1 disc requires a 90-degree rotation (or no cut on the key). In general, a key cut angle of θ degrees will result in a net rotation of the corresponding disc by $90 - \theta$ degrees since the cut surface does not contact the disc until the key has turned through an angle of θ degrees.

Discs for the Abloy Classic usually have a D-shaped hole, although there may also be a cut-out in the straight edge of the D (see Fig. 4.19). Abloy Profile discs have additional wards matched by profiling on the key blade. The discs have a stop lug on the rim that contacts the edge of the cut-out section of the shell at the extremes of rotation (0 and 90 degrees). Each disc also has a side-bar gate in its periphery. The disc in position 1 does not have a stop lug and can therefore turn freely. The reason for this is twofold: first, it prevents drilling [38] and second, it prevents tension from being applied naïvely to the first disc in the case of a manipulation attempt. The discs are reversible: for instance, a number 1 disc can be flipped over and used as a number 6 disc. This is a manufacturing convenience to reduce the number of components required for production.

When all discs have been correctly aligned by a quarter turn of the key in the clockwise direction (see Fig. 4.18), pressure exerted by a longitudinal bevel in the cylinder wall forces the side-bar radially inward into the channel formed by the discs. The shell is then free to turn (clockwise), releasing the locking balls in a padlock or turning the tail-piece in a cylinder lock. The key cannot be removed in the locked position since, with different angles of rotation on the various discs, the D-shaped cut-outs in the discs are not aligned. From the open position, if the key is turned anticlockwise, a position is reached at which the side-bar springs back out into the longitudinal channel in the cylinder wall. At this point, the uncut edge of the key blade immediately begins to rotate the discs back to their rest positions, relocking the side-bar. The key is then turned back to the point where the stop lugs

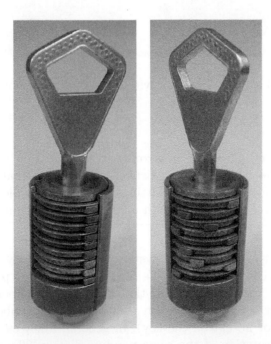

Figure 4.18: (Left) Key inserted with discs in rest positions. (Right) Key turned 90 degrees, leaving discs at various angles of rotation.

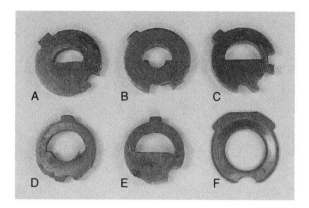

Figure 4.19: Assortment of single-cut and master-keyed Abloy discs: A—number 3 cut; B—number 5 and 6 cuts; C—number 1, 3, 4, and 6 cuts. B and D have cut-away sections. F is a separator. A, B, D, and E have false gates.

on the discs are all in angular alignment at 0 degrees rotation. The simultaneous alignment of the cut-outs in the discs then forms a keyway to permit removal of the key.

An Abloy lock with 11 active discs and six angular positions has theoretically $6^{11} = 362,797,056$ keying combinations, almost all of which are usable since the MACS is unrestricted; that is, a cut for a number 1 disc may be adjacent to a cut for a number 6 disc. One of the very few keying constraints is that at least one number 1 disc must be used so that, when the key is rotated to 90 degrees and the stop lug reaches the end of its travel, this disc is actually at the correct angle. If no number 1 disc were present, then it would be possible to turn the key more than a quarter turn, resulting in overrotation of the discs. To see this, note that each disc can rotate through 90 degrees, but, in the absence of a number 1 disc, the key would turn at least 18 degrees before contacting any of the discs. The requirement of having one number 1 disc reduces the number of theoretical codes to $6^n - 5^n$ where n is the number of discs, since all 5^n codes not containing a "1" must be excluded. Hence an 11-disc Abloy provides no more than 314 million combinations.

Master-keying is achieved by cutting more than one gate in one or more of the discs. Shallow (false-depth) gates are usually included to inhibit picking and impressioning, as illustrated in Fig. 4.20. Maneuvering of picking tools is made difficult by the geometry of the keyway and the relative positions of the discs when rotated. (For Abloy Profile locks this difficulty is particularly acute.) It is only feasible to apply tension to a number 1 disc, since it must be rotated by 90 degrees for its gate to register with the side-bar. In general, it is not known beforehand which discs are which, although the disc at position 11 (in an 11-disc lock) is often of the latter type. In addition, a fixed (nonrotating) disc can be used, so that the key blank must have a 90-degree cut in order for it to turn.

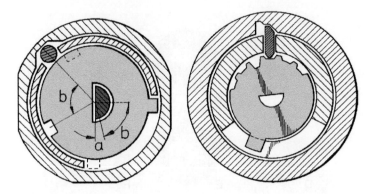

Figure 4.20: Operation of an Abloy lock with and without false gates (US 3,621,689 by R. Koskinen and K. H. Solitanner, US 3,948,065 by K. Martikainen).

It is generally agreed that it is possible to defeat the Abloy lock using a reader or jig similar to the Hobbs pick used to pick lever locks. Such a jig could be made using a coaxial rod fashioned to fit the keyway. The inner part of the rod tensions the core, while the outer part is free-sliding and can be rotated to test or align each of the discs in turn. In this way, the combination of the lock could be decoded and a key cut. If the discs had false gates, then this would only narrow down the number of possible keys that need to be cut and tried. The geometrical difficulties associated with this idea are: (i) once a disc is rotated by 90 degrees, there is only a quarter of a circle gap remaining in which to maneuver the jig forward or backward: (ii) the keyway is not centrally located in the face of the lock so that the jig will not lie along the axis of the discs. Two further problems must also be circumvented: (iii) cut-away discs may be included in the disc stack that are difficult to rotate during manual manipulation (see Fig. 4.19); (iv) false gate positions vary even for the same gate position. Thus it is not easy to locate the correct gate by "feel." However, for disc side-bar locks with symmetrically bitted keys and centered keyways (e.g., ABUS Plus and its numerous clones), the preceding idea is quite effective.

The Abloy lock has a most ingenious design in that it is simple to manufacture, contains relatively few moving parts, and yet is very hard to pick or impression, particularly when false gates are present. It is therefore a good choice for padlock mechanisms and outdoor environments. Early designs had a straight side-bar that could easily be removed by drilling a small hole in the face of the lock. This was rectified by making the side-bar L-shaped with an anchor point at the base of the slot in the shell (Fig. 4.15, center). It is then necessary to drill the side-bar along its entire length to defeat the lock.

We have already mentioned the close connection between the Abloy lock and Carl Kästner's safe lock from 1918. A number of high-security key-operated combination locks patented by Sargent & Greenleaf are also closely allied with the Abloy and Kästner locks. Diagrams from the relevant patents are given in Figs. 4.21 and 4.22.

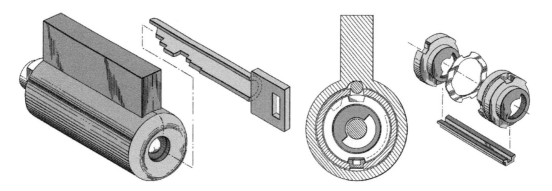

Figure 4.21: Sargent & Greenleaf's 1977 design for a rotary disc-tumbler side-bar lock (US patent 4,083,212 by P. R. Proefrock).

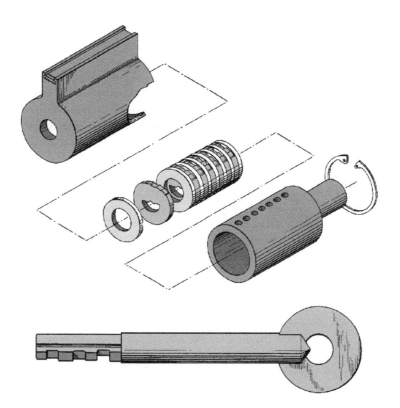

Figure 4.22: S&G's 1986 design for a disc-driven pin-tumbler lock (US patent 4,651,546 by W. R. Evans).

The La Gard 2200 safe lock is also based on a rotating disc mechansim, albeit with changeable discs. The Abloy principle has more recently been applied to some car locks (e.g., Ford Tibbe). The Abloy Classic and Abloy Profile designs have now been superseded by the Abloy DiskLock Pro, Protec, and Exec. Abloy locks are further discussed in two articles by Fey [38, 39].

Abloy DiskLock Pro

(FI) 11-disc + side-bar (4)

One problem associated with the original Abloy design was that it could only be opened in the one direction (clockwise) for a given handedness of key. A further problem was key breakage due to weakening of the key by maximum depth cuts. The Abloy DiskLock Pro (DLP) cleverly overcomes these drawbacks, also boasting a reversible key. The preliminary designs for a symmetrically bitted, bidirectional Abloy lock appeared in 1972 and 1977 (US patents 3,789,638 and 4,109,495), as shown in Figs. 4.23 and 4.24. They contain elements of both the Abloy DLP and its successor, the Protec. The production version of the DLP involved a number of minor modifications to the design of the discs and key bittings. In our description of the Abloy DLP, we measure angles clockwise from 12 o'clock as positive and anticlockwise angles as negative. As before, disc positions are numbered from the front of the cylinder.

The Abloy DLP cylinder, shown in Figs. 4.25–4.29, is somewhat more complicated than the earlier Abloy Classic design. The core comprises a shell or drum, a stack of 11 discs interleaved with 10 separators, one side-bar, two return bars, and a disc

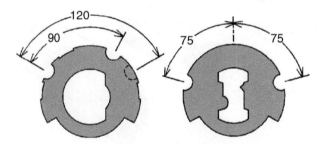

Figure 4.23: A 1977 design for bidirectional Abloy discs (US patent 4,109,495 by M. E. Roberts).

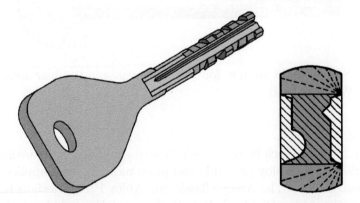

Figure 4.24: Design of a symmetrically bitted Abloy key (US patent 4,109,495).

Figure 4.25: Abloy DiskLock Pro oval cylinder and key.

Figure 4.26: (Left) Abloy DLP core. (Middle) Shell with disc stack removed and return bars visible. Side-bar in foreground. (Right) Disc controller.

Figure 4.27: Two views of Abloy DLP disc stack: twin return bars engaging driver discs (left); channel formed by separators for side-bar (right).

controller. A tweezer-like tension bar, attached to a profile plate inside the disc controller, provides support for the disc stack. There is also a drill protection plate at the back of the disc stack. The disc stack and controller subassembly are mounted in the shell, the back portion of which is fashioned into a tail-piece. The entire core assembly slots into the cylinder, which may be adapted to a variety of different formats.

The profile plate of the disc controller is constrained to turn ±90 degrees before its stop lug contacts the outer rim, limiting the rotation of the key. There is a spring-loaded locating ball that identifies the neutral (scrambled) alignment of the

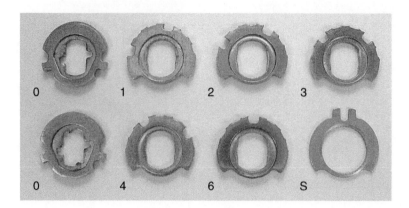

Figure 4.28: Discs from Abloy DLP cylinder: driver discs (0), code discs, and a separator (S).

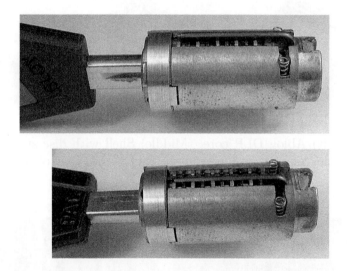

Figure 4.29: Abloy DLP core with key inserted (top). Key turned to 90 degrees and side-bar retracted (bottom).

profile plate within the disc controller, at which point the key may be inserted or withdrawn. The profile plate is flanked by two spring-loaded balls that provide positive location of the key blade when fully inserted (hence the dimple on the shoulder of the key blade).

Referring to Fig. 4.26, the shell contains a longitudinal cut-out at 12 o'clock for the side-bar, at the base of which is a spring-loaded triangular wedge that positively aligns the side-bar in the channel between the shell and the cylinder. The disc stack (Fig. 4.27) does not fully occupy the available space, but leaves just enough room for two return bars at 4 and 8 o'clock. Each of the return bars rests in a groove in the shell. The discs are normally scrambled so that their gates are out of alignment while their cut-outs are aligned, forming the keyway.

Of the 11 discs, the discs at positions 4 and 11 are driver or "0-discs," with a cut-out that matches the key blank section (see Fig. 4.28). The remaining nine discs, called code discs, are used to set the combination. All of these have the same broadly rectangular cut-out with contact points for the key surfaces in the top and bottom left- and right-hand edges.

The lower driver disc (in position 11) has an oddly shaped cut-out that provides extra key-profile control and also accepts the protrusions of the tension bar. The upper driver disc (in position 4) is a reduced-diameter disc with no side-bar gates that also has cuts for the tension bar. The tension bar itself has a long slot that rides in the narrow channel on the key stem. Turning the key tensions both discs 4 and 11 but none of the other discs. Discs 4 and 11 also contain gates for the two return bars, whereas all other active discs have a 135-degree bitting on the edge to allow limited travel past the return bars.

As the key is turned clockwise, the gates in the driver discs pick up the return bar at 4 o'clock and allow it to turn until it contacts the other return bar at 8 o'clock. The net effect of this is to limit rotation of the driver discs to 90 degrees. A similar argument applies when the key is turned anticlockwise. This clever mechanism therefore allows a maximum ±90-degree rotation of the discs in the shell.

During clockwise (CW) operation, the top right and bottom left contact points of the discs touch the bittings on either side of the key blade. Conversely, during counterclockwise (CCW) operation, the top left and bottom right contact points are active. Both sets of contact points are used in operating the lock since, once unlocked, the key cannot be withdrawn until the discs are scrambled again by turning the key in the opposite direction through 90 degrees. The control surfaces on the key determine at what angle the key first contacts the cut-outs in the discs, which determines their final angle of rotation. For instance, if the key bitting first contacts a disc at θ degrees of rotation, the disc will undergo a net rotation of $90 - \theta$ degrees once the key has completed its 90-degree rotation.

Each of the code discs may be one of six basic types. Code discs 1–5 are peripherally gated in two places, one for each direction of rotation. The gates are placed symmetrically in the case of a number 3 disc but asymmetrically for numbers 1, 2, 4, and 5. Discs 1 and 5 are mirror images, as are discs 2 and 4. Code disc 6, corresponding to the maximum cut on the key, has only a single gate: a number 6 disc does not need to be turned to align its gate. Discs also have false-depth gates to hamper decoding. Unlike the original Abloy, which has flat discs, Abloy DLP discs are embossed and must be inserted with the raised section toward the rear of the cylinder.

The gate positions on the discs are given in Table 4.1. This table also shows, for either CW or CCW operation, the angle at which the respective key bitting first contacts the disc (the contact angle) and the net rotation imparted to the disc.

Disc #	CW Gate	Contact Angle	Rotation	CCW Gate	Contact Angle	Rotation
1	−75	15	75	15	−75	−15
2	−60	30	60	30	−60	−30
3	−45	45	45	45	−45	−45
4	−30	60	30	60	−30	−60
5	−15	75	15	75	−15	−75
6	0	90	0	0	−90	0

Table 4.1: Gate locations, key contact angle, and net rotation for Abloy DiskLock Pro discs. All angles are in degrees.

There are six possible key-bitting angles in increments of 15 degrees from no-cut to 75 degrees. The set of bittings on the key determine the combination of disc angles presented to the side-bar. The bittings at positions 4 and 11 have to be no-cut (corresponding to maximum disc rotation) in order to align the driver discs. A correctly bitted key, when turned 90 degrees either CW or CCW, aligns the gates of all 11 discs at 12 o'clock underneath the side-bar channel.

Various key blank profiles are available giving more flexibility in key control. In terms of master-keying options, the six basic code disc types are supplemented by around 50 master discs that accept various multiple-code combinations. For instance, master discs exist for code combinations (2 3), (2 3 4), (2 4 6), and so on. There is even a "null" disc with code (1 2 3 4 5 6) that can be operated by any of the bitting angles on the key. Replacing a basic code disc with a master disc allows more than one key to operate the lock. For example, using master disc (2 3) in position 2 and (2 4 6) in position 5 would allow $2 \times 3 = 6$ different keys to operate the lock.

The Abloy DiskLock Pro system addresses several drawbacks of the older Abloy design: the disc controller prevents the key from being turned before it is fully inserted; key breakage is no longer a problem since the cuts do not weaken the key excessively. Two recent updates of the DiskLock Pro system are the Abloy Protec and Protec Industrial, which are easier to assemble. Apart from slight differences like disc numbering, cut-out profile, and gate offsets, the operating principle is identical to that of the Abloy DLP. The major differences can be appreciated by referring to Table 4.2, which gives the gate positions for the six code discs for the Abloy Protec and the contact angles for the key bittings. The ordering of gate positions for CCW operation is now out of sequence compared with Table 4.1. Note that disc number 4 has symmetric gate angles (instead of disc 3 in the Abloy DLP); numbers 3 and 5 and numbers 1 and 2 are mirror images. Disc number 6 is still gated at 12 o'clock, and driver discs are installed at positions 4 and 11. The key requires bittings with two different cut radiuses, which leads to enhanced copy protection.

A further feature of the Abloy Protec is the so-called disc-blocking system. This refers to a supplementary function of the return bars. The code discs for the Protec system contain a series of peripheral notches along the edge that contacts the return bars. The driver discs have a crescent-shaped cut-out that lodges one of the return

Disc #	CW Gate	Contact Angle	Rotation	CCW Gate	Contact Angle	Rotation
1	−75	15	75	60	−30	−60
2	−60	30	60	75	−15	−75
3	−45	45	45	15	−75	−15
4	−30	60	30	30	−60	−30
5	−15	75	15	45	−45	−45
6	0	90	0	0	−90	0

Table 4.2: Gate locations, key contact angle, and net rotation for Abloy Protec discs.

bars (depending on the direction of turning). As with the DLP, during operation only one of the return bars is picked up by the driver discs; the other remains in its channel. Rotation of the discs is stopped at the point where the two return bars come into contact. The difference with the Protec mechanism is that when the moving return bar contacts the stationary return bar, the moving bar is forced radially inward and into contact with the notches on the code discs. This action freezes the combination of the code discs, which must all be turned simultaneously to the correct angles to address the side-bar. A similar combination freezing principle is found in the Fichet-Bauche 787.

Both new systems enable extra master-keying possibilities via an increased set of master discs (around 55 different types in addition to the six basic code discs). A progressive indirect code is used for coding of the key cut sequence. The coding takes the form of a 7×10 look-up table for the seven disc numbers $(0, \ldots, 6)$ and 10 code discs (not counting the 11th disc). The code for each disc varies depending on its position in the code. For instance, a number 3 disc in position 1 is coded as a 9, but the same disc number in position 2 would be coded as a 5.

Electromechanical versions of the Abloy DiskLock Pro also exist, such as that described in UK patent 2,158,870. A transponder implanted in the key head transmits its unique code to the control electronics in the lock. Power for the key-top electronics is derived inductively from a source in the body of the lock.

ABUS Plus

(DE) 9-disc + side-bar (3)
(AU) **Chubb SMI** (equivalent)

The German company ABUS (standing for August Bremicker und Söhne) has been producing padlocks since 1924. The range of heavy-duty "Granit" padlocks, introduced around 1983, utilizes the ABUS Plus side-bar mechanism (Fig. 4.30). It is a close relative of the conventional Abloy but with a centrally located rectangular

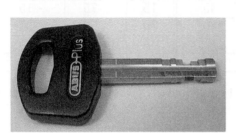

Figure 4.30: ABUS Plus key and Granit padlock.

Figure 4.31: (Left) Chubb SMI key. (Right) Chubb SMI padlock with disc stack and side-bar exposed.

keyway and a symmetric key that can be inserted either way round. The Chubb SMI heavy-duty 9-disc padlock, shown in Figs. 4.31–4.33, is another Abloy-type lock with a reversible key.

ABUS Plus locks typically have a stack of nine code discs, each corresponding to one of six possible rotation angles (see Fig. 4.32). The frontmost disc in the keyway is usually freely rotating, though not always, and the last disc may not be a number 1 disc (zero cut). Code discs may also include false-depth gates. The discs are interleaved with fixed separators to decouple their motion. Instead of a stop lug, the code discs have a reduced-diameter section; rotation is limited to the points of contact between the shoulders of the discs and the edges of a pillar in the shell.

When all the discs are rotated to the correct angles by the key, a longitudinal channel is formed by their gates. This allows the side-bar to drop into the channel as the core begins to rotate. The discs can be turned up to 180 degrees without fully inserting the key. If the key is cut quite deeply on both sides, it may be prone to shearing off in the lock when not inserted to the correct depth. Locks of this type are susceptible to decoding and picking with a specially designed coaxial tool resembling a 2-in-1 pick, although the process is not rapid enough to be attractive to thieves.

Figure 4.32: Abus Plus discs.

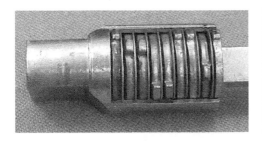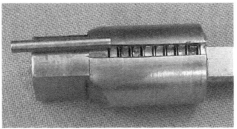

Figure 4.33: Chubb SMI core. (Left) Discs rotated by key. (Right) Gates aligned for side-bar.

DOM Diamant

(DE) 10-disc + 3 side-bar (4)

The DOM Diamant ("diamond") cylinder is a VdS class B approved lock representing the top level of security offered in DOM's cylinder lock range. Pictures of a DOM Diamant profile cylinder appear in Figs. 4.34–4.37. It is a variant of the Abloy lock with the principal difference lying in the design of the key and its contact surfaces with the discs. The initial concept was described in a 1993 German patent by H. P. Häuser of DOM Sicherheitstechnik (DE 43 14 208), filed in the United States in 1994 (US 5,613,389). The DOM design is similar to a 1982 Australian design by Ogden Industries shown in Fig. 4.38, published in a 1985 US patent. Whereas in a conventional Abloy side-bar lock, the discs are rotated manually by turning the key, the DOM and Ogden designs employ a specially shaped, asymmetric key that rotates the discs in *both* directions as it is inserted. How this is achieved will be clarified in the following description of the DOM Diamant lock. As usual, the lock cylinder is assumed to be viewed with the pin chamber portion at 12 o'clock.

The cylinder housing is of cast steel ("duracast") construction with a thick, hardened disc located at the front of the recessed keyway. This frontal disc provides vetting for the key profile as well as drill protection. The keyway is also very restrictive to impede the insertion of manipulation tools. The key has a steel shank broached by two channels to match the warding of the keyway. At the shoulder of the key, which is of round section, there is a reduced-diameter portion that bypasses a protrusion

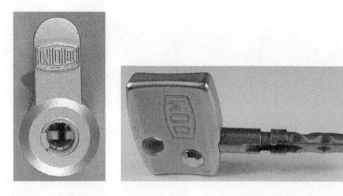

Figure 4.34: DOM Diamant high security rotating-disc cylinder and key.

Figure 4.35: (Left) DOM Diamant shell with side-bars and frontal cap. (Right) View down keyway with cap removed.

Figure 4.36: Sample of DOM Diamant left- and right-handed discs with separators.

on the housing just to the left of the keyway. The function of this protrusion is to ensure that the key is fully inserted before turning and to retain the key during operation.

A complicated set of transverse three-dimensional (3-D) bittings is apparent on both sides of the key blade. This 3-D pattern requires CNC milling to reproduce accurately, which greatly adds to the level of protection against unauthorized key

Figure 4.37: (Left and middle) DOM Diamant shell with key partially inserted. (Right) Full insertion of key aligns gates in discs.

duplication. The key is inserted with the larger part of the head at 12 o'clock. In this orientation, the deeper channel in the key blade is at 9 o'clock. One can discern four separate "quadrants" or tracks on the key surface. For later use, we refer to the four quadrants as top left (TL), bottom left (BL), top right (TR), and bottom right (BR).

The cylinder houses a brass shell (Fig. 4.35), similar to the Abloy lock, in which a number of stamped steel discs are mounted. The shell is blind at the rear and open at the front, capped by the hardened steel frontal disc. Two locating tabs ensure that the frontal disc is properly seated on the shell. The shell is longitudinally slotted at four places around its edge, the slots being at an angular spacing of 90 degrees. Three longitudinal channels are milled into the cylinder bore at 6, 9, and 12 o'clock. These three channels accommodate steel side-bars with pointed ends that register with the slots in the shell. The slot at 3 o'clock in the shell contains a nylon stop rod.

The cylinder contains 10 active discs and 11 separator discs (Fig. 4.36). The 11 separator discs, which are not involved in the locking function, each contain four tiny ball bearings to ensure smooth and independent operation of the rotating active discs. All active discs contain a system of three peripheral notches or gates spaced 90 degrees apart. Shallow, false-depth notches are also provided to thwart lockpicking and decoding attempts. The presence of the separator discs, one of which is made of rubber, also provides damping on the rotation of the discs.

Two types of active discs are employed, distinguished by the shape of their central B-shaped cut-out. Left-hand discs, acted on by the left-hand edge of the key blade, have a short central ward (the horizontal part of the B). Right-hand discs, acted on by the right-hand edge of the key blade, have a long central ward. In both cases, there are two contact points along the straight edge of the B in each disc. In the locked position, the left-hand discs are oriented with the B facing right,

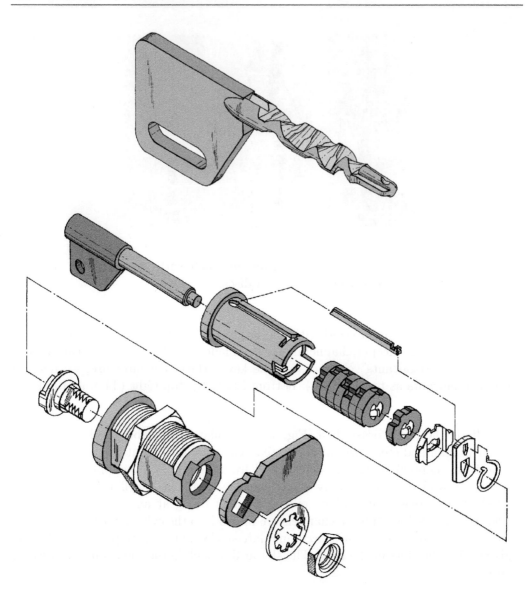

Figure 4.38: Ogden Industries' 1985 bidirectional side-bar lock design (US 4,512,166 by G. F. Dunphy and D. J. Newman).

whereas right-hand discs have their B facing to the left. The discs are arranged in an alternating sequence along the shell, with a right-hand disc at position 1 (at the front of the cylinder). The discs also feature a shallow peripheral recess at 3 o'clock spanning about 90 degrees. The nylon stop rod, seated between the recess and the cylinder bore, limits the angular travel of the discs: this is necessary to maintain their handedness and to ensure easy insertion of the key through the cut-outs in the discs.

The three gates in each disc are used to provide differs for the lock. While a 90-degree increment is always maintained in the spacing of the three gates, the angular offset of the gates may vary from one disc to the next. Focusing on the gate nearest the 9 o'clock position, there are five possible offsets, which we refer to as -2, -1, 0, 1, and 2. These offsets are identical for both left- and right-hand discs. The zero offset position is at 9 o'clock, corresponding to a zero-degree rotation of the disc from its locked position. Positive offsets are measured clockwise and negative offsets anticlockwise. The offset increment is approximately 15 degrees. For instance, a -2 disc has its gates at about 30 degrees past 6, 9, and 12 o'clock.

A left-hand disc with a negative offset must be rotated CW to align its gates with the channels in the shell at 6, 9, and 12 o'clock. Similarly, a right-hand disc with a negative offset must also be rotated CW to align its gates. Discs with positive offsets must be rotated counterclockwise for registration with the side-bars. Note that since the gates are at 90-degree increments, all three gates simultaneously register with the slots in the shell when the disc is correctly rotated.

As we mentioned before, the discs have bidirectional dynamics. This is achieved by the four bitting surfaces on the left and right sides of the key. As in an Abloy lock, the cuts are angled with respect to the plane of the key. It is convenient to visualize the bitting surface on each quadrant of the key as a sequence of peaks and troughs, joined in a smooth contour. At each of the 10 disc positions along the blade, a shallow cut produces a peak, while a deep cut produces a trough. The pair of tracks on either side of the key blade are cooperative: a peak in a given position on the top track requires a trough in the same position in the bottom track, and vice versa. There are five possible cut depths overall. In order to match the coding of the discs, we number these as -2, -1, 0, 1, and 2. Thus cuts 1 and 2 are peaks and -1 and -2 are troughs. A zero cut makes both top and bottom tracks equal in height.

The height of the peak corresponds to the angle of rotation imparted to the disc at that position. Thus a zero cut on a given side of the key causes both the top and bottom tracks to graze the contact surfaces of the cut-out in the disc, providing no rotation. Nonzero cuts impart either a CW or CCW rotation to a disc, depending on which side of the key blade the tracks are situated. A peak in either the TL or BR quadrants contacts the straight edge of the B in the discs to provide a CCW rotation. Conversely, a peak in the TR or BL quadrant provides a CW rotation. Opposite each peak, a trough is required to accommodate the matching contact surface of the disc as it rotates.

Since the discs alternate along the shell, cuts are made to pairs of tracks on alternate sides of the key. On each side there are five bitting positions. The bittings on a given side of the key blade operate the discs of the same handedness. The bitting sequence can be visualized as a matrix of cuts as illustrated in Table 4.3 for the key featured in the photographs. A dash in the table corresponds to a bridge between the cuts in the adjoining positions. The system is designed to allow a size 2 peak to be adjacent to a size 2 trough on either the same or the opposite side of the key blade. Thus there is no MACS constraint. Since each of the 10 discs has 5 possible gate offsets, there are theoretically $5^{10} = 9,765,625$ keying combinations. If only nonzero offsets are considered, there are $4^{10} = 1,048,576$ theoretical combinations.

Position	1	2	3	4	5	6	7	8	9	10
TL	-	−2	-	2	-	2	-	−1	-	−1
BL	-	2	-	−2	-	−2	-	1	-	1
TR	−2	-	1	-	−2	-	1	-	−1	-
BR	2	-	−1	-	2	-	−1	-	1	-

Table 4.3: Example bitting matrix for DOM Diamant key.

As the key is inserted (Fig. 4.37), the peaks on the tracks contact the cut-outs in the discs. With the key fully inserted, the angular positions of the 10 active discs are determined by the bittings in the four quadrants of the key. If the key correctly rotates all 10 discs, then their gate recesses register with the slots in the shell at 6, 9, and 12 o'clock. As the key is turned, the three side-bars are retracted into the channels formed by the gates. Once the shell is turned, the discs are held in position by the three side-bars.

The picking of the DOM Diamant mechanism, though theoretically achievable, is hampered by the presence of false-depth notches and the very limited space in the keyway. Since the frontal disc is fixed to the inner shell and not to the housing, it may be used to tension the lock. Interestingly, the presence of three side-bars has little effect on the manipulation difficulty, since alignment of the discs for one side-bar implies alignment for the other two. The extra side-bars do, however, increase the degree of drill resistance.

4.3   Lever Side-bar

Fichet-Bauche 484

(FR) 10-lever + 2 side-bar (4)

The Fichet Company was founded in 1825 in Paris by Alexandre Fichet. Fichet produced a fire-proof safe in 1840 and by 1879 was producing safe vaults with deposit boxes for banks. In 1967 Fichet merged with Bauche, another famous French lock and safe manufacturer, founded by Auguste Bauche in 1864. Fichet-Bauche was the leading supplier of high-security locks and safes in France up to the time of its acquisition by Gunnebo AB in 1999.

The Fichet-Bauche 484, shown in Figs. 4.39–4.41, is a classic French high-security lock, manufactured in a "2D monoblock" format with 26 mm cylinders. It was patented in 1949 in France, and a lock utilizing the 484 mechanism is described in UK patent 678,123 (1950). The principle is loosely based on much earlier designs described in US patents 408,147 (1889) by T. Taylor and 1,498,047 (1923)

Figure 4.39: Fichet 484 twin side-bar cylinder and H-profile key.

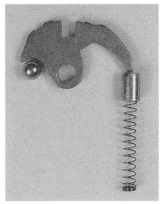

Figure 4.40: (Left) Fichet 484 core and one side-bar. (Right) Ball, rocker, and driver assembly.

by C. Ledin. Taylor's patent was for a lock with five blocking pins in which the two-sided key pivoted a set of rockers controlling the retraction of the pins. Ledin's patent suggested using a set of rockers to control the retraction of a side-bar. Enhancements of Fichet's original design are discussed in US patents 4,044,578 (1976) and 4,296,618 (1978); the second of these addresses the problem of key breakage. A drawing from the 1976 patent appears in Fig. 4.42.

The Fichet 484 lock is distinguished by its H-shaped keyway and elaborate key with two blades and four ramps. The key is not symmetric, with an extension of the cross of the H on one side serving to distinguish the left and right blades. The end of the key blade is forked and has the function of deploying a hinged tail-piece at the rear of the plug. The tail-piece is normally swung back and the clutch disengaged.[2]

---

[2]Fichet locks typically utilize a spring-loaded clutch rather than a fixed linkage: this is normally retracted until acted on by the key, causing it to engage the locking cam.

Figure 4.41: (Top) Fichet 484 core with key partially inserted. (Bottom) Key fully inserted to align rocker gates.

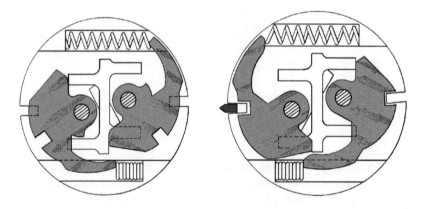

Figure 4.42: Operating principle of the Fichet-Bauche 484 from US patent 4,044,578 by F. Guiraud.

The plug (Fig. 4.40, left) contains two sets of five ball-driven counter-rotating rockers, one set located above and one below the H. The plug also accommodates a pair of side-bars sprung outward at each end, one side-bar across each set of five rockers. The rockers in each set share a common pivoting bar: the two bars being inserted just above and below the cross of the H (visible in Fig. 4.39).

The rockers in each set are mounted in an alternating sequence, three on one side and two on the other. Each rocker (Fig. 4.40, right) has a belly that rests on a ball bearing; the other side is elongated and pushes down on a capped driver spring. The ball is located in a bore that extends down to the keyway: the ramps on the key only ever contact the ball, which reduces friction. Each rocker also contains a gate in its outer edge, as well as false-depth gates to inhibit picking.

Each set of five rockers is actuated by two ramps on the key. The ramps on the top end of the H drive the rockers located at the top and vice versa for the ramps at the bottom end of the H. The lifting motion of the key blades is transformed into a pivoting motion of the rockers. For each side of the key (the top and bottom parts of the H), the three cuts in one ramp and the two cuts in the other ramp, which are staggered, must drive the rockers in alternating directions of rotation in order to align the gates (Fig. 4.41). When all five gates on the upper set of rockers and all five gates on the lower set are in alignment, the side-bars can be retracted into the channels formed by the gates. At this point the plug is free to rotate in the cylinder housing. The key also swings out the tail-piece, engaging the clutch through to the rest of the lock mechanism.

While the Fichet-Bauche 484 offers a high degree of pick resistance, it is susceptible to forced opening with a reinforced steel key called a *clé de force*. This is one reason why the 484 has been superseded by the Fichet-Bauche 787, which we encounter in Chapter 5.

Mottura

(IT) 6-lever + side-bar (3–4)

The Mottura push-key cylinder (Figs. 4.43–4.46), produced by the Italian company Mottura Serrature di Sicurezza S.p.A., is a one-star A2P-rated cylinder lock. It is typically installed on multiple-bolt deadlocks such as the Mottura model 30611. Judging from the shape of the end-bitted key, one might be tempted to conjecture that the mechanism is similar to that of the ISEO R6 or perhaps the Tover 27A (in Chapter 2), but this is not the case. The cylinder features six sliders or rods and a single side-bar. The design is not unlike the one from the 1970 patent diagram in Fig. 4.47. A mechanically reprogrammable version of the Mottura push-key lock was disclosed in a 1995 patent (US 5,791,181).

The key for the Mottura lock (Fig. 4.43) is flat with a groove to match a central ward in the keyway; it can only be inserted in one orientation. The key appears to have nine cut positions, but the outer two positions, which are uncut, function as shoulder stops. Of the remaining seven positions, the middle position (number 5) is a dummy, leaving positions 2, 3, 4, 6, 7, and 8 for the actual slider bittings. An attribute of the push-key design is that, in contrast with conventional lever locks, only axial force is applied through the key. This makes possible a fully unrestricted MACS, with the key exhibiting isolated prongs among its bittings.

Figure 4.43: (Left) Mottura side-bar cylinder with protection sleeve. (Right) End-bitted key.

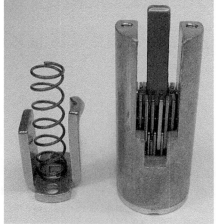

Figure 4.44: (Left) Mottura core with side-bar in foreground. (Right) Core with rear cover removed to show sliders and tail-piece.

The sliders are flat and heavily sprung from the rear by a T-shaped tail-piece (Fig. 4.44 right). Each slider has an intervening, fixed separator that allows it to move independently of the others. Sliders are supported in a cylindrical core flanked by a rear cover. On one side of the core, there is a cut-out into which is fitted a short side-bar shaped like a T (Fig. 4.44 left). The upper bar of the T rests against the edge of the sliders, also engaging a longitudinal groove in the cylinder that normally blocks rotation of the core. Each slider has a gate, the position of which may vary along the length of the slider (see Fig. 4.46). The relative positions of the gates provide differing in the lock. Furthermore, the contact point at the

Figure 4.45: Key depresses sliders to correct depths, permitting retraction of side-bar.

Figure 4.46: Slider pack from Mottura lock (left side faces front).

front of each slider is also subject to a variable-depth offset. The required depth of cut on the key is determined by both the gate position and the offset of the slider.

The key bittings must be such that all six sliders are simultaneously depressed to the correct depths, aligning their gates at the cross-piece of the T, as depicted in Fig. 4.45. As the key is turned, the side-bar retracts into the channel formed by the slider gates allowing the core to rotate. In general, only the slider that has been depressed the furthest will be in contact with the tail-piece, causing it to protrude from the rear of the cylinder and engage the boltwork of the lock.[3] The cylinder is protected by a toughened cylinder-guard and slotted front-piece. The entire guard assembly rotates freely to prevent sawing.

The Mottura 6-slider cylinder does not have standard dimensions, which is a poor formula for earning market share. So like many other lock manufacturers in Europe, Mottura are now producing Europrofile cylinders with horizontal keyways and reversible dimple keys. The "Champions" system, of which model C48 is the most recent addition, has five conventional pins acted on by a line of dimples on the key. An additional four driverless, rotating pins cooperate with a side-bar

---

[3]Mottura supplies a large range of heavy-gauge multipoint locks.

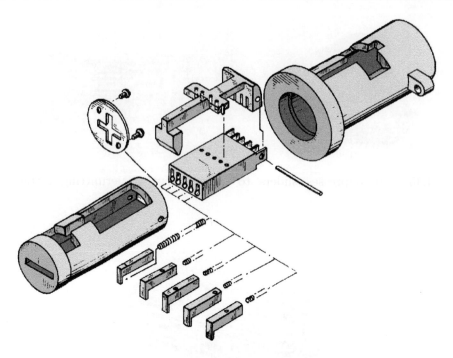

Figure 4.47: Design of a six-slider side-bar lock (US patent 3,604,231 by F. P. Buschi).

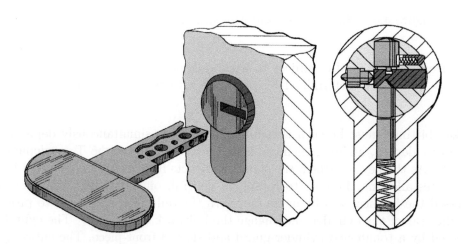

Figure 4.48: "Champions 48" side-bar lock design, with cammed rotating pins (US patent 6,490,898 by S. Mottura).

(like a Medeco pin without the lift dimension). The angle of rotation of the side-bar pins is set by the side-track milling on the key. There is also a floating ball, as in DOM iX locks, to prevent unauthorized key copying. For completeness, a diagram from the relevant patent has been included in Fig. 4.48 to show the design of this unusual lock. Further details may be obtained from Mottura's 2000 patent.

Ingersoll

(UK) 10-lever + side-bar (4–5)

Ingersoll locks, according to Evans [33], were originally produced by the Ingersoll Watch Company's lock manufacturing division around 1949. Ingersoll Locks were sold to the Yale lock company (Willenhall) in 1988. The flagship product is a 10-lever side-bar lock used in rim and mortice deadlocks for doors and in a range of rugged, 12 mm open- and close-shackle ball-locking padlocks. Our description is based on the Ingersoll HS712 "Impregnable" padlock, illustrated in Figs. 4.49–4.52. We assume that the cylinder is viewed with the keyway horizontal as in Fig. 4.49.

The Impregnable has a keyway shaped like a flattened M flanked by what appear to be ordinary wafer-tumblers. The padlock body is made of hardened steel laminations, incorporating a ceramic insert in the plug cap for protection against drilling.

Figure 4.49: (Left) Ingersoll "Impregnable" 10-lever padlock. (Right) Underside of padlock with cover removed.

Figure 4.50: (Left) Plug and side-bar from Ingersoll lock. (Right) Double-sided key.

Figure 4.51: Ingersoll lever pack and pivot rod.

Figure 4.52: Ingersoll plug with key inserted showing operation of alternating lever mechanism.

The plug is made of a die-cast zinc alloy (zamac). The key is double-sided with 10 nonsymmetric cuts that are staggered from top to bottom. The wafers are actually pivoting levers as in a Fichet-Bauche 484 lock and contain a gate in one or more places on their periphery.

Ten nickel silver levers are arranged in an alternating sequence along the plug, as shown in Fig. 4.50. Each C-shaped lever is sprung independently and shares a common pivoting axis located below the keyway at 6 o'clock (see Fig. 4.51). The edge of each lever opposite its pivot point has around 6.5 mm reserved for the 3-mm gate, allowing four gate positions in total. The side-bar, located above the keyway, is hinged at 2 o'clock and is strongly sprung in an outward radial direction with its knuckle-shaped outer edge resting in a channel milled into the housing at 12 o'clock. The inner edge of the side-bar does not contact the levers until the plug begins to turn and the side-bar is forced radially inward by the sloped sides of the channel.

When the correct key is inserted, its bittings displace the levers to the left or right in an alternating sequence, causing their circumferential gates to align (Fig. 4.52). This action creates a longitudinal channel that spans the length of the plug and allows the heavily sprung side-bar to enter, thus freeing the plug to turn. The principle is similar to the wafer side-bar design in Fig. 4.14. The keys are stamped with an indirect code from which the true cuts can only be inferred with reference to the appropriate code book. Keys are registered, and proof of ownership is required for duplicates to be made.

To gain an appreciation for the possible bitting codes in a 10-cut system like this, we assume that both sides of the key may be cut independently. This is reasonable since with only four possible cut depths and a wide key blade, a number 4 (deepest) cut can be made on each side of the key without weakening the blade excessively. We first note that the theoretical number of differs is $4^{10} = 1,048,576$ and that there are no symmetry constraints since the key is not reversible. To obtain the number of practically usable differs, we further assume that on each side of the key:

1. The MACS is 3.

2. Up to three cuts in a row may be identical.

We are thus ruling out single-side cut sequences like (1 1 1 1 2) since this violates constraint number 2. The MACS constraint is enforced by default since there are only four cut depths. Note also that by enforcing rule number 2, we automatically ensure that at least two cuts are different on each side of the key.

With these constraints we obtain 840 different single-side codes. Since the key is double-sided, the total number of usable combinations is the square: $840^2 = 705,600$. The two single-side codes are interleaved to make a double-sided code for the 10-cut key. For example, with even-numbered cuts (3 2 2 4 3) and odd-numbered cuts (1 2 2 4 3), the 10-cut code would be (1 3 2 2 2 2 4 4 3 3).

For master-keying, some levers are cut with more than a single gate to enable operation of the lock by multiple keys. For instance, a wider single gate can be made that accommodates two adjacent depths of cut such as (1 2), (2 3), or (3 4). If cut depths 1 and 4 must both operate the same lever, two separate gates result. Since there are 10 levers, a substantial number of MK options can be obtained in this way.

The rotation of the plug is heavily damped by a rubber bushing, so that there is very little tactile feedback to someone trying to pick the lock while it is being tensioned. This makes the Ingersoll a particularly difficult lockpicking challenge. As we mentioned in the introductory chapter, the median time to pick an Ingersoll padlock by an expert lockpicker, when it was possible, was in the vicinity of 30 minutes.

Miwa U9

(JP) 9-lever + side-bar (4)

Miwa's U9 lock builds on the design of their earlier 10-wafer cylinder, which was covered in Chapter 3. The lock was introduced around the year 2000 in response to a need for greater residential security in Japan. Pictures of the U9 lock appear in Figs. 4.53–4.55. The design is covered in a number of Japanese patents, a drawing from one of which (JP 2000-291300) is shown in Fig. 4.56. Since the U9 lock in many respects mimics the Ingersoll lock in the previous section, we give only a brief description of its operation here. The discussion assumes that the lock is positioned as in Fig. 4.56, with the side-bar at 12 o'clock.

The Miwa U9 lock is housed in a toughened steel cylinder with a polished metal front cap (see Fig. 4.53). The keyway of the lock has a slightly different profile to the earlier Miwa wafer lock, accepting a flat, nonreversible key with cuts on both

Figure 4.53: Miwa U9 9-lever side-bar cylinder and key.

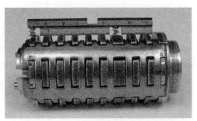

Figure 4.54: (Left) Miwa U9 barrel. (Right) Side-bar above laminated core.

Figure 4.55: (Top) Miwa U9 core with key partially inserted. (Bottom) Key fully inserted to align lever gates with channel.

Figure 4.56: Design cross-section for the Miwa U9 lock (from N. Ikuo's 1999 patent JP 2000-291300).

edges. The lock itself comprises a barrel and core, shown in Fig. 4.54. The barrel is mounted between two semicircular steel sleeves and secured by a Ç-clip at the rear of the cylinder. A steel front-piece contains a broaching for the keyway and provides a measure of drill protection.

The core itself is assembled from a number of preformed steel laminations held in place by two semicircular side plates. The plates are fastened to the front- and end-pieces of the core. The spaces between the laminations form a set of chambers for the tumblers, which closely resemble the levers in an Ingersoll lock. There is room for nine levers with integral flat springs. The levers are mounted on an axle at 6 o'clock that passes longitudinally through the laminations, and are limited in their pivoting motion by contact with either side of the core. Each lever contains one or more rectangular gates in its periphery (see Fig. 4.56) and may also contain false-depth notches to counter manipulation.

A steel side-bar is mounted in a channel at 12 o'clock formed by the cut-outs in the laminations. Two small springs, one at the end and one at a point two-thirds along the length of the side-bar, provide an outward radial bias. The side-bar normally rests with its apex in a longitudinal channel in the barrel, its retraction into the core being blocked by the levers. As in the Ingersoll lock, inserting the key causes the levers to pivot in proportion to the depth of cut on the side of the key that contacts them. A correctly bitted key is required to rotate all nine levers so that their gates are aligned at 12 o'clock, freeing the core to turn as the side-bar is displaced into the channel (see Fig. 4.55).

Whereas in an Ingersoll lock the levers are constrained to be mounted in an alternating sequence in their chambers, an innovative aspect of the U9 system is that the levers can be mounted either way round on the axle. This gives an extra degree of freedom when combinating the lock. Since there are nine levers, each of which may be mounted so that it acts either to the left (L) or the right (R) as the key is inserted, there are $2^9 = 512$ different mounting configurations. For instance, the lock in Fig. 4.53 has the following configuration from front to back: R, R, L, L, L, R, L, R, L. There are also four different depths of cut, corresponding to the four possible gate locations on a lever. Thus for each configuration there are $4^9 = 262,144$ possible lever combinations. The total number of theoretical system codes is the product of these two figures, or $2^9 \times 4^9 = 134,217,728$.

A high-security version of the U9 lock exists called the Miwa PR. This variant provides additional keying combinations and uses "closed" rather than horseshoe-shaped levers in addition to a dimple-bitted key. The PR design is covered in Japanese patents JP 2003-193715 and JP 2003-239577, the second of which discusses the inclusion of mobile elements in the key blade.

4.4   Driverless-Pin Side-bar

Medeco Cam-Lock

(US) 5-pin + side-bar (4)

The Medeco cam lock, pictured in Figs. 4.57–4.59, is a high-security driverless side-bar lock. It is available as a small-format (3/4″diameter) cylinder with either four or five inline pins. The lock is a popular choice for vending machines, cash boxes, coin-operated telephones, and other applications demanding a compact, high-security cam lock. The lock was designed by R. C. Spain and R. W. Oliver, as detailed in their 1971 patent (Fig. 4.60). Aspects of the pin design, such as the

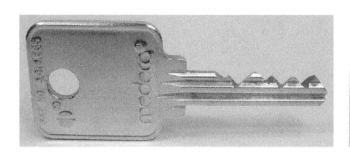

Figure 4.57: Medeco key and 5-pin cam lock cylinder.

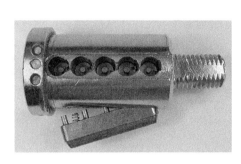
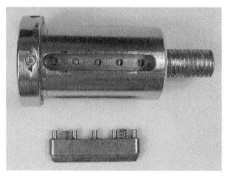

Figure 4.58: Two views of Medeco cam lock plug showing pin chambers and side-bar.

Figure 4.59: Medeco cam lock pins with either false-depth hole or antipick groove.

chisel-tip and locating tab were disclosed as early as 1891 in a patent by the Yale and Towne Manufacturing Company (US 457,753), where Spain previously worked.

Instead of the usual set of upper or driver pins present in a conventional pin-tumbler lock, the lower pins are sprung directly from within the plug, which has a retaining clip covering the pin chambers. Pins are limited in their angle of rotation by a stop tab that inhabits an enlarged-radius sector of the chambers facing the rear of the plug. A longitudinal milled groove at 3 o'clock in the plug houses the side-bar, which runs the length of the five pin chambers and is spring-biased radially outward in two places. The side-bar also has five posts that face inward toward the pin chambers, impinging on them through a set of holes in the side wall of the plug. The outward

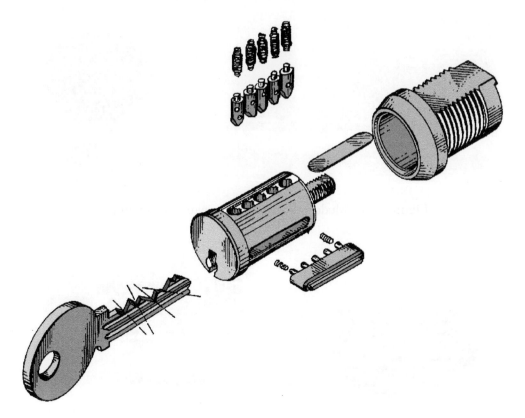

Figure 4.60: Medeco cam lock design (US patent 3,722,240 by R. C. Spain and R. N. Oliver).

face of the side-bar has a triangular apex that rests in a longitudinal channel in the barrel at 3 o'clock.

Each pin (Fig. 4.59) has one or more holes machined radially into its side. In addition, pins may be spooled near their top end or possess a false-depth (shallow) hole at another point in the side facing the side-bar. The pins are chisel-tipped to match the V-shaped cuts in the key. In the locked position, the posts of the side-bar are not in registration with the holes in the pins. Torque applied to the plug merely serves to bias the side-bar radially inward as its apex contacts the angled surface of the channel in the barrel.

To operate the lock, the correct depth hole in each of the five pins must be simultaneously raised and rotated by the correct amount to align it with the corresponding post in the side-bar. This is achieved by the insertion of a key whose bittings possess the correct depths and angles (0 or ±20 degrees). Once all the pins are correctly lifted and oriented, the side-bar can be retracted into the plug as the key is turned. This hole-in-pin principle is similar to that of the BiLock (considered next). The mechanism admits a large number of differs due to the dual functionality of the pins: the number of cut depths is effectively multiplied by three, being the number of possible cut angles. Master-keying is accomplished by equipping the pins with more than one correct-depth hole.

Despite its quite humble appearance and easy-access keyway, the Medeco cam lock is very difficult to manipulate due on the one hand to its very tight manufacturing tolerances and on the other to the presence of false-depth holes and spooled pins, which cause the side-bar posts to bind the pins in the wrong positions. Pin manipulation tools have been developed for decoding Medeco locks, but this remains a time-consuming task, requiring specialized equipment. The cylinder is also well endowed with drill-resistant inserts that qualify the lock for a UL 437 rating. Further details concerning Medeco locks may be found in the section on dual-action side-bar locks, as well as in the books by Roper [106] and Rathjen [102]. A variant of the Medeco cam lock by J. R. Smith of Shield Security Systems is described in a 1991 patent (Fig. 4.61). This modification called for two side-bars that engaged holes in opposite sides of the pins. Two differently bitted keys were applied in succession, each effecting half a turn of the plug.

BiLock

(AU) 12 pins in 2 rows + 2 side-bar (3–4)

The BiLock high-security lock, illustrated in Figs. 4.62–4.65, was invented in Australia by B. Preddey. As with other side-bar locks, the design was motivated by the need for a higher degree of resistance to picking, impressioning, and unauthorized key duplication than that provided by inline pin-tumbler locks. In 1981 Australian patent and design registration applications were lodged for the BiLock (see Fig. 4.66), which is now produced by the Australian Lock Company. We describe the operating principle of the lock and then go on to discuss some further developments in BiLock technology.

The twin-bladed BiLock key (Fig. 4.65) is formed by folding a steel blank, cut with a bitting pattern along two perpendicular edges, to form a U. The plastic key head

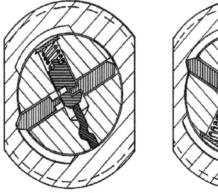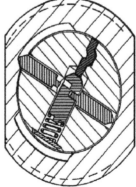

Figure 4.61: Twin-keyed, two side-bar cylinder proposed by Shield Security Systems (US patent 5,375,444 by J. R. Smith).

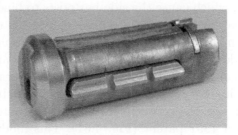

Figure 4.62: (Left) BiLock dual side-bar cylinder. (Right) BiLock plug.

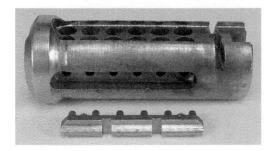

Figure 4.63: (Left) BiLock plug with pins removed, side-bar in foreground. (Right) Underside of plug.

Figure 4.64: BiLock pins showing the four sizes and antipicking features.

is held in place between the folded blades. Each blade has six bitting positions with four possible depths of cut and an unrestricted MACS. Thus there are theoretically $4^{12} = 16,777,216$ or over 16 million theoretical differs.

The lock cylinder accommodates a plug with two parrallel rows of six vertically oriented pin chambers located to the left and right of 12 o'clock (see Figs. 4.62 and 4.63). The plug is slotted at 3 and 9 o'clock to accept two side-bars. Each side-bar is sprung radially outward and is equipped with a 6-pronged fence. The cylinder does not use driver pins; instead, the 12 lower pins are biased from above by small springs retained by a copper slide. The pins are machined with a vertical channel on one side, with the side-bar prongs constraining them to move vertically in their chambers

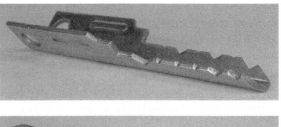

Figure 4.65: A BiLock blank is first cut and then folded to form a key.

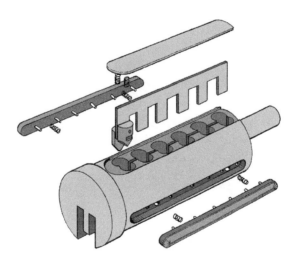

Figure 4.66: Early form BiLock design from US patent 4,478,061 (1982) by B. F. Preddey.

without twisting. The face of each pin may have one or more holes machined into it (see Fig. 4.64).

When a correctly bitted key is inserted, the six pins along each blade of the key are raised against the action of their driver springs such that the holes in the pins are in alignment with holes in the side wall of the plug. As the key is turned, the beveled edges of the twin side-bars ride out of the longitudinal channels in the cylinder housing, forcing the side-bars radially inward. The side-bar prongs protrude through side wall, impinging on the pins. Full retraction of the side-bars is allowed by the correct alignment of the holes in all 12 pins. The operation is similar in principle to the Medeco cam lock, minus the twist dimension.

The original 1982 patent suggested various security broachings for the keyway, but these did not appear in the production model and so the pins are easily accessible. Despite this fact, the lock has a high degree of manipulation resistance as long as the pins contain false-depth holes to partially engage the side-bar prongs. Without these pick-resistant pins, the security level of the cylinder is lessened considerably. The plug can also be fitted with hardened inserts as protection against drilling. In terms of master-keying, the system is extremely flexible. There are 11 different types of master pins that can be combined with the four basic depths of cut, making 15 different pin types in total. With 12 pin positions to choose from for each pin type, the system can be tailored to large-scale master-keyed systems.

Interestingly, around the same time of the BiLock patent, a patent was lodged by Strassmeir for what one might call a TriLock, pictured in Fig. 4.67. However it turns out that a two-bladed lock provides a more than adequate level of security and the TriLock design did not come to fruition. The middle blade would also make key manufacture and cutting rather difficult.

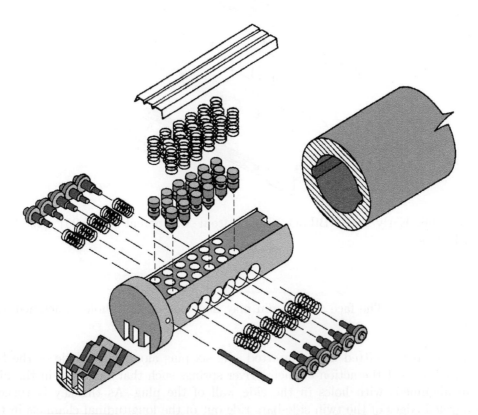

Figure 4.67: Driverless side pin cylinder with three-bladed key proposed as a successor to the BiLock (1983 US patent 4,603,565 by M. E. F. Strassmeir).

BiLock QCC

A more recent BiLock variant called QCC is shown in Figs. 4.68–4.71. The QCC has a removable core to facilitate rekeying, hence the initials, which stand for Quick Change Cylinder. The QCC system was first used in conjunction with the "First Generation" BiLock in the previous section and has now been adapted for use with

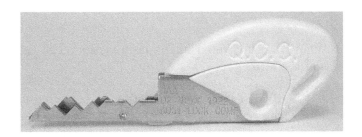

Figure 4.68: BiLock QCC control key.

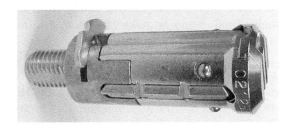

Figure 4.69: BiLock QCC plug with rear stub attached.

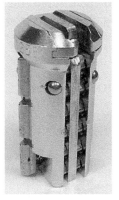

Figure 4.70: (Left) BiLock QCC cam lock barrel; (right) underside of QCC core.

Figure 4.71: Removal of BiLock QCC core.

the "New Generation" BiLock covered in the next section. The QCC cam lock in the photographs is of the latter type.

The plug (Fig. 4.69) is divided into a front section or core containing the active parts and a rear stub. The rear stub is rotatably anchored in the cylinder by a concealed spring clip. Regular keys for the First Generation QCC have a dimple on each side of the key blade between the second and third pin positions in order to distinguish them from conventional BiLock keys.[4] The front section of the core (Fig. 4.70, right) has two opposing pairs of ball bearings that protrude past the edge of the core: fixed at 2 and 10 o'clock and movable at 5 and 7 o'clock. The ball bearings ride in a milled-out circular groove at the front of the cylinder. There are also two short longitudinal channels, near 12 o'clock, running from the circular groove to the front of the cylinder (Fig. 4.70, left).

When the plug is turned to around 45 degrees (CW), one of the fixed ball bearings lines up with the short channel to the left of 12 o'clock, while the other lines up with the side-bar channel at 3 o'clock. Similarly, when the plug is turned to −45 degrees, the fixed ball bearings line up with the short channel to the right of 12 o'clock and the side-bar channel at 9 o'clock. If a regular operating key is inserted, the movable ball bearings remain in the circular groove, preventing the core from being extracted. On the other hand, when a control key, having profile dimples between pin positions 1 and 2, is inserted and turned to ±45 degrees, the movable ball bearings can be retracted into the dimples on each side of the key blade, enabling the front part of the core to be disengaged from the rear stub and removed from the cylinder (see Fig. 4.71). Repinning the plug and reinserting it is then a simple matter.

New Generation BiLock

(AU) 12 pins in 2 rows + 2 side-bar (4)

With the expiry of the original BiLock Patent, the Australian Lock Company introduced a new system called New Generation BiLock, or NG BiLock, as we will refer

---

[4]In the New Generation QCC BiLock, only the control keys have dimples on the side.

to it. The NG BiLock, shown in Figs. 4.72 and 4.73, is still clearly recognizable as a BiLock cylinder but has some additional features. A patent for the new design was filed by B. Preddey in 1998 in Australia and in the United States the following year (US 6,681,609). Keys and cores for the NG BiLock are supplied only through authorized agents. It is fully compatible with First Generation and QCC BiLock, and therefore the cores can be swapped over to upgrade the system.

The NG BiLock retains the 12-pin dual in-line twin side-bar construction of the original BiLock. Thus it still requires a U-shaped key with 12 cuts (6 per side). The keyway broaching, however, is different from the straight-sided U-profile: the NG keyway is curved inward at the bottom edges of the U. The key blade is shaped to match the new keyway and may also contain profile bullets. The keyway is actually cut into a removable insert that slots vertically into the front of the core and can thus be easily varied.

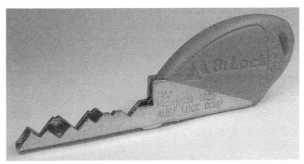

Figure 4.72: New Generation BiLock cut-away cylinder and key with movable element.

Figure 4.73: NG BiLock key aligns side-bar pins while movable element raises blocking rod.

The most important aspect from a design perspective is the inclusion of a movable element in the key blade (referred to in the marketing literature as the "13th locking dimension"). Just as in the DOM iX floating ball and Mul-T-Lock Interactive, the NG BiLock cannot be operated by a key without the active element, even if it has the right bittings.

In the NG BiLock (refer to Fig. 4.74), the movable element is in the form of a three-corner jack inserted through a hole in the bottom of the key blade, just before the first cut position, and secured by the plastic key head. The leading edge of this element is a scoop, which protrudes through the bottom of the key blade. The reciprocal of the movable element is a linkage mechanism of two pins placed in the front of the core. The function of the linkage is to deadlock the front of the side-bar mechanism. Since it is concealed behind the central ward of the keyway, the linkage cannot be operated by a key with a fixed blade.

The linkage comprises a pair of specially shaped rods. The first of these is a rod of similar diameter to the side-bar pins, with a forward-facing notch, mounted in a bore between the keyway insert and the front of the core. The line of action of this first rod is vertical, and it is spring-biased at 12 o'clock. A second rod is mounted in a horizontal bore running from 10 o'clock to 2 o'clock, traversing the core and held in place by the notch in the first rod. The second rod is equipped with a small disc on its right-hand end. The arrangement allows the linkage rods a modest amount of vertical travel. Since the first rod is spring-biased from the top, the default position for the linkage is down, with the disc on the horizontal rod obstructing the front of the right-hand side-bar channel. Unless the linkage is raised, the side-bar is effectively blocked.

When a correctly bitted NG BiLock key is presented to the lock, the movable element is the last part of the key to enter the keyway. The scoop end is initially in its rest

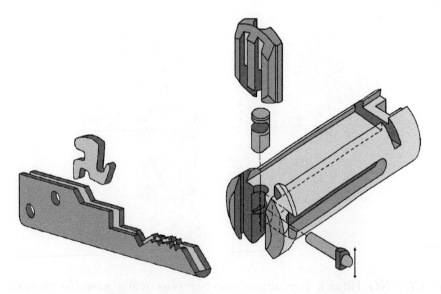

Figure 4.74: B. F. Preddey's design of the New Generation BiLock (US 6,681,609).

position in a trench in the key blade. As the element passes under the center ward, the hook on the element contacts the front of the keyway. The scoop is then pivoted up 0.050″ against the linkage rod, raising it along with the blocking disc and freeing the side-bar to operate in the normal manner.

The NG BiLock retains the advantages of the original design including the strong, compact key, and the durability of the mechanism. The movable element makes unauthorized key duplication impractical and also enhances the lock's already high level of manipulation resistance.

ASSA Desmo

(SE) 8-pin + 2 side-bar (4)

The ASSA Desmo, pictured in Figs. 4.75–4.78, is a miniature cam lock designed for high-traffic industrial environments. The 8-pin version of the cylinder contains two rows of four direct-drive pins and two side-bars. There are no top pins and, unlike the Medeco cam lock and BiLock, no driver springs. In other words, it really is a driverless side-bar lock. For this reason there is next to no resistance as the key is inserted, which makes the lock very durable and dirt-resistant.

The ASSA Desmo design (see Fig. 4.79) is explained in two 1992 patents by Häggström (US 5,517,840 and 5,582,050). It is tempting to think that this modern

Figure 4.75: ASSA Desmo direct-drive twin side-bar cylinder and key.

Figure 4.76: Different pin sizes used in the ASSA Desmo (shown inverted).

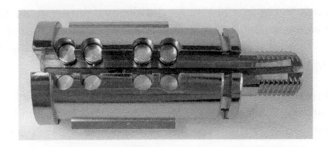

Figure 4.77: Top view of ASSA Desmo core.

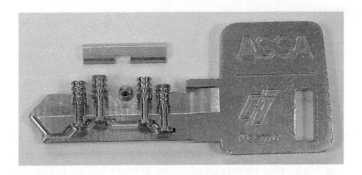

Figure 4.78: Pins aligned by ASSA Desmo key. Central pin retains key.

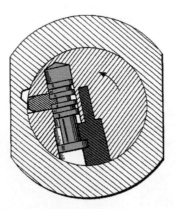

Figure 4.79: A. Häggström's ASSA Desmo design from US patent 5,517,840.

lock is an entirely original concept; however, as is often the case in the world of locks, this is not so. The design can in fact be traced to a 1933 patent by S. A. Liss of the Briggs and Stratton Corporation (US 2,070,233). This patent called for a 5-pin dual side-bar lock with a conventional flat key. The principal differences between this and the ASSA Desmo are in the presence of two rows of driverless pins and a two-track key. The similarities can be appreciated from Fig. 4.80. At the time, Liss's lock would have been uneconomical due to the high-precision machining required in its construction.

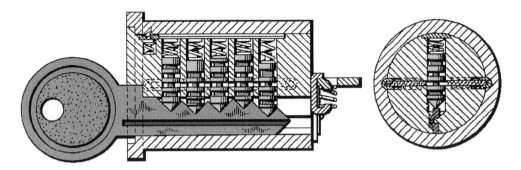

Figure 4.80: Briggs and Stratton's 1933 driverless side-bar lock (US patent 2,070,233 by S. A. Liss).

The ASSA Desmo has a nonreversible flat key that is wider at the bottom of the blade than at the top. There is a track milled into the bottom half of the blade on each side of the key (similar to a Bell or Dudley lock). The pins are chambered in vertical bores and are of spooled construction (refer to Fig. 4.76). They have a full-width base, a reduced-width midsection, and multiple spooling on the top end. A deep circumferential notch in the top half of the pin accommodates the side-bar fence. There is room for six different depths of cut, so with eight pins the theoretical number of differs is $6^8 = 1,679,616$. Taking into account MACS and other constraints, the practical figure is around 1.5 million.

As the key is inserted, the ramps at the end of the blade on both sides pick up the base of the pins and guide them into the edge milling. From this point on, the lifting of the pins is determined by the geometry of the two side tracks on the key. Insertion of the correct key causes all eight pins to be displaced such that their correct-depth notches form a channel at the height required for the side-bar fence. As the plug is turned (see Fig. 4.77), the side-bars are pushed out of a longitudinal groove on each side of the housing and their fences slot into the aforesaid channel. Each side-bar has an inward-facing pin that meshes with a single bitting on the top of the key-blade to retain the key during operation (see Fig. 4.78). Master-keying can be achieved through the use of pins with more than one correct-depth notch.

As expected in a high-security side-bar lock, the design is highly resistant to manipulation since false-depth notches are included on each of the eight pins. With twice as many false-depth notches as correct ones per pin, the odds are 2-to-1 against correctly setting each pin. This multiplies out to odds of 256-to-1 across all eight pins.

Tubar

(US) 8-pin + 2 side-bar (4)

In spite of the many design modifications proposed to axial pin-tumbler locks of the ACE/GEM type, they remain a much easier target than side-bar locks for lockpickers armed with the proper tools. But how can a side-bar mechanism be added to a tubular lock? One approach is to introduce a Bramah-style locking ring as in the JPM and Laperche locks, covered in Chapters 2 and 3. Another approach, detailed in a 1981 patent by the Chicago Lock Company (US 4,446,709), is to flatten the circle of pins into a rectangle, enabling a side-bar to be fitted on each of the longer edges. This greatly adds to the security and manipulation resistance of the lock. The new product, Tubar, is a twin side-bar 8-pin lock providing a high degree of resistance to unauthorized access for a lock of its size and construction. It is typically used in a cam lock or push-button format for vending machines.

The lock housing is made of sintered steel (see Fig. 4.81). A cylindrical brass plug and hardened front cap are mounted inside the housing. The cap contains the keyway broaching and, instead of being fixed in the housing, turns with the plug. A 2 × 4 matrix of axial borings in the front of the plug houses the eight steel pins, as shown in Fig. 4.82. On either side of the plug there are transverse and longitudinal channels that intersect in a cross. The side-bar comprises two parts that slot together. The first of these is a crescent with a flat edge that sits in the transverse channel. A straight bar is then inserted into a notch in the crescent, and this slots into the longitudinal channel. The two crescents impinge on the pin chambers through the sides of the plug, while the straight parts of the side-bars rest in grooves at 3 and 9 o'clock that run the length of the housing. Unless the side-bars are fully retracted, the plug assembly cannot be turned.

The pins are all of the same overall length including a spindle section at the bottom end that guides a driver spring (similar to the ACE system). Each pin possesses a reduced-girth section in one or more places along its length, forming a set of gates. All but one of the gates in each pin is shallow. The true gate is deeper than the others, and we simply refer to this as the gate. The offset of the gate from the front end of the pin varies from 0.175″(minimum cut) to 0.325″(maximum cut) in steps of 0.025″, yielding seven pin sizes. This system therefore supports in excess of five million (7^8) different key combinations.

The key comprises a flat alloy blade set in a plastic handle. The blade is rectangular with four end-bittings milled into each side. The blanks are restricted, and key

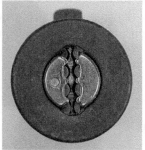

Figure 4.81: Tubar end-bitted key and twin side-bar cam lock.

Figure 4.82: (Left) Tubar core. (Middle) Core with cap removed. (Right) Tubar core with one side-bar.

copying requires proof of ownership. The key must be inserted the right way and pushed until its tip makes contact with the surface of the plug. If the key is correctly bitted, all eight pins will be depressed by the amounts required to align their gates with the transverse channels on either side of the plug. This provides a small gap to accommodate the side-bars as they move radially inward while the plug is turned.

During this time the cut-outs in the sides of the blade engage the rim of the housing, retaining the key in the plug. The action is smooth since the side-bars keep the pins at the correct depths without the need to maintain pressure on the key. If any one of the pins is not properly aligned, the crescent will not be fully recessed into the plug and the side-bar will block in its channel in the housing.

The presence of shallow gates on one or more (usually six out of eight) pins is to prevent the lock from being picked. Manipulation is hampered by the rounded edges on the crescents and gates together with the use of strong driver springs. Furthermore, the cylinder is highly resistant to attack by drills and hole-saws. The system is very effective and is listed by Underwriters Labs.

4.5 Wafer Side-bar

Lori

(US) 8-wafer + 2 side-bar (3)

The Lori side-bar lock is produced as a small-format cam lock for lockers, parking meters, and other applications involving the depositing and storage of coins. The particular model pictured in Figs. 4.83–4.85 comprises a cast zinc body and plug

Figure 4.83: Lori 8-wafer twin side-bar cam lock and key.

Figure 4.84: Lori barrel and plug.

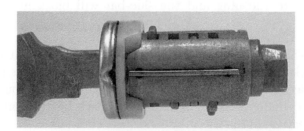
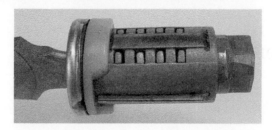

Figure 4.85: Operation of Lori lock.

with a two-track brass key. The idea is a simplification of a 1959 patent by F. J. Testa (US 3,035,433) that adds a side-bar to a dual-action Dudley wafer lock. The Lori lock is substantially similar to the Bell lock previously covered in Section 3.6, and so we give only a brief description of it here.

The plug houses eight bar-wafers arranged in two rows of four (see Fig. 4.84). A retaining clip runs along the middle of one side of the plug in between the two rows. The stubs on the bar-wafers protrude into the keyway through slots in the plug molding. The bar-wafers are not spring-biased but are instead guided into their positions by the tracks on either side of the key. Bar-wafers may be either underlifted or overlifted, in either case remaining proud of the shear line in one of the eight wafer chambers in the barrel. The bar-wafers are additionally provided with a V-shaped notch on their outer face. The plug has two slots, at 3 o'clock and 9 o'clock, that house two thin side-bars. In the rest position, the side-bars are outwardly spring-biased into longitudinal channels in the barrel.

Operation of the lock is assured by the insertion of a correctly milled two-track key. The tracks simultaneously raise or lower the eight bar-wafers to bring them to the shear line. At the same time the V-notches in the wafers come into registration with the twin side-bars, permitting the plug to rotate. The presence of the side-bars ensures that the wafers are maintained within the plug diameter while the plug is turned. The lock does not incorporate other security features such as false-depth notches or drill-resistant inserts. The key, however, requires specialized equipment for duplication, and the distribution of blanks is restricted.

Fichet-Bauche 666

(FR) 7-wafer + side-bar (3–4)

The Fichet-Bauche 666 has a similar principle of operation to the Fichet-Bauche 484 (covered earlier) but with one less side-bar and using wafers instead of rockers. Pictures of the lock are given in Figs. 4.86–4.88. The design was submitted in a 1964 French patent (FR 1,425,311). The key is a double-sided wafer type with a profile resembling a flattened M. Viewing the key in the orientation suggested by Fig. 4.86, there are four cuts on the lower edge and three cuts on the top edge. The blank

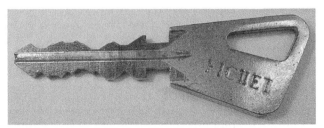

Figure 4.86: Fichet 666 7-wafer side-bar cylinder and key.

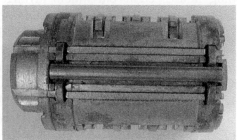

Figure 4.87: Two views of the Fichet 666 core.

Figure 4.88: Fichet 666 core with side-bar removed. Insertion of key aligns wafer gates with upper channel.

keys have seven cuts, all of the same depth, directly opposite the seven cuts just described. This gives the impression that the cylinder contains a large number of wafers; however, this is not the case. The plug (Fig. 4.87) holds only seven wafers that are alternately sprung: four from below and three from above. The wafers are much bigger than conventional wafer-tumblers, with a large cut-out and a narrow gate on the left-hand edge (near 9 o'clock). The function of the extra cuts in the key is to reduce the width of the blade to allow it to pass through the cut-outs in the wafers.

There are two longitudinal channels along the left-hand edge of the plug: a lower one at 8 o'clock and an upper one at 10 o'clock (see Fig. 4.88). Each of these contains a metallic strip spanning almost the length of the plug. One strip is fixed and the other movable. The strips are spring-biased radially outward from the plug, although only the strip in the upper channel can be retracted since the wafers are not gated for the lower channel. Two saddle pieces are mounted on the ends of the strips, with the side-bar rod balanced in the central notch of each saddle. The side-bar normally engages a shallow longitudinal groove at 9 o'clock in the cylinder housing. When the gates of all seven wafers are aligned with the upper channel by insertion of the correct key, the movable strip retracts into the channel as the side-bar is forced from its groove in the housing. Some of the wafers may have false-depth gates to inhibit picking.

EVVA 3KS

(AT) 12-wafer + 2 side-bar (3–4)
(IT) **Mottura KS** (equivalent)

The EVVA 3KS (3 Curve System[5]) is a close but somewhat more sophisticated relative of the Bell lock. It was patented in 1988 by K. Prunbauer of EVVA-Werk in Vienna, Austria. The 3KS lock and its internal components are pictured in Figs. 4.89–4.92. The distinctive flat key has six tracks (three per side) and is symmetrically bitted so as to be reversible. Looking at one end of the key, one can discern three pick-up slopes on each side: low, center, and high. It is convenient to call the low and high tracks "outer tracks" and the center track the "inner track." The shallow outer tracks are the same shape but have a constant vertical offset from each other. The inner track is deeper than the two outer tracks.

The plug (Fig. 4.90) houses 12 bar-wafers, 6 per side, in two parallel rows of chambers near 12 o'clock. The bar-wafers are not sprung and normally rest in their lowest positions (assuming vertical mounting of the cylinder). There are two distinct types of bar-wafer, guided by different tracks in the key (see Fig. 4.91). Each row of six

Figure 4.89: EVVA 3KS 6-track key and profile cylinder.

---

[5]The German word for curve is Kurve.

Figure 4.90: (Left) Underside of EVVA 3KS plug. (Right) Cylinder housing with internal ribbing.

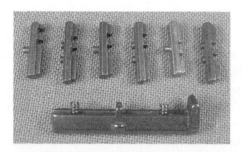

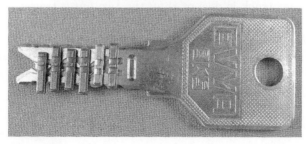

Figure 4.91: (Top) Bar-wafers and one of the side-bars. (Bottom) Alignment of gates by tracks on key.

bar-wafers has three wafers with two short stumps and three wafers with a single longer stump. The two types of wafers are installed in an alternating sequence. Short-stump wafers with two stumps are guided by the outer tracks in the key. Long single-stump wafers are guided by the inner track.

The inner track on each side of the key picks up long-stump wafers on both sides of the plug, displacing wafers on the left-hand side in the opposite direction to wafers on the right as the key is inserted. The high track on the left-hand side is equivalent to the low track on the right-hand side of the key. Thus, regardless of the orientation of the cylinder, both long- and short-stump wafers are always picked up by their respective ramps and guided into the correct tracks.

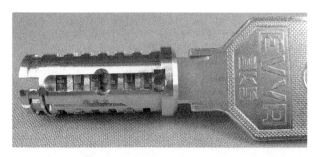

Figure 4.92: EVVA 3KS plug with side-bar removed showing displacement of bar-wafers by key.

The plug is flanked at 3 and 9 o'clock by twin side-bars that are spring-biased radially outward. Each side-bar has an apex on the outside edge to engage a longitudinal channel in the cylinder bore. The inside edge of each side-bar has a double ridge that matches a corresponding pair of gates in the outer edge of each wafer. All 12 wafers must be correctly in alignment to allow the twin side-bars to register with the gates (see Fig. 4.92).

For any particular choice of wafer sizes, there are six bitting positions or points per track at which the height of a wafer may be set. Remembering that the key is reversible, we see that three of these are needed to set three wafers on one side of the plug, with the other three being used for wafers on the opposite side of the plug when the key is inserted the other way. It follows that there are 18 bitting points per side of the key. This total is made up of six points for the inner track, six points for the low track, and six points at a constant offset above the low track for the high track. Figure 4.93 illustrates this idea for an inner track with four bitting points and outer tracks with three bitting points. The bitting points are joined up smoothly to form a track for computer-controlled milling of the key blank.

With three wafer sizes for the outer track and four sizes for the inner track, there are approximately $3^6 \times 4^6 = 2,985,984$ possible differs. The symmetry of the key, however, imposes additional constraints on the possible key codes. Furthermore, the top and bottom edges of the key have $2 \times 6 = 12$ angled profile bittings that must register with a profile bar (as in EVVA DPS/DPX locks). The profiling options increase the number of possible differs enormously. The original design featured an additional seven conventional pin-tumbler bittings along one edge of the key

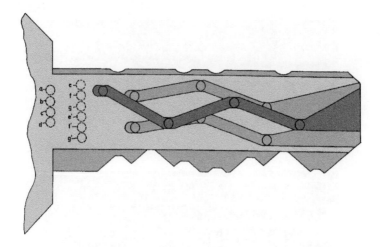

Figure 4.93: EVVA 3KS design from 1989 suggested additional pin-tumbler bittings (US patent 4,977,767 by K. Prunbauer). Letters refer to track-bitting heights.

blade, but these were suppressed in the production version in favor of having a reversible key.

The inside edge of the cylinder (Fig. 4.90) has a series of milled crenellations or ribs that allow the ends of the wafers to protrude past the normal shear line (as in Emhart and EVVA MCS locks). This applies even when the lock is in the open position. In other words, the wafers only serve to control the side-bar action.

A stub at the front of the side-bar on each side of the keyway impinges on a depression in the key blade to capture the key during rotation of the plug. The cylinder and plug contain hardened inserts to resist drilling. Note that there are no components where the pin chambers in a conventional lock are normally located. The lock also features a specially constructed cam that resists forced removal of the front part of the profile cylinder.

Cut keys can either be supplied by the factory or as a dealer permutation with only the edge profiling and outer tracks cut. An authorized locksmith then cuts the inner track on each side of the key according to a locally chosen combination of single-stump wafers [40].

More recent designs from EVVA include a 10-disc side-bar lock with a round key and a 9-wafer lock with a double-sided key like a flattened version of the Fichet 484. The disc side-bar lock (Fig. 4.94, top), described in a 2003 patent by K. Prunbauer (US 6,758,074), maps the 3KS key design onto the surface of a round key. The lock is a circular implementation of a Bell lock. The six tracks on the key pick up stubs on the discs, rotating their gates to align with a side-bar. The wafer lock (Fig. 4.94, bottom) is covered in a 2002 patent by K. Prunbauer and A. Reinhard (US 6,622,538). In this design, the lateral motion of the wafers is used to control the axial motion of a ribbed side-bar that threads a cut-out in the wafers. EVVA has also released a mechatronic version of the 3KS lock called the ELMO. Turning

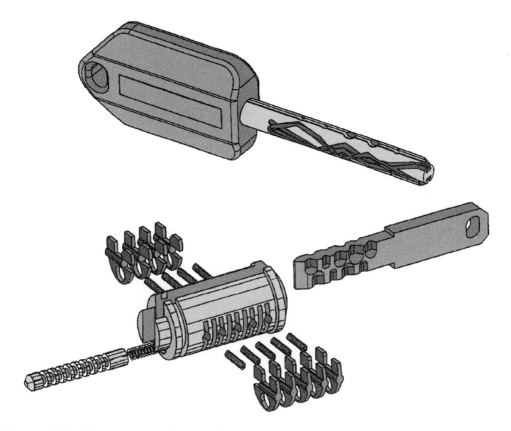

Figure 4.94: Two recent side-bar lock concepts from EVVA: (top) 10-disc "3KS" lock with round key (US 6,758,074); (bottom) 9-wafer axial lock with double-sided key (US 6,622,538).

the key completes a circuit that provides current to a miniature servo-motor, which operates a mechanical coupling to the locking cam.

4.6 Dual-action Side-bar

Medeco

(US) 6-pin + side-bar (4–5)

Medeco locks come in two basic application-dependent types: a rim or mortice cylinder, which uses six or seven pin-tumblers and a side-bar; and a smaller cam lock version with four or five driverless pins and a side-bar. The Medeco cam lock was described previously, and we assume some familiarity with its operating principles here. Both versions have a UL rating and are distinguished from most other pin-tumbler locks in their use of pins with two degrees of freedom: twist and lift.

Externally, Medeco cylinders look like ordinary pin-tumbler cylinders, except for the V-shaped bottoms on the lower pins. Both types are high-security locks, utilizing hardened inserts (crescents and rollers) to resist drilling. Medeco locks require special key duplication equipment, and the factory exercises control over distribution of the registered key blanks. The remainder of our discussion focuses on the Medeco dual-action cylinder lock illustrated in Figs. 4.95–4.98.

The Medeco lock uses lower pins with a single spline or slot along the edge and a distinctive chisel point. The lower pins have a locating tab that limits the range of rotation within the pin chamber; this ensures that the spline on the pin is always pointing to the right-hand half of the lock (3 o'clock) rather than to the left half. The side-bar is similar to the one used in a Medeco cam lock except that its posts are rectangular rather than round. The side-bar posts communicate with the pin chambers in the plug through slots milled into a longitudinal channel at 3 o'clock in the plug. The pins must simultaneously be raised to the shear line and rotated to one of three angles (center = 0, left = −20 and right = 20 degrees) in order for the splines to be brought in to registration with the rectangular side-bar posts. Angular alignment can occur for any degree of pin lifting. Although there are only six pin lengths, the extra degree of freedom provided by the angled cuts more than compensates for this by multiplying the number of basic pin types by three.

Figure 4.95: Medeco Biaxial 6-pin side-bar cylinder and key with angled cuts.

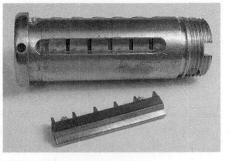

Figure 4.96: Two views of Medeco Biaxial plug and side-bar.

Figure 4.97: (Top) Medeco plug with key partially inserted. (Bottom) Key aligns pin slots with side-bar channel.

Figure 4.98: Medeco Biaxial lower pins with fore and aft offsets and false-depth slots.

As in other dual-action side-bar locks (like the ASSA Twin in the next section), it is not enough either to correctly raise the pins or to correctly align the side-bar slots; both conditions must be met simultaneously for all pins in the cylinder. It is this feature that bestows a very high level of security to the Medeco lock.

In a system that has pins with a single spline, master-keying can only be accomplished by the addition of master pins in the space above the side-bar fence. Thus a limitation arises since a master key with given bitting angles requires all the locks it operates to have pins with the same set of angles. In other words, the system cannot use the angular degree of freedom and reverts to being a conventional MK system. This was a recognized limitation of the 1968 Medeco patent (US 3,499,303). One way around this problem is to create additional splines on some of the pins to accept more than one cut angle on the key for each depth of cut. Another solution is the hole-in-pin approach adopted in the 1971 Medeco cam lock (US patent 3,722,240): this system may be master-keyed using both the twist and lift degrees of freedom. A further solution is provided by the Medeco Biaxial, described in the next section: this can have doubly-cut key bittings, with two different offsets for the same depth and angle.

Medeco Biaxial

The Medeco Biaxial, released in 1986, differs from the earlier model Medeco in a number of respects, the most significant of which is the skewed tips on the lower pins. The pins are also of a different length and have the locating tab sited differently [102]. The locking principle is unchanged, however: all pins must be simultaneously turned to the correct angles and raised to the shear line so that the side-bar posts can register with the splines in the pins, freeing the plug to rotate.

Regarding Biaxial pins, the tip of the pin is machined such that its lowest point is offset with respect to the pin's central axis. Taking the pin slot as a reference plane, the offset can be 0.031″ forward or aft. The alignment of the slot with respect to the flat edges of the tip can be at an angle of 0, −20, or +20 degrees. The diagrams in Figs. 4.99 and 4.100 show the specification for the key cuts. The original patent called for five possible angles of orientation including ±10 degrees.

Since there are two pin offsets (fore and aft) and three pin angles, it follows that there are six possible pin classes, each of which comes in six different lengths. The overall number of theoretical pinning combinations, ignoring MACS, is $6^6 \times 6^6 = 2{,}176{,}782{,}336$, or over two billion. Another interesting twist in the Biaxial design is that the MACS is variable since it depends on the pin offset. A MACS of 4 applies if a fore pin is placed next to an aft pin; the MACS is 3 for adjacent pins with the same offset and only 2 for an aft next to a fore pin. As mentioned before, Medeco Biaxial

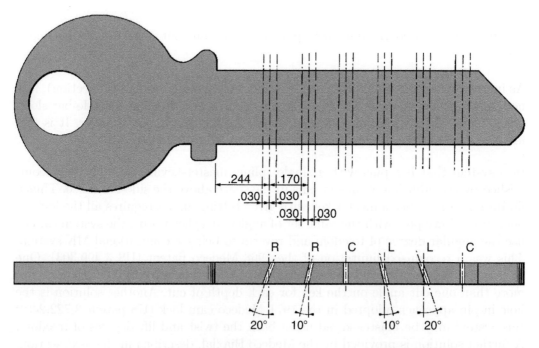

Figure 4.99: Spacing and bitting angle specification for Medeco Biaxial key from R. N. Oliver's 1985 patent (US 4,635,455).

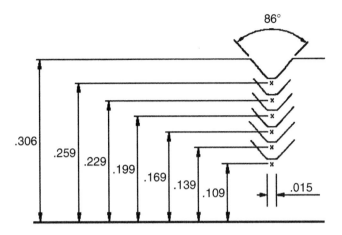

Figure 4.100: Cut specification for Medeco Biaxial key (US patent 4,635,455).

keys can also be "doubly-cut," meaning that at the same position, two differently offset cuts can be made on the same key: one for fore and one for aft pins. The cuts would normally be for the same depth and angle. This feature gives added flexibility in master-keying.

There have been reported pickings of the Medeco lock, in particular in response to a competition in the early 1970s offering a reward of up to US $10,000 for the picking of three Medeco cylinders. Since the pins can have a false-depth slot as well as mushroom drivers, it is unlikely that the lock can be reliably picked.

A decoding device for Medeco and its Biaxial version was proposed in a 1974 patent by G. V. Iaccino and R. A. Idoni (US 3,987,654). The decoder consisted of an extendible wire probe that could be maneuvered into the larger-radius portion of the lower pin chamber. The probe could be directed to sense the height and angular position of the locating tab with respect to the V-shaped end of the pin. This could then be converted into a bitting depth and angle to make a key. A further exploitable fact is that the spline runs all the way along the side of the lower pins and can therefore be probed from the keyway. In principle this allows a limited reading of the angular code of the lock. However, the presence of a false-depth slot on left- and right-angled pins or, alternatively, using pins with a spline of limited length, would effectively counter this type of decoding.

Patent protection for the Medeco Biaxial expired in 2005. The product has since been upgraded to the Medeco 3 [14]. The Biaxial cylinder, pins, and drivers remain compatible with the new model, which differs only in respect of the plug and side-bar. In addition, the key has a side-bitting on the lower right-hand side. A slider mechanism, spring-biased axially toward the front of the lock, is mounted in a slot in the underside of the plug. In the locked position, the slider engages the inner edge of the side-bar, blocking the action of the fence. The side-bitting on the key contacts the end of the slider, pushing it back to free the side-bar, while at the same time raising and rotating the pins in the usual manner.

A related lock, disclosed by J. M. Genakis in a 1981 patent (US 4,450,699), modified Medeco's rotating and lifting pin design in order to reduce wear on the pins. The pins proposed in the Genakis patent (Fig. 4.101) were of threaded construction with a V-shaped indentation in their tips, rather than the usual chisel point found in Medeco pins. The pins were provided with a pair of opposing slots cut perpendicularly into their sides. The key blade exhibited a set of triangular ridges at various orientations, designed to lift and rotate the pins. On full insertion, the indentations on the tips of the pins finished on top of the ridges of the key. This action brought the slots in the four or more pins into registration with the fence of the side-bar. Genakis also proposed other cylinder locks with rotating pins, one of which was discussed in the section on Emhart locks in Chapter 2.

ASSA Twin

(SE) 6-pin + 5-pin side-bar (4–5)

The ASSA Company, founded in Sweden in 1881, derives its name from its founder August Stenman. Originally a blacksmithing operation producing door hinges, ASSA entered the lock cylinder market in 1939. ASSA is now a subsidiary of ASSA Abloy AB–one of the largest lock-making companies in the world and the largest in the United States. The ASSA Twin 6000, shown in Figs. 4.102–4.105, is one of ASSA's most successful products. It was patented in 1980 by B. G. Widen (see Fig. 4.106). The original specification suggested a lock with seven ordinary pin-tumblers and either one or two rows of profile pins and side-bars in a number of configurations. The production version of the lock has six conventional pins and five side-bar pins with a single side-bar on the left-hand side of the keyway. The side-bar pins are simply referred to as side pins in what follows.

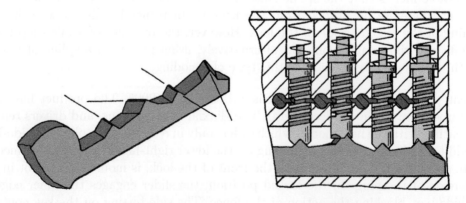

Figure 4.101: Side-bar lock with rotating hollow-tipped pins from J. M. Genakis's patent (US 4,450,699).

Figure 4.102: ASSA Twin 6000 cylinder and key.

Figure 4.103: Assa Twin plug with six top pins and five side pins. Inset shows side-bar.

Figure 4.104: (Left) Side pins resting on ridges of side-bar. (Right) Top pins and spooled drivers on key bittings.

The distinctive split-level key is made from nickel silver with a set of secondary millings on one side of the blade, lower down than the usual pin-tumbler bittings. The conventional pins are actuated by the top bittings of the key, while the side-millings address the five side pins at 7 o'clock in the keyway. The left side of the

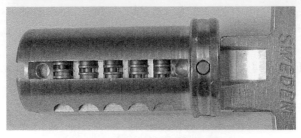

Figure 4.105: ASSA Twin with side-bar removed. (Top) Key partially inserted. (Bottom) Key fully inserted.

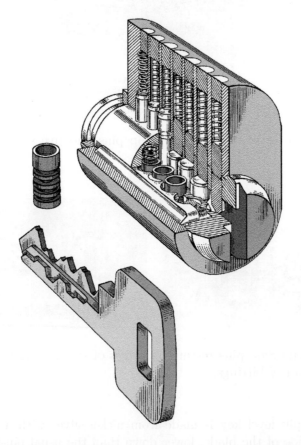

Figure 4.106: ASSA Twin cylinder and key design with side-bar pin detail (US patent 4,393,673 by B. G. Widen).

key has a ramp at the tip that picks up the round edge of the side pins as the key is inserted. The side-bar and conventional pin-tumbler mechanisms are totally independent: it is necessary for the key to possess the correct bittings for both top and side pins in order to operate the lock.

The plug has a specially constructed counter-milling along the left and right edges of the top pin chambers. The counter-milling is teamed with a spooled driver design (Fig. 4.104, right) that blocks if the top pins are underlifted while torque is applied to the plug. This type of antipicking device is traceable to Crousore's 1940 patent (US 2,283,489), and offers security against manipulation even in the absence of the side-bar, which we consider next.

The side-bar inhabits a longitudinal milling in the plug (Fig. 4.103) at 9 o'clock and is supported at each end by springs that provide an outward radial bias. The side-bar pins are internally spring-biased from above and are housed in an off-axis set of blind chambers in the plug, parallel to the borings for the top pins. These chambers intersect the longitudinal milling for the side-bar. Each side pin has several grooves around its girth, only one of which, the operating groove, is the correct depth for the side-bar fence. There are five possible ridge heights, and hence five possible heights for the operating groove or "sizes" of side pin, a particular combination of which forms what we will refer to as a side pin code.

The side-bar does not have a fence in the normal sense of the word. Instead, the fence is composed of five ridges milled into the inner wall of the side-bar. The important thing to note is that the side-bar ridges are in general at different levels of elevation. What this means is that the operating grooves of the side pins should not be aligned at the same level, but should instead be raised to differing levels corresponding to the heights of the side-bar ridges. The profile bittings on the key must therefore be such that this is achieved.

We refer to the set of side-bar ridge heights as the "ridge code." This introduces a second level of ambiguity into the unlocking of the side-bar: not only is the side pin code variable, but so is the ridge code. If both degrees of freedom are used to code the lock, it is impossible to determine the bittings for the side-bar from an examination of the side pins (which are accessible in the keyway and in theory could be probed).

In practice, the ridge code of the side-bar is taken to be identical to the profile bittings on the key. This simplifies matters by allowing all five side pins to have their operating grooves at the same height (corresponding to a "number 3" side pin, as in Fig. 4.104, left). Implementing the system in this way means that both the ridge code and the profile bittings can be referred to as the side-bar code, since they are the same. The theoretical number of side-bar codes is 3,125, stemming from the fact that there are five positions and five different profile bitting heights. The actual number of side-bar codes is 2,800, which can be realized using only 1,400 ridge codes since the side-bar may be installed either way round. An indirect 5-digit code is used to reference the actual side-bar code. The indirect code is stamped on the key blanks to identify them.

The side-bar code acts as a dealer permutation analogously to other dual-action locks such as Schlage Primus. The factory supplies side-bar-coded blanks and matching side-bars under various licensing agreements. For instance, a locksmith may have his or her own side-bar code for use on a local, regional, or national basis; a distributor may have its own exclusive set of side-bar codes. End-users may also have their own regional or national side-bar codes. There is no need for a multitude of different registered key broachings since the side-bar code fulfills this function. Another commercially winning aspect of the system is that the side-bar-coded blank keys may be treated as standard 6-pin key blanks from a local key-cutting perspective. Thus no special equipment is required to cut the bittings for the six top pin-tumblers. The locks are supplied in a number of formats including completely keyed (side-bar, top pins, and keys) and subassembled (side-bar mechanism included but uncombinated for top pins).

The terminology for ASSA locks differs from the standard used in this book for pin-tumbler locks. Position numbering runs from the tip of the key back to the shoulder (or tip to bow). There are nine depths of cut, with pin size 1 being longest and pin size 9 being shortest. The drivers for the top pins are spooled and come in four different sizes. This allows the pin stacks to be compensated (i.e., of roughly constant height), which reduces the susceptibility to decoding by feel since all pin-tumblers are under approximately the same tension. The drivers are made from stainless steel for drill resistance, and the other pins are nickel silver.

Master-keying is performed in the usual manner for inline pin-tumbler locks: master pins are inserted in the top pin stacks to introduce extra shear lines. There are six different sizes of master pin. The MACS is five depths, and over 160,000 usable differs are possible for a given side-bar profile. A MK system for a given application would normally use the same side-bar profile with differing achieved through the top cuts in the keys. The maximum top pin cut depth is not compatible with the highest side-bar cut, so depending on the side-bar bitting code, some of the conventional key codes may be excluded.

The ASSA Twin cylinder is UL 437 rated, having hardened pins inserted around the keyway and side-bar to resist drilling. In terms of manipulation resistance, everything hinges on the side-bar portion of the lock. There is already a very high level of pick resistance built into the lock on account of its dual-action mechanism and the use of spooled drivers on the top pins and false-depth grooves on the side pins. However, only a limited number of side-bar profiles are in use and these are allocated on a regional basis. If prior information or the sighting of a key can be used to determine the side-bar code, then a blank with the appropriate side-milling could in theory be prepared. The top pins could then be picked or impressioned using the ground-down blank to neutralize the side-bar. Without prior information on the side-bar code, the job of picking the ASSA Twin is very much in the "too hard basket."

ASSA also makes conventional pin-tumbler locks for high-security applications. These include the 600 series: a 6-pin cylinder with spooled drivers offering in excess of 250,000 usable differs. A more recent model is the ASSA Twin Combi, which has six top pins and five side pins. The Twin Combi is very similar to the Schlage Primus, having the same "finger pin" design. The presence of side pins gives the system a much higher degree of keying flexibility, copy protection, and resistance to picking than ordinary 6-pin cylinders. Since there are five different side pin elevations, the number of theoretical side-bar profiles is 5^5 or 3,125. Taking the number of top pin combinations as 600,000, the total number of keying possibilities is of the order of $3,125 \times 600,000$ or approximately 1.9 billion. We revisit the Twin Combi in the section on Schlage Primus.

In 1996 ASSA upgraded the ASSA Twin to revitalize the product via a new set of patents. The new models are called the Twin V-10, Twin Exclusive, and Twin Pro [3, 68]. The basic format of the lock is unchanged: six top pins and five side pins with a single side-bar on the left-hand side at 9 o'clock. The new system retains the geographical exclusivity of side-bar profiles (called keyways by ASSA) and the various distribution schemes that ensure strict control of blanks. The design and operation of the ASSA Twin V-10 are illustrated in Figs. 4.107–4.109.

The principal differences of the new ASSA products lie with the side pin design and the plug chambering. The finger pins are modeled on the Schlage Primus and ASSA Twin Combi, and must be lifted to the appropriate heights by the profile milling on the key. The pin elevations, of which there are five, provide $5^5 = 3,125$ theoretical side-bar codes. Approximately 2,800 of these are manufactured. Furthermore, the finger pin ends have two possible offsets or handednesses (left and right). The original finger pin design, shown in Fig. 4.110, included a third offset (center), which is not used in the V-10 system. The bores for the finger pins are eccentric in cross-section, with a portion of larger radius to accommodate the end of the finger pin that contacts the profile milling on the key. The finger pins can thus be lifted but not rotated.

Whereas the finger pin bittings for the original ASSA Twin are regularly spaced along the secondary milling of the key, the offsets in the V-10 call for a nonuniform

Figure 4.107: (Left) ASSA Twin V-10 cylinder and key. (Right) Key with cuts for both left- and right-handed finger pins.

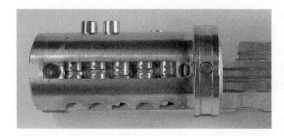
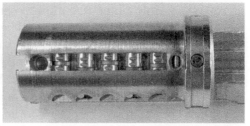

Figure 4.108: Operation of ASSA Twin V-10 plug with side-bar removed to show positioning of finger pins.

Figure 4.109: (Left) Left-handed finger pins and coded side-bar. (Right) Twin V-10 driver pins and plug with counter-millings on pin chambers.

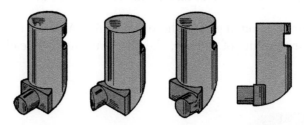

Figure 4.110: ASSA Twin V-10 finger pin design from B. G. Widen's 1988 patent (US 5,067,335).

or irregularly spaced bitting pattern in general. With two offsets and five pins, there are $2^5 = 32$ offset combinations. The product of the elevation and offset degrees of freedom yields roughly 89,600 overall profile bitting variations. An extra advantage is that the secondary milling can accommodate cuts for several offsets at once, as shown in Fig. 4.107. This adds master-keying flexibility to the system since change keys can be cut for a given offset combination, with the MK bittings covering several combinations in order to operate more than one differently keyed lock. The handed side pin concept has also recently been applied to upgrade the ASSA Twin 6000 to the Twin Global 7000 model. The new ASSA designs are summed up in US patents 5,067,335 (1988) and 5,640,865 (1994).

Lockwood Twin

(AU) 6-pin + 3-pin side-bar (4–5)

The resemblance of the Lockwood Twin to the ASSA Twin 6000 is more than coincidental. ASSA Abloy AB now wholly owns the Lockwood Company, previously the largest supplier of locks in Australia, following their 50% acquisition in 1999. The Lockwood Twin 6200, appearing in Figs. 4.111–4.113, is a modified version of the ASSA Twin designed to Australian standards.

Like the Assa Twin, the Lockwood Twin is a dual-action lock, having six conventional pin-tumblers and three profile pins controlling a side-bar. The lock comprises a cylinder and core of standard diameter that retrofits most existing Lockwood locks.

The side pins are chambered parallel to the six inline pins, but are offset to the left, in between pins 2–5. The side pins are of special ribbed construction, and are inserted from the bottom of the plug. Their top end is hollow to allow space for a light-gauge driver spring, while their bottom end is flat with a slight chamfer. Each side pin has several grooves around its girth with only one groove deep enough to accommodate the side-bar fence.

A short side-bar is mounted in a slot at 9 o'clock in the plug, straddling the three side pins (see Fig. 4.112). The side-bar has a fence along the top of its inner face and an apex along its outer face. The apex of the side-bar normally sits in a longitudinal

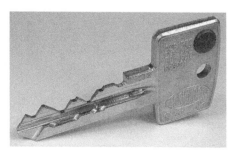

Figure 4.111: Lockwood Twin 6200 cylinder and key.

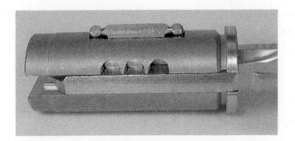

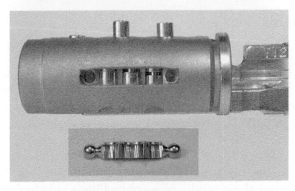

Figure 4.112: (Top) Side pins are fitted from underside of plug. (Middle & bottom) Operation of Lockwood Twin; inset shows side-bar.

Figure 4.113: Set of top, bottom, and side pins from Lockwood Twin.

channel at 9 o'clock in the cylinder body. The ends of the side-bar are shaped to seat small driver springs that supply outward radial tension.

The rest of the cylinder is of conventional design: there is a line of six chambers containing the driver pins, which are usually spooled to increase the manipulation and impressioning resistance. The cylinder also incorporates drill protection in the form of a hardened ball embedded in the plug at 12 o'clock just below the shear line and a hardened rod mounted horizontally across the body in front of the pin chamber.

The key profile, which is part of the registered design, has side bittings on the lower left of the blade and conventional V cuts on the top of the blade. The leading edge of the left side of the key is angled and beveled to ensure smooth pick up of the side pins. Unlike the Assa Twin, where the side-bar ridge heights match the side milling on the key, it is the side pins that are matched to the side-bar code. The key can only be issued by Lockwood, having a side-bar profile precut at the factory. The conventional bittings for the six pin-tumblers can then be made at the dealer's shop according to the requirements of the particular system being supplied.

The design specifies nine pin sizes, numbered from 1 to 9, with a MACS of 5. There are eight master pin sizes in increments of 0.6 mm and four sizes of (spooled) driver pins, which are used to compensate the pin stacks. It follows that the number of differs for the top key bittings is of the order of 220,000 (refer to Table 2.2 in Chapter 2). Since there are five sizes of profile pin, there are theoretically $5^3 = 125$ different side-bar profiles. Each of these supports the full range of top pin-tumbler differs. The system is therefore suitable for multilevel master-keyed suites. As in the ASSA Twin, the side-bar profile is not generally used for master-keying but rather as a dealer permutation for control over the distribution of blanks. This translates to a negligible risk of key interchange and unauthorized key duplication due to the tight tolerances and special side-bitting on the key.

The operating principle, which is the same as in the ASSA Twin 6000, is now briefly described. As the key is inserted, the ramp on its left edge picks up the three side pins and locates them in their respective bittings. At the same time, the top bittings of the key address the six conventional pins. If the side-profile of the key is correct, the deep groove in each profile pin will line up with the top edge of the side-bar slot at 9 o'clock, creating a channel for the side-bar fence. Assuming that the top bittings have brought all six pin-tumblers to the shear line, the side-bar rides out of the longitudinal channel in the cylinder as the key begins to turn and engages the side-bar pins. This principle has been in use for more than 20 years and has proven to be highly reliable. The key and pins are made of nickel silver to reduce wear and resist corrosion.

The manipulation resistance of this type of lock to picks and pick guns is very high[6] because it is not feasible to pick the inline pins unless the side pins are also at the

---

[6]Similar comments apply to other dual-action mechanisms like Yale 5000, Banham, and Schlage Primus.

right height. The shallow grooves on the side pins provide protection against picking and decoding the side-bar profile. Thus one could only expect to narrow down the number of possible side-profiles that would need to be tried, bearing in mind that the side-bitting requires specialized key-cutting equipment.

Banham

(UK) 6-pin + 5-disc side-bar (4–5)
(US) **Yale 5000** (equivalent)

In addition to conventional 6 pin-tumbler and 7-lever mortice locks, Banham in the United Kingdom produces a 6-pin high-security cylinder lock equivalent to the Yale 5000 [69] with side-millings on the key. Yale also produces a 7-pin version of the lock. The Banham side-bar lock is pictured in Figs. 4.114 and 4.115. The cylinder features drill-resistant pins and an optional hardened cylinder guard-ring. The five side-bittings on the key are on the right-hand side and are positioned halfway between each of the pin-tumbler cuts.

The relevant patent reference is US 4,638,651 (1985) by W. Surko on behalf of Yale Security Inc. The mechanism bears a resemblance in its operating principle to two 1980 patents, the first by H. Wolter of DOM-Sicherheitstechnik (US 4,377,082) and the second by Prunbauer of EVVA-Werk (US 4,434,636). Wolter's patent discusses a system of active profile rockers driven by floating balls embedded in the key blade. Prunbauer's patent proposed a number of active profile pin designs and blocking side-bar mechanisms actuated by secondary bittings on the edge of the key. Prunbauer's patent also formed the basis for the ABUS TS 5000 profile pins illustrated in Chapter 2 (Fig. 2.27).

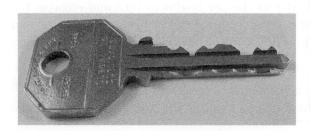

Figure 4.114: Banham key with side-bar bittings and 6-pin high-security cylinder.

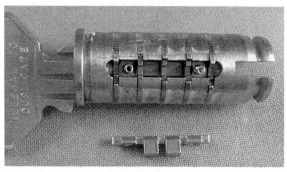

Figure 4.115: Three views of Banham plug with key inserted and sliders aligned for side-bar.

The Banham side-bar mechanism consists of five crescent-shaped rocker arms or sliders that are sprung at one end and slide freely on the inner edge of the cylinder bore (see Fig. 4.115). The side-bar elements are clearly visible in the keyway. Each slider has a gate to accept the side-bar, with extra notching to counter manipulation. The reader is probably already aware of the operating principle: the correct key aligns all the slider gates and at the same time raises all the pins to the shear line to allow the plug to rotate.

Since the side-millings are positioned at the bottom of the key blade on one side only, the bitting depth may vary across the whole width of the key. The implication is that the side-bar bittings can be almost as deep as the thickness of the key. There are three different bitting depths, making a total of $3^5 = 243$ side-bar profiles. These are issued by the factory as dealer permutations on blank keys. The top bittings on the keys are then cut locally on standard equipment. As with ASSA Twin and Schlage Primus, the presence of two independent locking mechanisms (pins and side-bar) gives this lock a high degree of manipulation resistance as well as a very large number of keying combinations.

Schlage Primus

(US) 6-pin + 5-pin side-bar (4–5)

The Schlage Primus is an enhancement of the standard Schlage 6-pin cylinder supplemented by a side-milling on the key blade that operates a side-bar. Pictures of the lock appear in Figs. 4.116–4.118. The idea is credited to B. Widen, formerly of the ASSA Company and inventor of the ASSA Twin covered previously. The Primus design was enunciated in US patents 4,756,177 and 4,815,307, stemming from a 1986 Swedish patent. The original idea is traceable to F. Testa's 1959 patent mentioned earlier in connection with Bell/Dudley locks.

Starting with a conventional wafer or pin-tumbler lock, a supplementary set of tumblers or bar-wafers is added that are actuated by a side-milling on the key. The bar-wafers, which must be lifted to the correct elevations by a lateral track in the key, authorize the retraction of a side-bar. Widen's patent added the dimension of rotation to this locking concept. The original design considered improvements to Medeco locks through the inclusion of side pins with both rotational and elevational degrees of freedom. These enhancements would increase the pick and impression resistance while also yielding a truly huge number of permutations. The side-bar design for this lock had a set of alternating posts for both the main pins and the side pins. The commercial embodiment of the patent was applied to conventional rather than twisting pin locks; however, the twisting side pin idea was retained.

In a Schlage Primus cylinder there are five profile or finger pins whose tips are visible at 6 o'clock in the keyway, as seen in Fig. 4.116. These secondary pins inhabit bores that are parallel to the main row of six pin-tumbler chambers, but are longitudinally offset from them (see Fig. 4.117). As in the ASSA Twin, the finger pins are spring-biased in a downward direction, with their lower outward edge resting against the cylinder bore. The finger pins are shaped like a golf club at the bottom, having a

Figure 4.116: Schlage Primus side-bar cylinder and key with finger-pin cuts on bottom edge.

Figure 4.117: Underside of Schlage Primus plug with finger pins and side-bar in foreground.

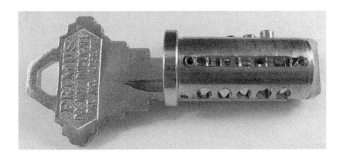

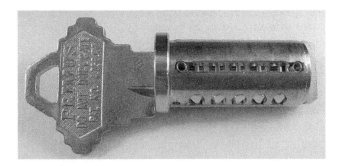

Figure 4.118: Operation of Schlage Primus core with side-bar removed: key aligns finger pins.

straight stem with a hollow top to accommodate a spring. The club end or foot faces inward toward the center of the keyway. The pins are identical in shape and overall size, with a pinched section on their outer edge. We will refer to the pinched section as the ridge. (The original patent specified pins with a circular hole to accommodate a side-bar post as in the Medeco cam lock.)

The ridges are not required to be at the same height on each finger pin or to be in the same angular alignment with respect to the foot. There are two independent degrees of freedom for orienting the finger pins: rotation and elevation. The embodiment of the lock suggested in Widen's 1987 patent admitted three angles of rotation (0, ±15 degrees) and three different elevations, yielding $3^5 \times 3^5 = 59,059$ finger pin permutations. In practice, three rotations and two elevations are used.

The side-bar sits in a longitudinal groove at 3 o'clock in the plug and is spring-biased radially outward into a channel of triangular section in the cylinder. The side-bar can be described as of "female" type: instead of the usual fence protrusions seen in the ASSA Twin and Medeco locks, it is regularly slotted in five places, with the slots facing radially inward. The side pin bores intersect the side-bar channel transversely so that the ridges of the finger pins form an obstruction to the inward radial movement of the side-bar.

The key is a modification of the standard Schlage 6-pin blank, having a wavy side-milling on the lower right side of the blade (viewing the key as it is inserted in the lock, as in Fig. 4.116). The side-milling is along the very bottom of the blade to allow room for the usual pin-tumbler bittings. Furthermore, the broaching of the key blank is such that the motion of the finger pins is unobstructed. The interesting thing about the side-bit milling (or SBM, as it is called) is that the cut centers are generally not equally spaced. This is necessary to cause the finger pins to rotate in the forward or aft direction as their feet are guided by the SBM. If the cut centers were evenly spaced at the halfway points between the main tumbler bittings, the finger pins would end up in a transverse orientation.

Considerable attention is paid to the fabrication of the SBM in Widen's patent. For instance, the cutting angle of the CNC milling machine must be inclined in the forward and aft directions (toward the bow or tip of the key blade) while cutting the two slopes on the side-milling. The boundary between these two cut surfaces is adjusted so as to minimize key wear on the finger pins.

With the insertion of a key with the correct top- and side-bitting profiles, the six conventional pin-tumblers are raised to their respective shear lines. The frontal ramp of the secondary milling on the key picks up the finger pins, which are lifted and pivoted as the SBM slides under their feet. At full insertion of the key, the side-milling imparts a minimal lift to each finger pin while twisting it either to the left or right, or leaving it centered, so that the ridges are in registration with the slots in the side-bar (see Fig. 4.118). As the plug is turned, the side-bar moves out of its channel in the cylinder, and its slots mesh with the ridges in the finger pins. Retraction of the side-bar is not possible unless all finger pins are correctly lifted and twisted.

In common with other dual-action side-bar locks such as ASSA Twin and Yale 5000, the side-bit milling on the key blade is controlled by the factory. Blanks with preassigned SBMs are supplied to authorized locksmiths or agents who then combinate the key by adding the pin-tumbler cuts according to the particular job specification. There are various levels of key control in operation. For instance, a four-level system includes [48, 95]: (i) locally stocked key blanks with a standard "open" side-milling; (ii) locally stocked key blanks with factory side-milling; (iii) factory-controlled key blanks with randomly selected side-milling; and (iv) factory-controlled key blanks with restricted side-milling. The current key control system used by Schlage has nine levels and is described in [108].

The factory control of the side-milling minimizes the risk of a key operating a lock for which it was not intended, as well as restricting the availability of blanks that could be used to impression the lock. Blanks with the same dealer permutation may be used as in a conventional MK system by adding master pins to the pin stacks. Primus is also supplied with conventional key profile variations ("obverse keyways") for a number of multiplex systems.

There is no possibility of a key of one dealer permutation opening a lock in a system with a different dealer permutation since it will not release the side-bar. In terms of keying possibilities, with three different finger pin angles (left, center, and right) and two elevations (low and high), it follows that the number of theoretical side-bar profiles is $6^5 = 7,776$. This multiplies the number of 6-pin differs so that the overall number of keying combinations is around 7,776 × 600,000 or 4.6 billion. It should be remembered that for a given supplier it is the top cuts that are varied and not the side-bar profiles. In addition, several different keyway profiles are available, with this number of combinations applying equally to each key section.

An advantage of this type of design is that it represents a quite minor variation on the conventional pin-tumbler system—cylinders only need a channel to be broached for the side-bar. This makes it easy to retrofit the Primus into existing installations already using Schlage locks, thereby enhancing security in an economical way. The Schlage Primus is also made in an interchangeable-core format. When drill-resistant inserts are present in the plug and cylinder, the lock satisfies the UL 437 standard. Without drill protection the cylinder pins and side-bar could be drilled in a matter of minutes.

An additional feature of the Primus side-bar mechanism is that it can be integrated with the Schlage Everest check pin and security profile (covered in Chapter 2). The combination of these two technologies is the subject of Widen's 1993 Swedish patent, submitted in the United States in 1996 (patents 5,715,717 and 5,809,816). The undercut groove in the side-milling can be included to enhance the already high level of copy protection. The check pin is included in the last side pin position (at the rear of the plug). An Everest Primus key operates both the Everest Primus and Everest locks. In the latter case, the side-bit milling is only used to actuate the check pin. Further information is contained in [109].

The Schlage Primus design has been redeployed in a 5-pin cylinder by IKON AG. This cylinder incorporates a 4-finger pin side-bar with lift and twist degrees of freedom. Spooled driver pins and finger pins with antipicking notches are included [40].

Figure 4.119: ASSA Twin Combi key with undercut groove for finger pins.

Another close relative of the Schlage Primus is the ASSA Twin Combi. The key, shown in Fig. 4.119, possesses an undercut groove on the side-milling similar to the Schlage Everest. The only difference in operating principle between these two locks is that, while Schlage Primus uses fore and aft twisting finger pins, the ASSA Twin Combi uses vertical-lift side pins. Like the Schlage Primus, side pins for the Twin Combi have a pinch mark at a certain height, and they must all be raised by the side-milling on the key to the height of the side-bar fence. Recent variants of the Schlage Primus and ASSA Twin Combi exist that utilize finger pins with both twist and lift dimensions. For instance, finger pins are made for two heights as well as three angles, yielding $2^5 \times 3^5 = 7,776$ possible side-bitting profiles, all of which can accept the usual range of top pin-tumbler bittings. The tight tolerances and the presence of the side-bar ensure that both of these systems are, for all practical purposes, immune from manipulation by conventional lock-picks or pick guns.

The importance of tight control over the supply of key blanks for the entire range of Schlage and ASSA dual-action locks can be appreciated from the following observation: given a blank with the correct side-milling, one can easily construct a tensioning key that unlocks the side-bar and reduces the lock to a conventional pin-tumbler lock from a picking perspective [12].

Scorpion CX-5

(CA) 6-pin + 5-pin side-bar (4–5)

The Scorpion CX-5, pictured in Figs. 4.120–4.122, is one of the most recent high-security side-bar locks reported in this book. Introduced in 2003, it is distributed by Can-Am Door Hardware Inc., based in Canada. The lock is produced in a variety of formats suitable for retrofitting to existing knob-sets, padlocks, and rim- and mortice-cylinder locks including interchangeable cores. The cylinder, which is UL 437 rated, includes significant drill protection in the form of both vertical and transverse hardened rods around the keyway and side-bar (see Fig. 4.121). The relevant patent

Figure 4.120: Scorpion CX-5 key and dual-action side-bar lock cylinder.

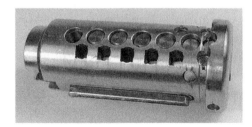

Figure 4.121: (Left) Top view of plug from Scorpion lock. (Right) Side view of plug with side-bar removed and key partially inserted.

Figure 4.122: (Left) Finger pins from Scorpion lock. (Right) Finger pins on key blade, side-bar at top.

is US 6,477,876 (2000) by J. K. Kim, first filed in Korea in 1999. Both single- and twin-side-bar implementations of the lock are given in the patent.

The CX-5 incorporates inline pin-tumblers as well as a side-bar. Its operating principles are similar to those of the ASSA Twin and Schlage Primus, with the distinction that the finger pins are springless. Instead of a ramp on the key blade, the CX-5 utilizes a track in the left-hand side of the key to pick up and guide the finger pins (see Fig. 4.122). In this respect the lock is similar to the EVVA 3KS.

The lock consists of a brass cylinder and plug with chambering for six conventional pin-tumblers. Spooled driver pins are installed to increase pick resistance. The plug contains a slot at 9 o'clock that accommodates a steel side-bar, radially spring-biased at its front and rear ends. In the locked position, the apex of the side-bar engages a longitudinal channel in the barrel. The mechanism is dual-action, requiring the simultaneous retraction of the side-bar together with the alignment of the pin-tumblers.

The five finger pins are chambered in a row of vertical bores offset to the left of the main pin chambers in the plug. The finger pin bores are staggered with respect to the regular pins, a compact arrangement that provides an acceptable amount of travel for the finger pins allowing them to protrude beyond the edge of the plug (as occurs in the EVVA 3KS). The inside edges of the bores are open to the keyway, while their outside edges intersect with the side-bar slot. The finger pins are of square section with an inwardly facing stump that impinges on the lower left-hand side of the keyway. The outward edge of each finger pin contains a gate, the vertical offset of which is varied to yield four different finger pin sizes. All finger pins are equipped with a secondary, false-depth gate to thwart manipulation. The travel of the finger pins is limited by contact with the cylinder bore at the lower extreme and by the stump contacting a longitudinal ward on the left-hand side of the keyway at the upper extreme.

The key contains six bittings along the top of the blade and a milled track on the left-hand side, which addresses the finger pins. Since there are four different finger pin sizes, the theoretical maximum number of side-bar configurations is 4^5, or 1,024. Practical constraints, such as the elimination of repeated entries, reduce this to around 900. In much the same way as in the ASSA series of side-bar locks, the side-bar permutations can be treated as different key profiles allowing the factory to exercise control over the distribution of key blanks. Each profile may be assigned to a specific dealer, institution, or geographical region. The top bittings on the key are then left to the discretion of the local supplier.

Operation is as follows, assuming that a correctly bitted key also possessing the correct side-bar code is presented to the lock. Regardless of the orientation of the cylinder, the pick-up slopes on the left-hand side of the key direct the stumps of the finger pins into the track. There is no need for spring-biasing on the finger pins, which reduces wear on the stumps and on the side-milling of the key. The finger pins are guided by the track as the key is inserted, finishing at the heights required to align their gates with the side-bar slot. At the same time the top bittings of the key raise the bottom pins to the shear line of the plug. Turning the key has the effect of applying inward radial force to the side-bar, which is retracted into the plug as it begins to rotate.

Despite the apparently limited vertical space between the upper edge of the track and top of the key blade, there is still room for 10 depths of cut for the regular pins. These vary in length up to approximately 0.31″ with a depth increment of 15 thousandths of an inch. Thus even a bitting for a number 10 pin, the deepest cut, does not interfere with the highest bitting point on the track (corresponding to a finger pin with its gate in the lowest position). The conclusion is that for each possible permutation of side-bar pins, the full range of 6-pin differs is available to combinate the lock. Assuming a MACS of 6 with typical bitting rules (see Chapter 2), the number of combinations is in excess of 400,000 per side-bar profile.

Manipulation of this type of mechanism, as mentioned in connection with the ASSA Twin, depends to a large extent on having prior information on the side-bar permutation. The lock is machined to very tight tolerances so that it is next to impossible to manipulate the side pins independently of the regular pins. Even if the side-bar can be neutralized by an appropriate "skeleton key," manipulation of the conventional pin-tumblers in the remaining keyway space, coupled with the presence of spooled drivers, would require a high degree of finesse.

Chapter 5

Lever Locks

The successes of lockpickers have always been a constant impetus to an improved construction, since these served to expose the weaknesses and technical deficiencies of supposed security locks. *V. J. M. Eras, c. 1957*

5.1   Introduction

The subject of lever locks has received much attention, in large measure due to their use in safes and vaults. While the lever lock has been superseded by keyless electronic and combination locks in true high-security applications, it is still widely used in smaller safes. Lever locks of varying levels of quality are equally used as door locks in many countries. Due to the strength of materials employed in their construction, lever locks are fundamentally more secure and robust than pin-tumbler locks, but owing to their size, cost, and lack of modularity they have lost market share.

Since their invention in late 17th-century Europe, thousands of modifications and patents have appeared. It is not possible to do justice to these in the space of a single book chapter. Nonetheless, in what follows we will try to give the reader a representative sample covering a wide range of lever locks, most of which are still in use today. The operation of a modern 6-lever lock, produced by Ross Security Locks,[1] is explained pictorially in the series of Figs. 5.1–5.5. This particular type of lever lock is referred to as a single-entry rim lock, meaning that it is mounted on the back of the door rather than mortised into it, and has a keyhole in the front face of the lock only. Double-entry or double-sided locks, which are often of the mortice variety, have keyholes in both the front and back faces and may be operated from either side of the door.

---

[1]The Ross 100 has been superseded by the 102 model.

Figure 5.1: Ross 100 lever lock with levers removed. Key contacts talon of bolt in either direction.

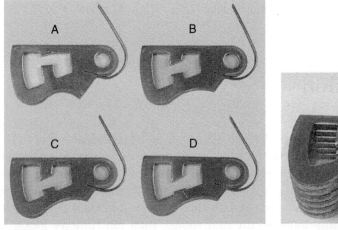

Figure 5.2: (Left) Levers from Ross 100: (A) low-lift; (B, C) midlift; (D) high-lift. (Right) Lever pack: bellies of different-sized levers are not identical.

Historical Perspective on the Lever Lock

The medieval warded lock held sway in continental Europe for approximately seven centuries, finally being replaced at the lower end of the market by lever and pin-tumbler locks in the late 19th century. While blacksmiths and lock makers produced intricately embellished keys and locks, some of them works of art,[2] the basic principle of the warded lock remained largely unchanged until the late 18th century, when the

---

[2]See for instance [9, 25, 29, 64, 125]. Extracts of some 17th- and 18th-century works on warded locks are reprinted in [75].

Figure 5.3: Lever lock with single lever installed: underlifting (left) and correct lifting (right) of lever by key bit.

Figure 5.4: (Left) Lever lock in locked position. (Right) Correct key lifts all six levers to align gates with bolt stump.

industrial revolution in England began to transform the production of manufactured goods. The consequent spread of urban zones around big cities like London also led to an increase in crime.

During this time, locksmiths were beginning to propose alternatives to the warded lock, which was easy to bypass with a skeleton key and required little skill to impression by waxing or marking a blank key with soot from a candle flame. Examples of keys for warded locks and warded lever locks are exhibited in Figs. 5.6–5.9 (see also Fig. 1.1 in Chapter 1). All of these keys were designed for rim locks mounted on the inside surface of the door, hence requiring a long key shank.

Early versions of the lever lock in England and continental Europe were of the single-acting type: a single spring-loaded lever with a stump entered a gate in the top edge of the bolt, preventing its lateral motion. The correct key matched the fixed wards

Figure 5.5: (Left) Key contacts bolt talon, pushing stump through gates. (Right) With bolt retracted, stump finishes in right-hand pocket.

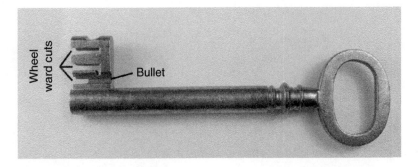

Figure 5.6: Hook and ward pipe key for one-sided rim lock.

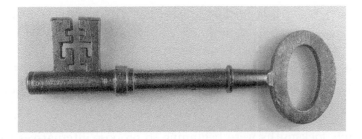

Figure 5.7: Key for double-entry rim lock with bridge wards.

in the lock and raised the lever sufficiently to allow its stump to clear the gate in the bolt. This arrangement was, however, just as easy to defeat by impressioning as a warded lock without the lever.

The Barron patent of 1778 in England (UK patent 1,200) marked the introduction of the double-acting lever, that is, a lever that had to be raised by the key bit

Figure 5.8: A skeleton key bypasses fixed obstructions in a warded lock. Keys shown above would work the same locks as the keys in Figs. 5.6 and 5.7.

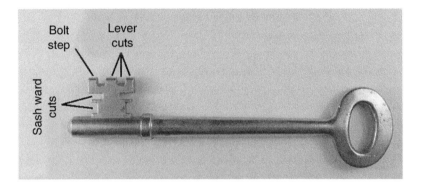

Figure 5.9: Key for 3-lever double-entry rim lock with sash wards.

adequately to clear a lower gate in the bolt, but not overraised or it would be blocked by the upper part of the gate. The embodiment shown in Fig. 5.10 contained two such levers acting in parallel with a 3-pocketed gate in the bolt for double-throw operation. Barron locks with as many as four levers were produced.

Different-sized bellies on the levers required different cuts on the bit of the key to ensure that both lever stumps could pass freely through the gates. Such a lock could not be impressioned "in one go" since the two cuts, being of unknown depths initially, had to be incrementally approached lest they be filed too deeply for the key to work. However, as expressed by John Chubb in his 1850 paper to the Institution of Civil Engineers in London:

> On account of only two tumblers being used in these locks, it is obvious
> that no great changes or permutations, can be made in the combinations,
> so as to prevent the evil of keys passing a lock for which they were not
> made.

The evolution of the English lever lock can be traced through a series of patents over the 40 years following Robert Barron's invention to the 6-lever "detector" lock of Jeremiah Chubb, which was patented in February 1818 (UK 4,219). Whereas earlier lever locks had gates in the bolt and stumps on the levers, the Chubb lock had the

Figure 5.10: Early 19th-century 2-lever Barron night latch (London Science Museum display).

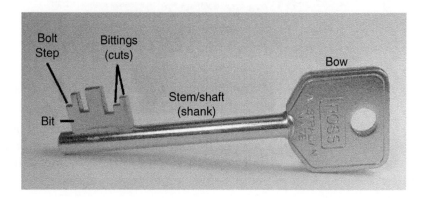

Figure 5.11: Naming conventions for double- and single-entry lever lock keys.

familiar form that we recognize today: a bolt having a single stump and a set of levers with two-chambered gates through which the stump must pass. Terminology for a modern Chubb-type lever lock is explained in Figs. 5.11–5.13.

The manipulation resistance of Chubb's detector lock was tested in earnest by a convicted locksmith cum lockpicker who stood to gain a free pardon and a cash reward of £100 from the Chubb Company, which was then based in Portsea [91]. Imagine the man's desperation when, after 10 weeks of trying to pick the lock, he admitted defeat and was sent back to jail to serve out his term! As a result of this invention, it was the Chubb Company that received a reward of 100 guineas from the U.K. Crown for having developed an "unpickable lock." The reward was used to found Chubb Lock Manufacturers in Wolverhampton in June 1818 [49], which at the time was the heartland of English locksmithing.

Figure 5.12: Naming conventions for bolt used in Chubb-type lever locks.

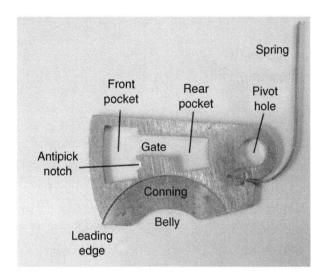

Figure 5.13: Terminology for English or Chubb-type levers.

Further improvements were brought in by Charles Chubb, Jeremiah's brother, and by Charles's son John Chubb and Ebenezer Hunter in a series of patents, the last of which (UK 11,523) was in 1847 [22]. By this time, Chubb's was well established in London, selling its locks to the English aristocracy. The detector lock, pictured in its modern form in Figs. 5.14 and 5.15, contained a seventh "detector" lever that was triggered if any of the other six levers was overraised by picking or by an incorrect key. The original detector lock employed a "regulating key" to reset it to its operating state [99]; the regulating key could not be used to open the lock. In later versions, the operating key was required to reset the lock by turning it in the locking direction. The Chubb Company was duly proud of its locks, which

Figure 5.14: Pipe key for Chubb 7-lever detector lock.

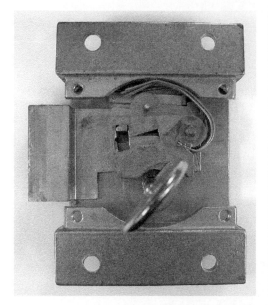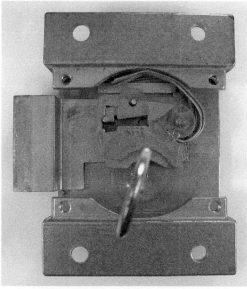

Figure 5.15: Chubb-type detector lock by John Tann with curtain and front levers removed: in triggered state (left); with detector released by turning key CCW (right).

offered excellent security, having as many as 30 changes for each lever. We provide a description of a modern detector lever lock in a later section.

The Chubb detector lock withstood attempts to open it by manipulation and impressioning until the 1851 Great Exhibition, during which A. C. Hobbs picked the lock open in 25 minutes in front of 11 witnesses. Hobbs's technique involved applying heavy tension to the talon of the bolt while incrementally adjusting only levers that were binding. This resulted in a controlled convergence to the correct configuration without the need to overlift any of the levers. This display, while undoubtedly causing alarm to the Chubb Company and some of its prestigious customers, provided an impetus to the industry for finding ways to improve the security of lever locks. Over the five-year period following the 1851 Exhibition, many new designs were patented, as evidenced in George Price's 1856 book [99]. According to Price, Hobbs also supported mechanization in the lock-making industry, which even in the mid-1800s was still dominated by manual labor, much of it performed by children working in deplorable conditions. We revisit Hobbs's accomplishments in the section on changeable lever locks.

Double-acting lever locks all possess a single-bolt stump that must pass into or through a gate formed by the levers. Although lever lock designs exist with two or even three stumps, the locking principle remains the same, with the extra stumps merely providing added security against manipulation and drilling. An alternative *balance lever* principle is espoused in T. Parsons's patent of 1832 (UK 6,350) [5, 22, 91]. The innovative aspect is that a pair of gates are placed in the underside of the bolt rather than in the levers (see Fig. 5.16). The levers are shaped like a see-saw with upturned ends, pivoted at their midpoint. The left or right end of the lever impinges on the left- or right-hand gate in the bolt, depending on which end of the see-saw is raised. The key acts on the levers via a shallow belly in between the pivot point and their right-hand end. When the bolt-step of the key contacts the talon of the bolt, the key bittings must be of the precise height required to bring the balance levers into a horizontal position. Underlifting causes the left end of the lever to block the sliding motion of the bolt, while overlifting causes the right end to do likewise. One advantage of the construction is that the levers bear no significant load and can therefore be made quite thin. This results in a compact lock that can have many levers.

Parsons offered a reward of 1,000 guineas for the picking of a three-inch padlock with 26 balance levers in 1834 [99]. The challenge was taken up by three well-qualified candidates, all of whom failed to pick the lock. The Parsons lock, unlike the Chubb detector and Bramah locks, was, however, not singled out for special attention by A. C. Hobbs at the Great Exhibition in 1851 and presumably remained unpicked. Although the original linear balance lever design is now no longer in use, the circular variant of the balance lever is the basis for the German CAWI lock described in a later section.

Figure 5.16: (Left) Parsons 5-lever drawer lock in locked position. (Right) Key raises balance levers to disengage bolt (note: bolt talon removed).

It is interesting to note that as early as 1846, John Chubb [22] produced a quadruple detector lever lock for banks having four sets of six levers operated by a 4-bitted key (see Fig. 5.17). Other examples of lever locks with triple- or quadruple-bitted keys are furnished by the German safe manufacturer Panzer (later Bode-Panzer), whose products included the Cerberus and the Tangential [26]. The Cerberus lock, named after a mythical dog-like creature with three heads, had a 3-bitted key with 18 levers employing the balance lever principle mentioned previously.

The Bode-Panzer Tangential safe lock, based on designs issued prior to World War II, employed a dual-dial mechanism requiring a 4-bitted key and combination for opening (see Fig. 5.18). Of the two dials, only one was actually a wheel-pack combination lock; the other was a plunger, called a lafette, for a key-operated lock. The key was inserted through a hole in the spindle revealed by pulling out and then pivoting the false combination dial (refer to Fig. 5.116). Returning the dial to its front plate had the effect of transporting the key to the rear of the mechanism, where it operated a four-directional slider lock as the dial was turned. The system was an enhancement of an earlier quadruple-bitted key lock by Panzer from 1907, pictured in Fig. 5.19. The Panzer lock featured a system of 12 interleaved levers and sliders. Two of the

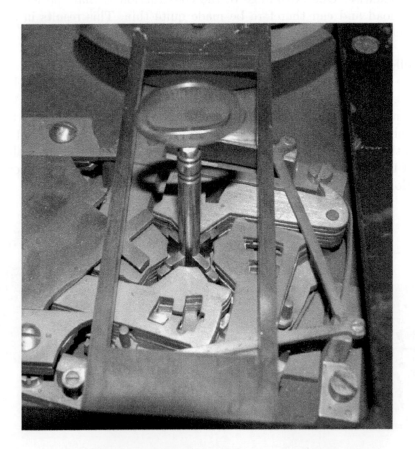

Figure 5.17: Internal view of Chubb's quadruple 6-lever bank lock, held at the London Science Museum. (Courtesy R. Hopkins).

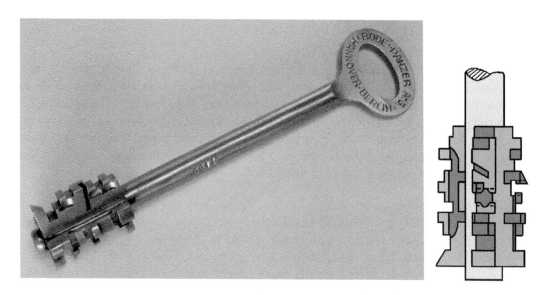

Figure 5.18: Bode-Panzer quadruple-bitted key with millings, bevels, and angled cuts with a diagram from its 1933 patent (DE 642,131 by H. Bode).

Figure 5.19: Panzer 12-lever lock from 1907 taking a 4-bitted key.

four key bits first raised the six levers to free a blocking stump that permitted further rotation of the core so that the other two bits of the key could actuate the six opposing sliders. Further details of these intricate mechanisms are contained in German patents 214691 (1907), 418982 (1924), 642131 (1933), and 646623 (1933).

Double-bitted Key Lever Locks

Until the 1870s, lock makers in the United Kingdom such as Chubb, Milner, and Chatwood concentrated on lever locks with single-bitted keys [96]. Meanwhile in Europe the focus had shifted to lever locks with double-bitted keys. Depending on the mechanism employed, double-bitted keys can offer a higher level of security and are more difficult to copy than their single-bitted counterparts. The double-bitted lever lock has a number of embodiments, including the Italian and German lever designs with constant-width double-bitted keys.

The German lever design is typified by the Novum lock depicted in Figs. 5.20 and 5.21. It was originally produced by the Theodor Kromer Company in the late 1800s (see, for example, 1909 German patent DE 214,693). A similar design principle was espoused in the Max Zahn "Federlos" or springless lock. The obvious simplicity of locks such as these belie their high degree of effectiveness and reliability. Inside the lock case there is little more than a bolt with integral stump and a stack of levers with identical cut-outs.

Compared with conventional Chubb levers, the gate in a German lever lock is shifted to the edge of the lever opposite its hinge point. The lever cut-out has two opposing circular arcs and clearance for the keyway (refer to Fig. 5.45). A consequence of this construction is that a double-bitted key is required to operate the lock. The offset of the constant-width bittings varies along the key stem according to the positions of the gates in the levers. Since the bittings of the key contact the upper and lower arches in each lever, the levers do not actually need to be spring-biased. However, a spring assembly may still be used to return the levers to the locked position on withdrawal of the key. The Mauer Variator lock presented in a later section of this chapter features this type of cut-out lever construction.

Figure 5.20: Kromer Novum double-bitted key with constant-width bittings.

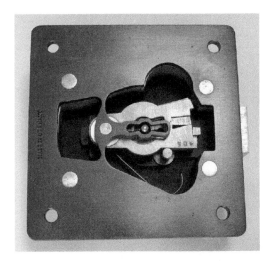

Figure 5.21: Kromer Novum 8-lever safe lock in locked (left) and unlocked positions (right).

German levers may also be configured to act in opposing directions without changing the design of the key. An example is the Kromer Reling lock from the 1880s. The lock shown in Fig. 5.22 was produced after World War II and employs a set of nine levers, spring-biased from above and below in an alternating sequence. A tenth lever with a circular outline was also included in the pack: this lever prevented the bolt stump from contacting the gates of the other levers until the key was turned. The Ross 700 lock operates on a very similar principle to this.

The Italian double-throw lever, or "mandata," is employed in locks made by CISA, Cerruti, Fiam, Mottura, Potent, and many other companies. These locks require a double-bitted key with constant-width bittings except for the bolt-step. The Silca 201 key blank catalogue [111] lists dozens of brands of locks of this type.

Italian lever locks of the kind depicted in Fig. 5.23 typically have multiple turn operation during which the bolt stump must pass through a series of lever gates. In double-entry locks, the center step of the key operates the bolt, whereas in single-entry locks the bolt is usually operated by the last step (see Fig. 5.24). Naturally, a symmetric double-bitted key is required if the lock is required to be operated from either side of the door.

Instead of a single pivot point, Italian levers are slotted so as to slide in a direction perpendicular to the throw of the bolt. The motion of the levers is therefore linearly constrained rather than pivoting as in Chubb locks. In a double-turn lock, there are three rectangular pockets in each lever, corresponding to the unlocked, single-locked, and double-locked positions of the bolt. There are thus two gates in each lever. While the gate width is constant and sufficient to allow passage of the bolt stump, its vertical offset is variable to allow for different cut depths.

Figure 5.22: Post–World War II Kromer Reling 10-lever lock and double-bitted key.

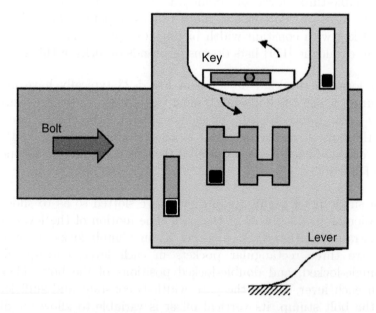

Figure 5.23: Type of lever typically found in Italian double-bitted lever locks.

Figure 5.24: Constant-width double-bitted key from an Italian 8-lever lock.

In each lever, the height of the gate between the first and second positions is generally different from that of the gate between the second and third positions. This vertical offset in gate heights is required so that if one side of the constant-width key bitting raises the lever to the correct height to pass from first to second position, then the other side of the same bitting will allow the bolt to pass from second to third position.

The double-throw system applies equally well to lever locks with single-bitted keys and Chubb-type levers. This system, popular in France and other European countries, utilizes two gates at equal heights with three pockets through which the bolt stump must pass. The level of physical security is increased since the bolt has greater penetration into the strike.

The German and Italian lever lock designs we have so far seen are operated by a double-bitted key with a constant bitting width. A further class exists called twin-lever locks, which have a double-bitted key with truly independent bittings. Such a lock incorporates either Chubb-type levers, arranged in two separate stacks with the bellies facing each other, or a single stack of alternating levers with cut-outs. Bittings on each side of the double-bitted key address both sets of levers in order to open the lock with a single turn. Usually these locks are intended for use in safes and safe deposit boxes (see Fig. 5.25), although they have been produced for domestic purposes like mortice door locks, for example, by Lips [30]. By doubling the number of levers from 6 to 12, or 7 to 14, for instance, the number of key permutations is squared. Locks of this type, though obviously expensive to make due to the number of components and mechanical precision required, do provide a very high level of security, especially if serrations or notches are provided on the surfaces where the lever gates meet the bolt stumps. An example is provided by the Chubb-Lips 6K207 14-lever two-stump lock shown in Fig. 5.26. We present a twin-lever lock with a symmetric key (Ross 700) later in this chapter.

Another system involving a double-bitted key is employed in the Kromer Protector lock. The locking principle, on which the Chubb Ava lock is based, was developed around 1870 [30]. Early patent references for the Protector lock include DE 3,523 (1878) and DE 17,157 (1881). Kromer held the firm belief that double-bitted locks

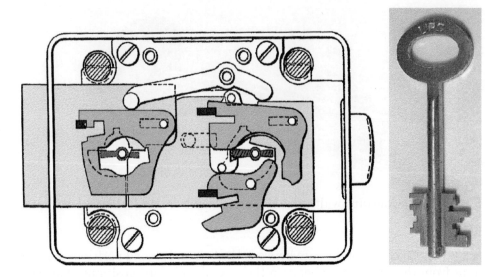

Figure 5.25: (Left) Eras and Lips's 1924 safe deposit lock design with twin packs of 7 levers (UK patent 224,175 by V. J. M. Eras and H. J. J. M. Lips). (Right) Lips double-bitted renter's key.

Figure 5.26: Chubb-Lips 6K207 14-lever two-stump safe lock.

offered greater security than the conventional high-security locks of the time, including those of Chubb and Bramah. He stated [116] that even Bramah-Chubb locks (of the type covered in Chapter 3):

> owe their frequent use only to the low price and the ignorance of the public.

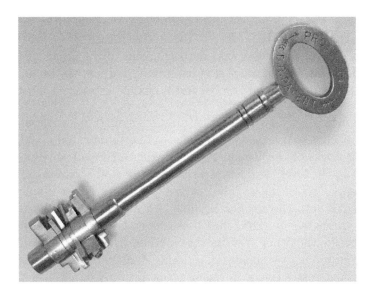

Figure 5.27: Kromer Differential Protector double-bitted key with pivoting key steps, patented in 1932 (DE 620,053).

The Kromer Protector design incorporates elements of the wafer-tumbler system, but it is really a cylindrical lever lock with a double-bitted key. The lock comprises a rotatable core and stator. The core contains a stack of 10 or 11 slideably-mounted levers that alternate in their directions of action. The push-pull action of the key brings all sliders so that their ends are flush with the core. Until 1931, most Protector locks were fitted with one-part, horseshoe-shaped sliders with open bellies. Split-levers or lever pairs were introduced by Otto Sellin in 1931 in order to enhance the level of copy protection by doubling the number of contact points on the key. Kromer went as far as producing a system called the Differential Protector, which boasted a double-bitted key with pivoting key steps as shown in Fig. 5.27.

One hundred years after the invention of the Bramah lock in 1784, Mr. Kromer felt confident enough to say:

> I believe that I have created a lock which no expert will ever be able to open without force.

However, as we will see later in this chapter, like many other great makers of locks, he would be proved wrong.

Master-Keying in Lever Locks

We have not yet spoken of master-keying in the context of lever locks. The traditional method of master-keying for this type of lock is to use sash wards and keyway variations. This method, while producing considerable flexibility, is all too easily bypassed. If we restrict ourselves to Chubb-type levers, then the conventional technique involves widening one or more of the lever gates in order to let more than one bitting height operate the lock. An alternative measure is to include a stud or pin riveted onto one of the levers that actuates some of the adjacent levers. Neither of these methods is attractive since it substantially reduces the security of the lock, rendering it easier to manipulate or convert a change key into a master key.

In systems requiring only two levels of master-keying, (e.g., servant and master keys), one approach is to use the two-keyhole system (illustrated in [30]). The system employs sliding levers of the Italian type, utilizing a servant keyhole that operates the lower gates on the levers and a master keyhole located above the servant keyhole, operating an upper set of gates. The servant key has a larger diameter stem so it cannot be inserted into the master keyhole. An alternative is the British master-keying system, patented in 1898, that uses longer levers with two bellies and two keyholes, one beside the other. Yet another method from 1922 described in [21] uses a compound lever having two bellies of differing diameters. All of these systems are, however, somewhat restrictive in terms of their master-keying potential.

A more flexible system, shown in Fig. 5.28, was patented in 1949 by F. J. Butter who was the chief designer at Chubb's and later at Josiah Parkes & Sons Ltd., Union Works, England [89]. The Butter's system requires only a single keyway and makes use of a type of edge-gated or peripheral lever, as depicted in Fig. 5.29. This type of construction has inherently more leverage than in a conventional Chubb lock because the gate moves at a larger radius from the pivot than the fulcrum of the lever, where the key acts. The operation of the lock will be described in detail later, but the important point is that many narrow gates may be cut in any one lever. This substantially increases the master-keying possibilities without requiring fixed wards, which are sometimes used to increase the number of key blank profiles. Added security is provided, as with other lever locks, in the form of notches or teeth on either side of the gate that hinder manipulation. The Butter's system is incorporated in the Chubb 3G110 lever lock covered later in this chapter.

Despite the improvements in master-keying brought about by the Butter's system, the high cost, size and relative inflexibility of lever locks outweigh their security advantages in many applications. Consequently, lever locks are often not seen as a viable option for large-scale master-keyed systems where low unit cost and flexibility are paramount.

Changeable Lever and Combination Locks

Of the many high-security lever lock designs that have been proposed, the most intricate are changeable lever or permutation locks. These include both key-operated and keyless combination locks. Many such locks were produced in the late 19th and early 20th centuries by Milner, Tann, Ratner, Chatwood, and other noteworthy British lock and safe manufacturers [21, 34]. Hopkins's book on the J. M. Mossman collection [57] and its recent extension [31] are a good source of North American examples of these locks.

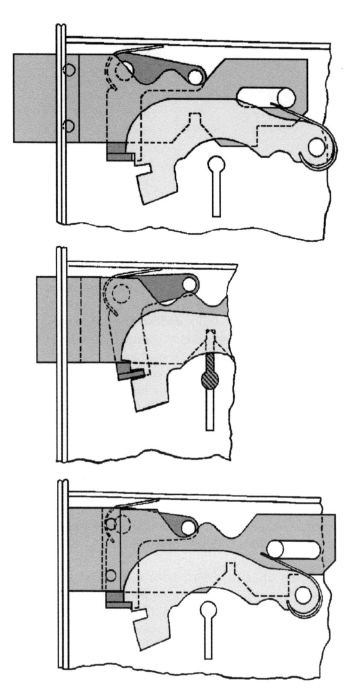

Figure 5.28: Opening sequence of the Butter's system lever lock (UK patent 661,501 by F. J. Butter).

The motivation for changeable lever locks is easy to understand. Since most lever and safe locks are not modular, it is necessary to dismantle the lock in order to change its combination. This is an onerous task and one that is totally unsuited to applications where rapid or frequent changing may be required, such as in banks and

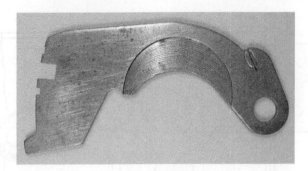

Figure 5.29: Lever from a Chubb 3G110 Butter's system lock.

financial institutions. Key-changeable lever locks permit the user to recombinate the lock for operation by another key without the need for disassembly.

Key-changeable lever locks are generally based on meshing levers, gears, or rack and pinion mechanisms, as illustrated in Figs. 5.30 and 5.31. A further method, described in Watson's 1909 patent (US 1,136,067), endows the levers with several pivot points that can be engaged by a slideable secondary stump. Examples of this type include the 7-lever Bauer Sphinx and the LeFebure dual-control lock, used in the United States for safe deposit boxes [122]. Other implementations employ identical levers with either (i) an array of variable bolt stumps or (ii) gates whose position with respect to the bolt stump may be varied. In the first case, this is achieved through use of a change key that unclamps the bolt-stump array so that it may be adjusted to suit a new key (see Fig. 5.48). In the second case, applicable to German levers with cut-outs, the recombination is effected by a mechanical switch that provides temporary disengagement of the gate array. Examples of both of these concepts are provided in the sections on the S&G 6804 and Mauer locks later in the chapter.

Another type of mechanism, depicted in Fig. 5.32, employs key-operated changeable discs. Locks of this type include the Sargent & Greenleaf model 6860 (7-wheel) and the La Gard model 2200 (4-wheel) combination locks. The idea of using a double-bitted key to rotate the discs in a lock with a fixed combination is traceable to Kromer's Central lock from the 1880s, subsequently renamed the Integral. Romer's Scandinavian padlock in Chapter 4 (Fig. 4.3) also used discs with inner steps of differing radiuses. Fig. 5.33 provides an illustration of a pre-1920 model Central lock from the Swiss firm Bauer, inventors of the Kaba lock covered in Chapter 2. The La Gard 2200, shown in Figs. 5.34 and 5.35, is a close relative of the Carl Kästner lock covered in Chapter 4 that replaces fixed discs with changeable discs.

Recombination of a lock with changeable discs is usually achieved by turning all wheels to align their gates and then inserting a change key from the rear of the lock into the wheel-pack. The change key releases a clutch mechanism between the inner and outer parts of the wheels that allows them to be recombined to a new key. This is analogous to the method used to set a new combination in a keyless combination lock such as the S&G 6700 series (see Figs. 5.36 and 5.37). For further details on keyless combination locks the reader may wish to consult [122].

Figure 5.30: Pinion-type changeable 5-lever lock by John Tann (top to bottom): (1) locked position; (2) key raises lever to align gate in pinion with stump; (3) bolt is withrawn, sliding pinion toward fixed stump.

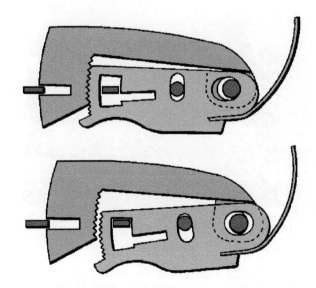

Figure 5.31: Meshing levers for changeable lever lock from US patent 666,697 (1900) by J. Roche.

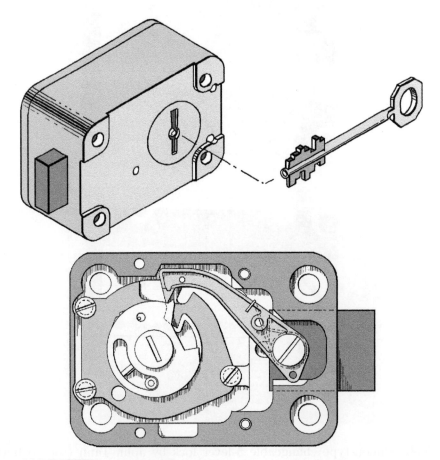

Figure 5.32: In S&G's model 6860 changeable combination lock, the key contacts steps in the disc edges at various angles of rotation (1981 US patent 4,375,159 by C. G. Bechtiger and J. Peyronnet).

Figure 5.33: Core from Bauer Central 7-disc lock with double-bitted key.

Figure 5.34: La Gard 2200 4-wheel safe lock with back cover and cam wheel removed. (Top) Locked position. (Bottom) Key aligns gates to engage fence of drop arm and withdraw bolt.

Figure 5.35: Detachable key tip from La Gard 2200 safe lock.

Figure 5.36: S&G 6700 series 3-wheel combination lock with rear cover removed.

Figure 5.37: Combination change in S&G wheel-pack combination lock performed by releasing clutch linking inner and gate wheels (indicated by arrows) via square change keyhole at top. Released position on right.

A particularly famous specimen from the Day & Newell Company in New York, of which A. C. Hobbs was a representative, was called the parautoptic lock, meaning "concealed from view."[3] The Newell lock, or Hobbs's lock, as it was also known, was

---

[3]In those days, the study of ancient Greek was much more prevalent than it is now.

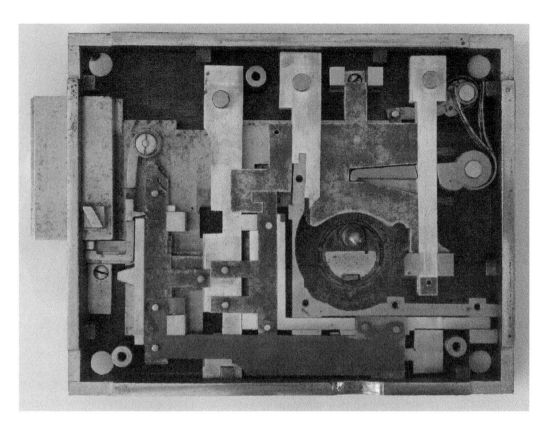

Figure 5.38: (Top) Hobbs 6-lever parautoptic lock with cover plate and curtain removed (Courtesy P. A. Prescott, www.antique-locks.com). (Bottom) Key for a 15-lever Day and Newell parautoptic lock with rearrangeable steps (London Science Museum display).

patented by R. Newell in 1838 (US 944). It was a key-changeable lock with a key whose bit sequence could be rearranged in any order. Newell managed to pick the original version of his own lock, a fact that induced him to improve the design in two further patents, issued in 1844 and 1851 (US 3,747 and 8,145). A 15-lever example of the parautoptic lock was displayed at the Great Exhibition in 1851. A 6-lever parautoptic lock made in 1863 by Hobbs Hart and Co. is shown in Fig. 5.38.

Permutation locks taking a key with removeable bits had been invented earlier by MacKinnon in 1835 [99] and by Dr. S. Andrews in 1836 [50], although these locks required the order of the levers to be changed manually. The Newell lock avoided this problem by including in its mechanism a system of compound levers with serrated edges forming an array of variable stumps that meshed with a V-shaped projection.

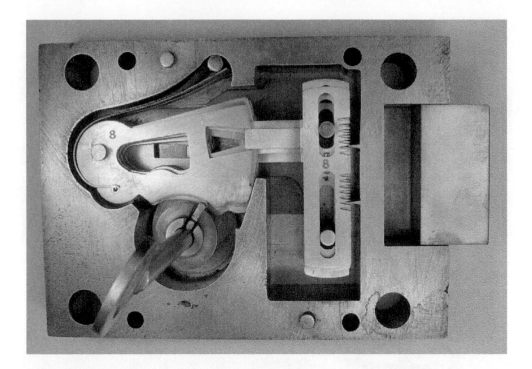

Figure 5.39: Operation of 9-lever Hobbs banker's solid change-key lock: notches in stump array, raised according to the key profile, engage V-shaped projections in case.

Once disengaged by the correct key, the stumps could be shifted and reengaged, thus recombining the lock to any desired key. As a precaution against inspecting the lever-tumblers, the parautoptic lock also included a shutter that blocked the keyhole while the curtain was being turned. The automatic setting action of the bolt-stump array by the key is more easily appreciated by reference to Fig. 5.39, which shows a 9-lever Hobbs "Protector" change-key bank lock from around 1880.

Since the number of rearrangements or permutations of N objects is N factorial $= N \times (N-1) \times ... \times 3 \times 2 \times 1$, a 10-lever Newell lock with 10 different steps would have 3,628,800 possible key changes. If M of the steps were identical, then this number would be divided by M factorial.[4] The rearrangeable-bit key is still in use today in the form of a make-up key or pin and cam tool. Once a lever lock has been decoded (a topic we consider next), a working key can quickly be assembled by mounting bits of the required length onto the supporting stem.

The 1851 Newell lock was offered as a picking challenge with a reward of £200 at the Great Exhibition. A Mr. Garbutt accepted the challenge and spent a total of 30 days in an unsuccessful picking attempt [99]. Newell's formidable lock consequently appeared to be unpickable. By 1856, Linus Yale Junior had developed a soft-key impressioning technique for it in the course of promoting his own brand of

---

[4]Refer to Appendix A for further details on permutations.

Figure 5.40: The key for Linus Yale Junior's Magic Infallible bank lock had a removeable pod of 8 bits that could be rearranged by shuffling or flipping.

"unpickable" bank locks. Yale Junior described his technique for the 10-lever Newell lock in the third person [135]:

> his method is so exceedingly simple that any smart lad of sixteen can in a short time make a wooden key, the exact transcript of the owner's, which will open these locks, and relock them either on the same or any other combination he may choose, in an incredibly short space of time.

According to witnesses' reports of Yale's technique, an "incredibly short space of time" was of the order of one to three hours.

Yale's own lock, the Magic Infallible bank lock, was a springless recombinatable lock sporting a key with a detachable pod holding eight bits, as shown in Fig. 5.40. On insertion and turning of the key, the pod was released from its dovetail on the key stem and transported to a guarded section of the lock behind an intervening steel plate where its bittings were pressed rather than swept against the levers to release the bolt. The closed-off keyway of Yale's lock also made it effectively powder proof. Like Hobbs's lock, the key bits could be permuted to change the combination; the lock adapted itself to the new combination through a set of slideable fences. For an appreciation of the internal mechanism, it is best to refer to Yale's 1856 dissertation [135] or to Yale's subsequent patents (US 28,710 and US 32,331).

An extra twist was added in Yale's lock by endowing the key bits with cuts on either side. Only the cuts along one side were active in operating the lock; the other cuts provided an extra degree of freedom for recombination. Whereas a key with eight rearrangeable steps would have 40,320 (8 factorial) different permutations due to shuffling, the Yale Magic bank lock multiplied this by the number of possible flippings of the bits, being 2^8 or 256, to yield a total of 10,321,920 key changes.

Yale's confidence in this product was such that he set out a challenge to the public, offering \$3,000[5] to anyone who could pick an installed Magic Infallible bank lock without damaging it. By the time Yale died in 1868, no one had claimed the reward.

Differing, Decoding, and Security Features

Lock and safe manufacturers have given much thought to protecting their lever locks against forms of attack such as sawing, cutting, drilling, punching, and the use of gunpowder and nitroglycerin. Without digressing into the specialized materials and technology of safes, we mention some aspects relevant to door locks. Good-quality lever locks typically have toughened steel front plates to protect the stump, gates, and pivot from drilling. Hardened pins or rollers are inserted into brass bolts to prevent them from being sawed through with power tools. Alternatively, the bolt may be of composite construction, containing ceramic or other tool-resistant materials. Another popular formula is to equip the lock with several round bolts instead of a single latchbolt of rectangular section. The round bolts should include freely rotating steel sleeves for protection against sawing.

Naturally, the security offered by a lever lock is not solely a function of its physical strength. A good-quality lever lock should also offer a large number of combinations and be difficult to pick, impression, and decode. The number of combinations offered by a lever lock is determined by the number of levers and the number of sizes (or changes). Theoretically, a lock with M levers, each of which may be of N different sizes, provides N^M (N to the power of M) combinations. Further details on lever lock combinations are given in the section on Chubb locks.

We have already mentioned features like notches around the gate and/or on the bolt stump that impede manual picking and impressioning. The inclusion of false gates also makes it more difficult to pick the lock using 2-in-1 picks (or "curtain picks," if the lock has a curtain). The process of decoding, in contrast to manipulation, tries to infer the key cuts from inspection of the lock. We now focus on factors that hamper decoding by inspection of the lever bellies.

Two lever locks of the same type are said to differ if they require keys with different cuts to open them. Let us assume that the lock is made to differ by changing the height of the gate from one lever to the next. As explained by Eras [30], the diameter of the lever belly required to maintain the gate at a constant height while the key is turned is a function of the height of the gate. It is important to minimize the vertical width of the gate so that the stump passes exactly through it, subject to the tolerance of the key bitting. For this reason the bellies of levers with different gate positions are generally of differing dimensions (refer to Fig. 5.41).

Whenever levers are made to differ in proportion to the size of the lever belly, it is easy to derive information about the required key cuts by inspection of the lever

---

[5]A hefty sum at a time when the salary for unskilled workers in the United States was around \$3 a week.

Figure 5.41: Levers from a Chubb 3G114 lock differ in respect of gate position and belly radius.

Figure 5.42: The curtain surrounds the key bit and restricts access via the keyhole.

bellies, either by optical or mechanical means. Most locks are susceptible to decoding in one form or another, but for lever locks the problem is quite acute since their physical size allows more access for tools to inspect the levers via the space below and in between them.

In early lever locks without fixed wards, it was relatively easy to inspect the bellies of the levers, thereby allowing someone to read or decode the lock and make a working key. An American locksmith by the name of Mr. Hodge, present during John Chubb's 1850 dissertation in London, described how a heated mixture of glue and molasses could be injected into a lock and cut out with a thin-bladed instrument to impression the range and curve of the lever bellies. Chubb consequently included a rotating skirt, called a *curtain*, or barrel and curtain, that surrounded the keyway (see Fig. 5.42). The curtain flanks the key as it turned, restricting access to the levers. This makes it much more difficult to compromise the lock by inspection or manipulation. The idea of a curtain had apparently been around for some time, with one of Chubb's interlocutors referring to a Swiss lock of 1762 that included such a device.

A first step in preventing the combination of a lever lock from being read is to use "belly groupings." In this approach, gates of similar height are cut in levers of the same belly shape so inspection will only narrow down the key combination. This is a feature of modern Chubb locks like the 3G114 where there are eight gate positions but only three belly groupings.

A second way of providing differing that is hard to decode was adopted by Chubb for its 6K75 safe lock (see Figs. 5.43 and 5.44). The method entails reshaping the

Figure 5.43: Chubb 6K75 8-lever safe lock with identical cut-outs in levers (curtain removed).

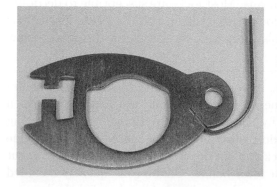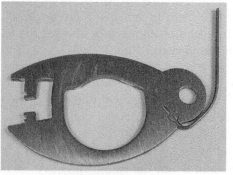

Figure 5.44: Chubb 6K75 levers with differing gates but the same belly cut-outs. Lever on right has antipicking notches on either side of gate.

levers so that their gate is moved from its position above the belly to the edge of the lever furthest from the pivot point. This construction also dispenses with one of the lever pockets. The gate and stump are then made thinner horizontally, and the remaining pocket is enlarged to allow for the extra travel of the lever end (since it is further from the fulcrum). This implementation allows all the lever bellies to be cut to the same size. The construction is such that the gate is raised to the exact height required to accept the bolt stump just as the bolt step of the key bit engages the talon of the bolt. From the keyhole, no differences in the rest positions of the levers are discernible. This type of lever is closely related to a much earlier "closed-lever" design from 1860 due to R. W. Parkin, a partner of Samuel Chatwood [43]. The closed-lever construction, suited to both single- and double-bitted keys, reduces the space around the keyway as a means of protecting against both manipulation and the use of gunpowder.

Another way to ensure constant radiusing on the lever bellies, preferred by German lock manufacturers, is to use a differently shaped, closed lever that has a cut-out in the middle through which the key enters. Examples of this type of lock were given in the section on lever locks with double-bitted keys. Referring to Fig. 5.45, if the size of the cut-out (the dimension L) is kept constant and only the position of the gate is varied, then the key has the property that, with the exception of the bolt-step, all of its bittings are of the same overall width. Moreover, all levers have identical cut-outs so that the key cuts are not related to the cut-outs in the levers. An added advantage is that the lock cannot be impressioned with the usual technique of incrementally deepening the cuts on the key since a shallow cut on one side of the bit must be matched by a deep cut on the other side.

The design of belly cut-outs for lever locks is still an area where improvements are being made. Even when all levers have identical bellies or cut-outs, it is still possible to decode the lock from inspection of the wear patterns on the surfaces of the levers where the steps of the key contact them. This stems from the simple observation that, since the motion of the key bit is circular, steps of different size correspond to different radiuses that contact the camming surfaces of the levers at different points. Measuring the distance from the axis of the keyway to the extremities of the wear pattern on a lever establishes the height of the key bitting for that lever. A modern

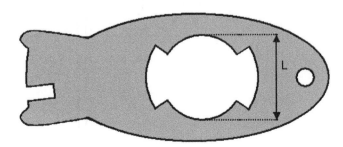

Figure 5.45: Type of lever typically found in German double-bitted lever locks.

technique for decoding a key for a lever lock involves the insertion of a fiber-optic borescope to locate and measure the wear patterns.

Countermeasures for this technique include contouring both sides of the lever belly to prevent, as far as possible, marking of its surface due to wear. Another method, evidenced in a 1988 patent by a Swedish firm (US 4,836,000), applies to Italian-type levers. The design calls for a specially shaped curtain with a triangular base. Two vertices of the triangle are normally in contact with the lever cut-outs. The dimensions are such that the curtain leaves a broad wear pattern that effectively hides the wear pattern made by the key bit.

Further Examples

The "floating cam" lock was developed in 1966 by P. E. Schweizer and R. K. Thompson of Bell Labs for use in U.S. public telephones. The internal details of this highly manipulation-resistant lock are shown in Fig. 5.46. The lock utilizes a system of up to eight levers with peripheral gates, not unlike Butter's system levers. The central idea in the design is that the floating cam decouples the motion of the bolt stump from the lever gating. Tensioning the cam does not help in determining the correct positions of the levers. Instead, the levers must simultaneously present the correct combination

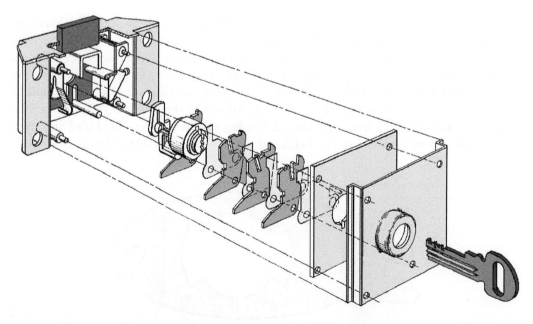

Figure 5.46: Bell Labs "floating cam" lever lock (US patent 3,402,581 by P. E. Schweizer and R. K. Thompson Jr.).

when the cam is turned. A similar idea implemented with rotating gears is used in the Fichet-Bauche 787 lock, which is covered later in this chapter. Another attribute of the Bell Labs lock is the presence of a detector mechanism that requires resetting if a lever is overlifted. The detector may also be used to trip a microswitch to signal that a picking attempt is in progress.

A further category of lever locks that we consider is that of safe deposit box locks, which are used in bank vaults and secure storage facilities. Many locks of this type are known [30, 57], and essentially any key-operated high security lock can double as a safe deposit lock. A particular class is that of dual-control safe deposit locks. This is a lock accepting two different keys, both of which must be operated, usually sequentially, to open the lock. There are several possible implementations of the dual-control mechanism, the two major variants being single-keyway multiple-gate lever locks and locks with two separate keyways. A further implementation, which we discuss later, is a single-keyway lock with a two-part or split key. Single-keyway multiple-gate lever locks were made by Milner and Chubb, among others, in the latter half of the 19th century. George Price produced a triple-control lever lock with three sets of gates and four pockets in each lever. A different key was required to move the bolt stump through each set of gates.

Various implementations of triple-control "key and combination" safe deposit locks were developed by a number of European manufacturers, including Lips and Fichet, around 1900 [41]. These locks typically had two keyways, requiring single- or double-bitted keys to operate two lever locks and an additional set of three or four combination dials operating a "click" or "clicker" mechanism. The naming of the lock is based on the French word "cliquet," meaning ratchet, which also describes the noise made by the lock. The Lips safe deposit lock incorporated two lever locks, one of them using twin levers (as in Fig. 5.25) and four alphabetic combination dials. In the Fichet triple-control lock, shown in Fig. 5.47, the upper lock had five Chubb-type levers, whereas the lower lock employed five compound levers with gating in their outer edges to increase the available differs. The bolt could not, however, be released without also adjusting the combination wheels. Each wheel had 26 positions and was operated by a knurled knob. The wheel positions were set by counting the number of clicks of the ratchets or, in the case of the Lips lock, dialing the letters of the combination. This would align the gate of each toothed wheel with a system of stumps controlling the motion of the bolt. Other types of click lock were key driven. Letter combination and multiple-dial locks of this sort were popular in Europe, whereas U.S. manufacturers focused more on refinements to the single-spindle/multiple-wheel combination lock developed by Yale, Sargent & Greenleaf, and others. An unfortunate aspect of the click lock was that someone listening could ascertain the combination.

In a two-keyway dual-control lock, one keyway accommodates the preparatory or guard key, kept by the bank manager, while the other accepts the renter or client's key. The guard key is inserted first and turned to enable the mechanism. The renter's key is then inserted and turned, opening the deposit box. Usually the renter's key is retained in the lock when it is in the unlocked position. A number of

Figure 5.47: Triple-control Fichet safe deposit lock with 3-knob "click" mechanism. First, the three knobs must be set at the correct positions. The guard key is then inserted in the upper keyhole and turned to allow access for the renter's key (top), which is inserted in the lower keyhole to operate the compound 5-lever lock. The fretting is ornamental—the original lock would have had a plain steel front plate.

distinct embodiments are possible for this type of dual-control lock. For lever locks, we can make a distinction depending on whether one or two stumps are attached to the bolt. In the case of two stumps, there are two independent lever packs. The gates on both sets of levers must be aligned by the correct keys to permit retraction of the bolt. For single-stump dual-control locks, the levers for both keyways are interleaved and share the same pivot point. The renter's key is inserted into a

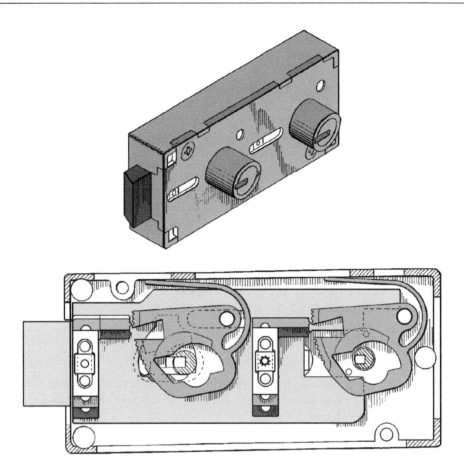

Figure 5.48: S&G's 1982 design of the 4500 series dual-control safe deposit lock utilizing slideable clamped fences to allow recombination (US patent 4,462,230 by W. R. Evans).

keyhole near the forend of the lock, while the guard keyhole is situated further back. The guard levers in this case have an extended tail portion. Examples of both types of lock are presented later in this chapter. An illustration from a 1982 patent by Sargent and Greenleaf is given in Fig. 5.48 in which variable lever locks are harnessed to facilitate rekeying of both the renter and guard sections of the lock.

Lever Lock Classification

In the following sections we present a number of different types of lever locks. Some of these have already been mentioned in this introduction, others will be covered subsequently. The classification we have chosen for what constitutes a lever lock has more to do with the shape and action of the key than the actual internal operating principle. Lever locks generally have keys that must be turned to set the positions of the tumblers. For this reason, locks such as the Chubb Ava and Kromer Protector, despite their obvious similarities to wafer locks, are considered to be lever locks. Master-keying aspects are discussed for locks that are intended for use on normal (residential and office) doors. Master-keying is not required for lever locks designed for use on safe and vault doors.

Locks from the following categories have been included.

1. Conventional (English) lever locks with single-bitted keys: these include the Chubb 5-lever and 7-lever, Chubb 5-detainer (Butter's system) and Chubb detector locks. The Fichet-Bauche "sans souci" with its twin-bitted key is also in this category. We have also included the Ross 102 and Ross 600 lever locks, the last of which uses vertical-lift levers.

2. Italian double-throw vertical-lift lever locks with double-bitted keys having a constant bitting width. Examples: Mottura and Nova Acytra, among others.

3. Axial or push-lever locks with end-bitted keys. Examples: NS Fichet, Muel, Miller.

4. Radial lever locks having circular symmetry in the arrangement of their levers and fluted keys. Examples: Fichet-Bauche Monopole, Fichet-Bauche M2B, Cotterill "climax detector."

5. Cylindrical lever locks having either alternating sliding levers or balance levers stacked in a cylindrical core and operated by a double-bitted key. Examples: Chubb Ava, Kromer Protector, Kühne Panzer, CAWI.

6. Geared lever locks whose levers have serrated edges that mesh with toothed wheels containing a gate. The Fichet-Bauche 787 is an example of such a lock having an end-bitted key.

7. Trap-door locks: turning the key operates a trap-door at the rear of the cylinder. The key is then pushed through to the rear chamber to operate the lock mechanism. The Dény 3-lever lock with its double-bitted key is the only example presented here.

8. Twin-lever locks having two sets of opposing levers operated by a double-bitted key with independent bittings. Examples: Ross 700, Chubb-Lips 6K207.

9. Changeable lever locks with a variable bolt stump mechanism. Examples: Sargent & Greenleaf 6804, Mauer Variator.

10. Dual-control safe deposit box locks: requiring two keys and having two lever packs with either shared or different pivots and one or two bolt stumps, respectively. Examples include Mosler 5700 (single-stump), S&G 4400, and Diebold 175 (both double-stump).

5.2 Conventional

Chubb

(UK) 5–9 lever (3–4)

The Chubb lock is the original English lever lock, traceable to inventions by Barron and Chubb (late 18th century) that sought to make the warded lock more secure. Some of the early history of the Chubb Company was covered in the introduction to this chapter. We take up the story again here, drawing on material from Evans [34], Gunn [49], and the references therein.

In the first quarter of the 19th century, the Chubb Company was based in Wolverhampton, England. The safe business was established in London in 1837 following the development of a burglar-resistant safe, but moved to Wolverhampton in the early 1900s. The company diversified its lock and safe business to include fire protection after World War II. Chubb took over Hobbs Hart and Co. in 1956, Chatwood-Milner Ltd. (Liverpool) in 1959, then Josiah Parkes and Sons (Willenhall, makers of Union locks) in 1965 and Lips (Netherlands) in 1973. Chubb's Lock Security Group underwent a number of acquisitions through the 1980s and 1990s, being acquired by ASSA Abloy in 2000, which retained the security lock division while selling the safe-making division to Gunnebo. ASSA Abloy continues to market locks from Chubb, Yale, and Union.

Chubb produces a large range of lever and cylinder locks for residential, commercial, and high-security applications. In terms of architectural lever locks, the three principal types are: (1) the 5-lever mortice deadlock (3G114); (2) the 7-lever mortice deadlock (3G117, 3G227); and (3) the 5-detainer deadlock (3G110, 3G135). There are many variations on these basic types depending on the locking function, overall case dimensions, bolt throw, and bolt type. For example, the 3G114 is also produced as a sashlock—a two-bolt format with both deadlocking and latchbolt operation. We describe a 7-lever sashbolt (the 3K277) later in this section. All Chubb locks conform to British standard BS 3621 in terms of their resistance to sawing, forcing, drilling, and picking (see [78]). The quoted number of differs is 1,000 for the 3G114 and 6,000 for the 3G117.

The operating principle of a Chubb lever lock was explained pictorially in the chapter introduction. We give a more thorough coverage here, based on the Chubb 3G114 illustrated in Figs. 5.49–5.52. The lock is assumed to be viewed with the direction of bolt throw to the left and the cover plate on top. Lever positions are numbered from 1 to 5, starting from the front of the lock (bow to tip on the key).

The lock case (Fig. 5.51) is constructed from folded steel with studs for the attachment of the front cover plate and one stud, called a pivot, on which the

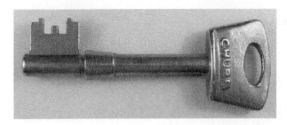

Figure 5.49: Key for Chubb 5-lever double-entry lock.

Figure 5.50: Chubb model 3G114 5-lever mortice deadlock.

Figure 5.51: Chubb 5-lever lock with bolt in locked position. Note antipicking notches near lever gates.

Figure 5.52: Two views of bolt stump passing through aligned gates of Chubb 5-lever lock.

levers are mounted. The thick steel forend is spot-welded onto the case and contains a rectangular hole for the bolt, which is made of cast brass with an integral stump. The bolt is slideably mounted on a small stud. The stump has a rectangular section with a V-notch facing the pivot. A steel hardplate protects the casing against drilling; this is particularly important in a lever lock, since one small hole suffices to align the levers. Furthermore, without drill protection, it is an easy matter to drill out either the stump or the lever pivot in order to neutralize the locking mechanism. In addition, the bolt has hardened roller inserts to prevent sawing.

The only other components are the levers, the curtain, and its locating spring. The curtain has a sleeve or barrel, forming the keyway, and a flange or skirt with a camming surface to contact the talon of the bolt (see also Fig. 5.42). The sleeve of the curtain is sandwiched between the rear of the lock case and the front cover, and can rotate freely through a full turn. The curtain restricts access to the interior of the lock, increasing the difficulty of manipulation and reading of the levers from the wear patterns on their bellies.

The key (Fig. 5.49) is a classic single-bitted type with a stop collar around the stem just in front of the bit. The bit is flat in section, although in high-security applications keyway wards may be applied to vary the key profile. The key has bittings to address each lever plus a bolt-step with a width roughly equal to two cuts. Cuts are often radiused to reduce wear at the contact surfaces of the levers. The bolt-step of the key drives the bolt indirectly via a cut-out in the base of the curtain, which is a more durable arrangement than having the key actually bearing on the bolt.

The levers are of stamped brass with a single leaf spring anchored in a slot near the pivot hole. Levers have a broadly rectangular cut-out with a forward and rear pocket, identical on all levers. The height of the gate between these two pockets is variable from one lever to the next (see Fig. 5.41). There is also a single antipicking notch above or below the outer edge of the gate (further from the pivot). The notch on the lever may engage with the notch in the stump if the levers are incorrectly lifted. During operation, there is a clearance of about 16 thousandths of an inch between the edges of the stump and the gates.

The lever bellies are not identical, differing in respect to their radiused outer portion. The inner edge of the belly, near the pivot, is linear. The function of the contouring around the belly curve (or "conning") is to reduce the thickness of the edge of the lever. This ensures reliable contact with the correct step in the key, preventing levers from being actuated by adjacent high steps. On each side, the belly ends in a horizontal edge. The construction is such that, for either direction of turning, the key bit first encounters the horizontal edge.

Consider first the operation of a lock containing only a single lever. For unlocking (clockwise from the front of the lock), the side of the key bit contacts the horizontal section of the lever belly nearly tangentially. As the key is turned, the key cut next encounters the radiused part of the belly (see Fig. 5.52). It is at this point that the talon of the bolt is engaged, advancing the stump toward the gate. If the key bitting is of the correct height, the gate recess will be in alignment with the bolt stump and the passage of the bolt will continue uninhibited as the key is turned. Since the radiused edge of the belly describes a circle centered on the key stem, the bit maintains the lever at a constant height, with its gate horizontal while the bolt stump is in sliding motion. By the time the key bit reaches the linear segment of the belly, the bolt stump has passed through the gate and further rotation of the key allows the lever to pivot downward under spring action to its rest position. The lever's motion is stopped by contact between the top of the rear pocket and the upper edge of the bolt stump.

On double-entry mortice locks that can be operated from either side of the door, like the one in Fig. 5.50, the key is symmetrically bitted with the outermost steps on the bit used for throwing the bolt. Since the bolt-step is roughly twice as wide as a regular bitting, a 5-lever Chubb mortice lock requires a key with seven cuts, whereas as 7-lever lock takes a key with nine or ten cuts. When the key is inserted from the outside, the first five bittings (in the 5-lever case) raise the five levers to align their gates at the correct height while cuts 6 and 7 drive the curtain, which in turn drives the bolt so that its stump passes through the aligned gates.

Conversely, when the key is inserted from the inside of the door, bittings 7 through to 3 work levers 1 to 5, in that order, with the step consisting of bittings 1 and 2 driving the curtain. The requirement for double-entry operation therefore means that only the first four cuts on the key are independent, with bittings 3 and 5 being equal. Bitting 4 always operates the fourth lever, regardless of which side the key is inserted from. Key symmetry substantially reduces the number of differs compared with a 5-lever lock with independent bitting, such as that used for a rim lock or safe lock requiring key access from only one side.

The original (pre-1950) model of the Chubb 3G114 5-lever lock had seven regular lever sizes, all with different gate heights, belly radiuses, and trailing edges. In addition, a monitor lever, similar to the one shown on the right side of Fig. 5.53, was sometimes used to block the keyhole and prevent removal of the key until the bolt was fully thrown. This is important since when the bolt-step of the key and the talon are very worn, it is possible for the bolt to be left in a half-open/half-closed

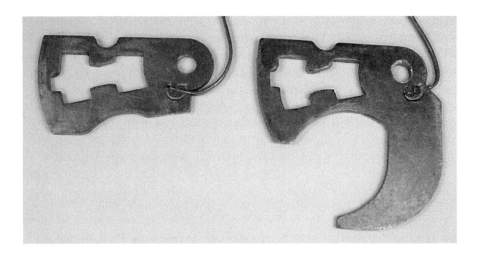

Figure 5.53: Levers from a Chubb detector lock: (left) regular lever; (right) monitor lever.

state. The monitor lever gives an early indication of this fault by partially blocking the keyhole.

The new (post–1988) 3G114 model has eight lever sizes, all of which have one antipicking notch above or below the gate (depending on the gate position). Thin bellies are used to lessen the problem of one key cut contacting two adjacent levers. The eight gate positions give eight possible depths of cut in the key bit. A high gate requires a deep cut in the key, a low gate requires a shallow cut. Said another way, a lever with a high gate (a low-lift lever) requires less lift than a lever with a low gate (a high-lift lever). Lever size numbering starts from 1, corresponding to the highest lift and the shallowest key cut. The depth increment in the key bittings is roughly 45 thousandths of an inch.

For symmetric operation, the bittings on the key must be of the form ABCDCBA, where each of A, B, C, and D is one of the eight possible cut depths. Since there are only four independent key bittings out of the seven, the theoretical total number of differs is $8^4 = 4,096$. (Note that for a 5-lever lock with one-sided operation, this would be $8^5 = 32,768$.) However, not all of these are usable: codes with repeated entries such as (1 1 1 2 1 1 1) are too easy to duplicate; codes like (1 8 8 1 8 8 1) excessively weaken the key bit.

In a practical key series it is necessary to impose some restrictions on the bitting codes. Typically, the following bitting rules are applied to the first four cuts (ABCD):

1. The code is excluded if three or more of the first four cuts are identical.

2. At least two of the first four cuts must be different.

3. At least one adjacent cut difference must be two or more.

The first constraint also results in there being at least two different depths in the first four cuts. In addition, a MACS constraint may be imposed so that codes like (1 8 1 8 1 8 1) are excluded. Note that the MACS constraint is not due to undercutting of adjacent bittings, as occurs in a pin-tumbler lock, so a very large or possibly unrestricted MACS may be acceptable.

The number of key codes subject to these constraints has been computed for different values of MACS in Table 5.1, which covers double-entry 5-lever locks with 7, 8, and 9 depths of cut, respectively. Program listings, contained in Appendix F, allow the user to print out the full list of usable codes. For the 3G114, which has eight depths of cut, a MACS of 3 gives only 1,446 usable differs, the first of which is (1 1 2 4 2 1 1) and the last of which is (7 7 6 4 6 7 7).

We have assumed that the MACS constraint is *symmetric*; in other words it is applied without regard to whether a peak or a trough is left in the key bit. In practice, it is quite acceptable to have a deep trough in the bit as long as it does not result in any single large peaks that excessively weaken the key. For instance, the code (1 2 1 7 1 2 1) has a deep cut in the middle of the bit, while (7 2 1 3 1 2 7) has deep cuts at either end of the bit. In order to generate codes such as these, an *asymmetric MACS* must be introduced. This means that in going from position 3 to position 4 (the midpoint in a symmetric 7-cut key), we allow a large or possibly unrestricted MACS, whereas a smaller MACS is applied in going the opposite way. We also need to ensure that there are no isolated "peaks," which is equivalent to saying that a deep cut in position 3 (say) is compensated by a shallow cut in position 1. One possible set of asymmetric MACS rules can be summarized as follows:

1. The height difference from position 1 to 2 satisfies the MACS.

2. The height difference from position 3 to 2 satisfies the MACS.

3. The height difference from position 4 to 3 satisfies the MACS.

MACS	7 depths	8 depths	9 depths
2	354	444	534
3	890	1,182	1,480
4	1,434	2,026	2,664
5	1,876	2,804	3,852
6	2,132	3,410	4,900
7	-	3,752	5,694
8	-	-	6,134

Table 5.1: Number of usable codes as a function of MACS for 5-lever double-entry Chubb-type locks with 7, 8, and 9 depths of cut (subject to constraints listed in text).

4. The sum of the cut depths at positions 1 and 3 does not exceed $L + 1$ where L is the number of cut depths (including zero).

Allowing the MACS to be asymmetric in this manner yields the figures listed in Table 5.2. It can be seen that the use of an asymmetric MACS gives more usable combinations than a symmetric MACS of 2, but the overall number of codes is less when the MACS is greater than 2. The number of differs quoted by Chubb for the 3G114 corresponds roughly to an asymmetric MACS of 3. For this value of MACS and with eight depths of cut, the code series runs from (1 1 2 4 2 1 1) to (8 4 1 8 1 4 8). Some examples of the bitting patterns from this series are displayed in Fig. 5.54. We stress that the code series generated according to the above rules is provided only as an example. The rules may be relaxed to provide an increased number of codes.

Master-keying of this type of lever lock is accomplished by widening the lever gates, with a commensurate decrease in security. A more secure and flexible method is the Butter's system, described later in this chapter.

The commercial Chubb lock is difficult to pick, especially when notched gatings are present, but, with practice, the task may be accomplished with a specially adapted 2-in-1 pick (called a curtain pick) that provides tension to the curtain while allowing manipulation of the levers. The sleeve of the curtain also prevents the key from skewing in the keyway, so, unlike wafer locks, jiggling a key with approximately the right cuts in a Chubb lever lock is much less effective. A difference of one depth increment in a given lever shows up as an overlap of one-third of the stump width—enough to engage the antipicking notches.

A vulnerability of the old model 3G114 lock is that the levers have different belly sizes and trailing edges, and these are in direct relation to the gate positions. Even though the keyhole is obstructed by the curtain, it is still possible to insert a reading tool to decode the levers and make a working key. In the newer model 3G114, levers with similar gate offsets have the same belly radius and trailing edge. Although it is impossible to decode the lock by inspection, using three belly groupings allows the code of the lock to be narrowed down to a workable number of keys that must

MACS	7 depths	8 depths	9 depths
2	544	855	1,254
3	759	1,173	1,703
4	963	1,491	2,153
5	1,127	1,776	2,589
6	1,220	1,997	2,967
7	-	2,120	3,253
8	-	-	3,410

Table 5.2: Number of usable codes for 5-lever symmetric Chubb-type locks with asymmetric MACS.

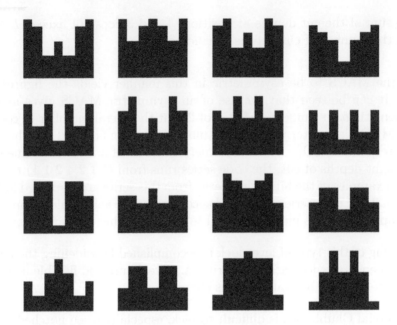

Figure 5.54: Examples of symmetric 5-lever key-bitting patterns generated with asymmetric MACS of 3 and 8 depths of cut. Code in top left is (1 2 6 4 6 2 1). Code in bottom right is (8 4 1 4 1 4 8).

be tried (similar to the progressioning of a pin-tumbler or wafer lock, described in Chapter 7).

A Chubb 3K277 7-lever double-entry mortice sashlock is shown in Figs. 5.55–5.58. The lock includes a latchbolt operated by a handle from both sides of the door as well as a key-operated deadlocking bolt with 2 cm throw. Other 7-lever Chubb mortice locks like the 3G117 and 3G227 have the same type of deadlock mechanism as the 3K277. The lock is also supplied with a roller bolt instead of a latchbolt (model 3K77).

Instead of discussing the operating principles of the 7-lever Chubb lock, which are the same as those of the 5-lever and should by now be familiar to the reader, we mention a number of minor differences in construction. Unlike Chubb 3G114 levers, the levers for the 3K277 (see Fig. 5.57) are open-ended with a single antipick notch. As before, the different-sized levers are arranged into a small number of belly groupings to protect against decoding by inspection. The belly is more streamlined with a curved leading edge to reduce marking of its surface due to wear at the initial point of contact with the key. The edge of the conning on the levers is straight instead of curved, which results in easier manufacturing.

The key for the 7-lever Chubb lock (Fig. 5.55) has 10 bitting positions and seven depths of cut. Since the lock is of the double-entry type, the key must be symmetric. This means that the 10 cuts must be of the form ABCDEEDCBA, where each of A, B, C, D, and E is one of the seven possible cut depths. The symmetry constraint

Figure 5.55: Symmetric single-bitted key for Chubb 3K277 7-lever lock.

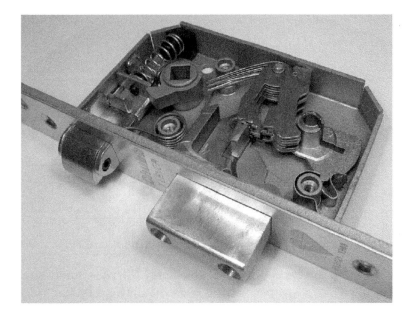

Figure 5.56: Chubb 3K277 7-lever double-entry sashlock with front cover removed.

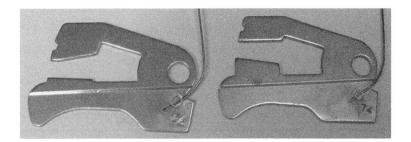

Figure 5.57: Number 1 (left) and number 4 (right) open-ended levers from Chubb 3K277 lock.

results in only five independent cuts. When the key is inserted from the front of the lock, the first seven bittings raise the levers while the last three cuts are covered by the rear of the curtain. For operation from the back, it is the last seven cuts that contact the levers. Table 5.3 gives the number of key codes subject to the bitting rules

Figure 5.58: Chubb 3K277 7-lever sashlock being operated by correct key.

MACS	7 depths	with rules	8 depths	with rules
2	2,363	1,434	2,986	1,852
3	6,083	4,454	8,300	6,214
4	10,483	8,172	15,500	12,370
5	14,407	11,554	22,914	18,844
6	16,807	13,644	29,114	24,322
7	-	-	32,768	27,572

Table 5.3: Number of usable codes for double-entry 7-lever Chubb-type locks with 7 and 8 depths of cut with symmetric MACS constraint and bitting rules taken into account.

in the symmetric MACS case described previously for the Chubb 3G114. Results for both seven and eight depths of cut have been provided. Note that applying an asymmetric MACS would generally reduce the numbers shown in the table. We stress that for a lever lock the MACS is not an essential constraint—large values of adjacent cut difference are quite acceptable in practice.

The 7-lever Chubb lock is highly resistant to manipulation with conventional lever lock-picks. However, a curtain pick could be applied first to determine the likely cuts on the key and second to assemble a make-up key on this basis. Alternatively, a pin and cam tool could be applied to decode and pick the lock: this tool functions like a Sputnik decoder for a pin-tumbler lock with adjustable pin heights on a lever-key bit [122].

An even more secure system is used in the Chubb 6K75 8-lever safe lock illustrated in Figs. 5.59–5.61 (see also Figs. 5.43 and 5.44). The system, which is in effect a single-bitted version of a German lever lock, derives from an earlier double-bitted key lock produced by Chatwood called the Impregnable [43]. The Chubb model 6K75

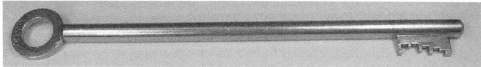

Figure 5.59: (Top) Chubb 6K75 8-lever safe lock with cover removed. (Bottom) Chubb 6K75 key (stem length $5\frac{1}{2}''$).

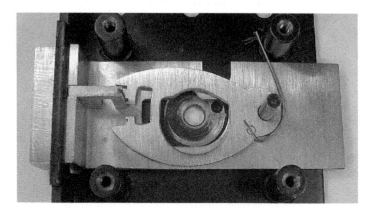

Figure 5.60: Curtain and all but one lever removed to show action of antipicking notches on lever and stump.

employs levers with identical belly cut-outs and peripheral gates with antipicking notches. The stump has a groove on its edge to match the notches in the levers (see Fig. 5.60). The lock includes a curtain with a locating notch, which engages a curtain lever at 12 o'clock, as shown in Fig. 5.59.

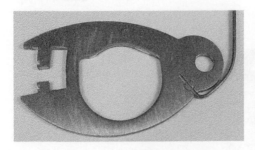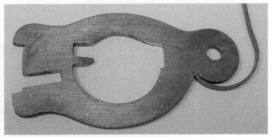

Figure 5.61: Chubb 6K75 levers. (Left) Regular lever with antipicking notches. (Right) Monitor lever.

A monitor lever may also be installed (see Fig. 5.61). This lever has a slotted gate with no rear pocket and a stop lug situated just beneath the lower lip of the keyway. When the key is inserted with its bit in the 9 o'clock position and turned clockwise, it raises the monitor lever to the correct height, allowing the bolt stump to enter its gate while at the same time positioning the stop lug just above the top lip of the keyway. On completion of a full 360 degree turn, the key may be removed from the lock, leaving it in the open position. On reinsertion, the monitor lever prevents the key from being turned further in the unlocking direction by contact of the key bit against the stop lug. The lock could potentially be decoded by minute examination (e.g., with a borescope) of the wear pattern left by the key bittings on the bellies of the levers, although this would require a means of accurately measuring the markings.

As one might expect of one of the world's oldest surviving lock companies, Chubb produces many other types of lever locks, including models with nine or more levers as well as locks for safe deposit boxes (see section on dual-control locks). A further example is the Chubb-Lips 6K207—a VdS Class 2 rated twin-lever lock with two packs of seven levers and two stumps. A picture of this lock is given in Fig. 5.26 in the chapter introduction. The advantage of having more levers is additional keying combinations as well as enhanced resistance to picking and impressioning. Other locks from Chubb include the Biaxial, which is equivalent to the Medeco Biaxial; a magnetic-tumbler lock called the 3G222, described briefly in Chapter 6; and an electronic cylinder lock called Eloctro that has a rectangular keyway. The Eloctro employs inductive code transmission so that no electrical contacts are required. The idea is further described in UK patents 2,252,356 (1992) and 2,273,128 (1994).

Ross 102

(AU) 6-lever (3–4)

The Australian company K. J. Ross Security Locks specializes in the manufacture of lever locks for safes and vaults. The range of Ross locks includes models 102, 600, and 700, all of which are described in this chapter. As the Ross 102 lock differs only marginally from the 6-lever Ross 100 lock in the chapter introduction, we cover it only briefly here.

The Ross 102 lever lock, pictured in Figs. 5.62–5.64, has a folded steel body and cast brass bolt. The lock is designed for rim mounting onto the rear face of a safe door. The bolt is slideably supported by a rectangular cut-out in the forend of the lock and by a stud in the case. A pressed brass stump is attached to the bolt. The lock is produced in left- and right-handed versions, depending on the opening direction of the safe door. The lock accepts a single-bitted steel key with six cuts and a bolt-step at the tip end of the bit. Since the lock is designed for one-sided operation, there is no symmetry requirement on the key bittings. The number of components has been kept to a minimum so as to reduce manufacturing costs.

Figure 5.62: Ross 102 6-lever lock and single-bitted key.

Figure 5.63: Ross 102 with cover removed showing guard underneath lever pack.

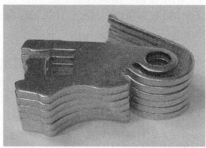

Figure 5.64: Set of six Ross 102 levers arranged in order. (Left) Gate positions. (Right) Belly curves, differing on trailing edge.

Unlike Chubb locks that use a separate leaf spring, levers for the Ross 102 have an integral spring that is stamped from a single piece of brass, which facilitates assembly. There are also a number of functional differences in the design of the levers. First, levers have only a single (inner) pocket, with the remnant of the outer pocket serving to stop the levers against the bolt stump in the locked position. The bolt stump has no antipicking notches, but the gates in the levers do contain this feature. The levers differ in respect of the vertical position of the gate and also in the radius of the belly. The differences in the lever bellies are most noticeable at the trailing edge of the levers, near their pivot points.

An interesting feature is the presence of a knuckle between the straight, leading edge and the curved radius of each lever (see Fig. 5.64). The reason for the knuckle can be appreciated by reference to the Chubb 3G114 lock, in which there is no knuckle. Normally, during unlocking, the first point of contact of the key bit with the lever occurs on the leading or outer edge of the lever belly: cuts of different radiuses contact this edge at different points, leaving marks that can be used to decode the lock. In the Ross 102, the first point of contact is the knuckle. The flat edge of the key bit bears on the knuckle up to a certain angle of rotation, at which point there is contact between the key cut and the knuckle. Now although different cuts contact the knuckle at slightly different points, the differences are minute since they are mapped onto the tightly curved surface of the knuckle. This makes it virtually impossible to decode the lock via the wear pattern left by the key.

The reader will have noticed the presence of the guard plate situated on the lower edge of the bolt tail near the center of the keyhole (Fig. 5.63). The function of the guard plate is threefold. First, it operates in a similar manner to the curtain in a Chubb lock, restricting access to the levers in the case of a manipulation attempt. Second, it prevents the impressioning of the trailing edges of the levers since the key cannot be turned the wrong way to raise them. Lastly, it covers the trailing edges of the levers when the bolt is in the locked position, hampering attempts to read the wear pattern left at the inner edge of the bellies.

With six lever sizes, corresponding to different depths of cut on the key, a 6-lever Ross 102 lock has a maxmimum of 6^6 or 46,656 differs. Naturally, this is reduced by practical keying constraints. Although there is no explicit drill protection, the inclusion of a steel bolt stump and a shoulder on the bolt that bears on the inner edge of the forend considerably inceases the resistance of the lock to forcing of the bolt. A sintered steel insert can also be added to the front of the lock to protect the keyhole.

Ross 600

(AU) 6-lever (4)

The Ross 600 series, pictured in Figs. 5.65–5.67 was introduced around 1991. It contains a number of innovative features, the most important of which is the use of

Figure 5.65: Ross 600 series 6-lever safe lock and key.

Figure 5.66: Ross 600 with cover removed to show vertical-lift levers.

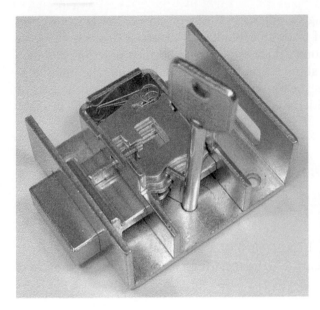

Figure 5.67: (Left) Ross 600 operated by correct key: center step of key drives bolt. (Right) Levers are identical except for gate position.

vertical-lift levers. The design was patented in Australia in 1993 by K. J., G. H., and B. A. Ross (AU 641,024B); a US patent was filed in 1995 (US 5,560,234). As in the Ross 102, the lock has six levers and takes a single-bitted, nonsymmetric key. Unlike Chubb locks, there is no curtain. The number of components has been kept to a minimum so as to reduce manufacturing costs.

The lock is constructed with a heavy-gauge folded steel case and moving parts made of brass, except for the bolt stump, which is made of steel. Rather than the conventional pivoting type, the levers have straight forward and aft sides and are slideably mounted inside a compartment formed by a U-shaped piece of folded steel; the side-walls of the compartment guide the levers as they are lifted. The leaf springs bear on the top edge of the compartment, biasing the levers toward the keyhole. Since there is no pivot point, the vertical-lift lever mechanism is naturally more drill resistant. This is further reinforced by the inclusion of ball bearings in the mounting posts of the lock case.

The stamped brass levers (see Fig. 5.66) have a single pocket and, due to the vertical lift design, have an identical outline. In particular, the belly profile is identical from one lever to the next. The use of identical lever bellies prevents decoding and impressioning of the lock using conventional techniques. The lever bellies are contoured away from their radiused central portion on both sides. The contouring is designed to minimize the wear pattern left on the lever bellies by the key, making it more difficult to determine the key bittings from visual inspection or impressioning. The only differentiating feature in the levers is the position of the gate. The lever gates have serrated edges: below the gate on low-lift levers, above the gate on high-lift levers, and both above and below on mid-lift levers. The serrations can engage a notch in the bolt stump (not shown), greatly increasing the difficulty of manipulating the lock.

The six levers are disposed in two groups of three: one below the bolt and one above. Counting eight positions along the key bit, the levers are operated by the cuts at positions 1–3 and 6–8, with positions 4 and 5 reserved for the bolt-step (a zero-depth cut). There are 10 cut depths and correspondingly 10 different gate positions. The approximate depth increment in the cuts is 0.75 mm. Cuts are numbered from 0 (no cut or maximum lift) to 9 (deepest cut or minimum lift). The theoretical maximum number of key combinations is $10^6 = 1,000,000$. Typical keying constraints allow for a MACS of 4 between adjacent lever cuts and 8 adjacent to the bolt-step.

From a lockpicking perspective there are a number of challenges. These stem from the serrated edges of the lever gates and the fact that tension must be applied at a position in between the two packs of three levers. A normal 2-in-1 pick cannot be applied since this would only give access to either the first or last three levers, depending on the tensioning arrangement. A number of companies have developed tools for decoding/picking the Ross 600 lock. The quoted time for nondestructive opening of the lock using the tool from Prescott's site [97] is 20 to 40 minutes.

Detector Lock

(UK) 7-lever (4)

The original Chubb detector lever lock was patented by Charles and Jeremiah Chubb in 1818. The essence of the idea, which is highly effective, is to block the operation of the lock in the event of a manipulation attempt by picking or use of incorrect keys. Whereas previous designs had required the lock to be dismantled in order to reset it for normal operation, the detector lever lock could be reset by inserting a regulating key and turning it the wrong way (i.e., counterclockwise in a lock with the bolt on the left).

In this section we present a more contemporary version of the Chubb detector lock produced by Tann in the 1960s. John Tann, a traditional British safe-making company, was established in 1795. It became part of Rosengrens AB in 1990 and is now owned by the Swedish Gunnebo Group of companies that also acquired Chubb's safe division in August 2000. The vault doors on the Tower of London that protect the Crown Jewels were produced by Rosengrens-Tann. Two double-bitted keys are required to operate the locks on these doors.

The Tann detector lock shown in Figs. 5.68–5.72 has seven levers and is designed for single-sided operation by a single-bitted key. The lock, including its case, is made almost exclusively from brass components. The bolt has a narrow stump and

Figure 5.68: Tann 7-lever detector lock (with cover removed) and key.

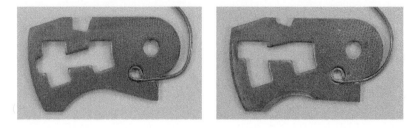

Figure 5.69: Detector lock levers have constant-width section at top of gate.

two talons: one for normal operation and one for resetting the detector lever. Our description assumes left-handed operation, namely, clockwise turning of the key to open with the bolt on the left and the keyhole below the bolt. Numbering of levers proceeds from front to back.

The pack of seven levers comprises three different lever types. The front lever, in position 1, is a regular Chubb-type lever with a notch in the belly to accommodate a lug on the periphery of the curtain (see Fig. 5.68). This serves to provide positive location for the curtain at the point where its slot aligns with the keyhole, permitting the key to be inserted or withdrawn.

The next five levers in positions 2–6 resemble Chubb levers but also contain two additional notches (see Fig. 5.69). One notch is positioned opposite the gate in the left-hand edge of the outer pocket. We will refer to this notch as the resetting notch. The second notch is positioned in the top edge of the lever above the gate and will be referred to as the detector notch. Levers with the same gate cut may have

Figure 5.70: (Left) Overlifting of lever triggers detector latch. (Right) Bolt stump stopped below gate of detector lever.

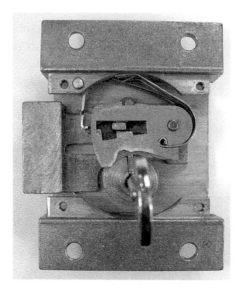
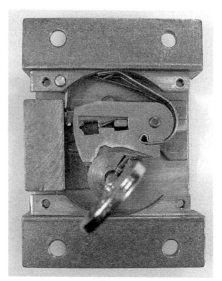

Figure 5.71: Levers raised to correct height by key for passage of bolt stump.

differing belly sizes, as in (new model) Chubb 3G114 locks. This makes it impossible to determine the key cuts by reading the bellies of the levers.

The lever in position 7 is the detector lever, differing in a number of respects from a standard Chubb lever (see Fig. 5.70). First, the detector lever possesses an arm or stump anchored in its top edge at the gate radius. The lever has a regular outer

Figure 5.72: Turning the key the wrong way resets the detector lever.

pocket with an enlarged-width portion in its left edge. The inner pocket has been replaced by a slot at the height of the gate. Finally, the outer edge of the detector lever, furthest from the pivot point, has an overhang or nose. A detector latch, consisting of a steel hook pivoted from above, is also present. The detector latch is spring-biased so that its hooked end is in light contact with the outer edge of the detector lever.

In the locked position, with the levers at rest as in Fig. 5.68, the arm of the detector lever traverses levers 2 to 5, stopping just short of the first lever. The notches in levers 2–5 register with the detector lever arm, but there is a small gap that allows a limited amount of lift for each lever. The detector notch of each lever is adjusted to maintain a constant distance from the notch to the top of the gate, regardless of the height of the gate. This ensures that a lever of any size can be raised by the same amount before its detector notch first contacts the detector arm. The top lever, which also locates the curtain, functions independently of the detector lever.

In normal operation, the key is inserted and turned clockwise to unlock the lock. The key must correctly address the first six levers in order to register their gates with the bolt stump. In addition, the key must have a seventh cut to raise the detector lever so that its gate channel is aligned with the bolt stump. At this point the detector notches on levers 2–6 are also aligned slightly below the detector arm. The bolt-step of the key then drives the stump through the lever gates by contact with the left-hand talon in the bolt. The first six levers are used to combinate the lock, although the detector lever can also be used for differing. The function of the slot in the detector lever is to ensure that, while in the unlocked position, the detector lever is held at constant height by the bolt stump and cannot therefore be overlifted. This ensures that the dectector lever cannot be accidentally triggered when the lock is unlocked.

Consider what happens if one or more of the regular levers 2–6 is overlifted, as might occur in a picking attempt or if an incorrectly bitted key were tried. As a regular lever is lifted, a point is reached where its detector notch contacts the detector lever arm. Further lifting of the lever then also raises the detector lever by mechanical coupling through the arm. If the lever in question is lifted by more than about 5 mm, the nose of the detector lever reaches the hook end of the detector latch. This triggers the detector latch, holding the detector lever in an overraised position and effectively blocks passage of the bolt stump. In addition, the detector lever may itself be overlifted to trigger the latch. It should be noted that a small amount of overlifting is tolerated before the detector lever is triggered.

Once the detector is triggered, the bolt cannot be displaced to the right by manipulation of the levers, even by the correct key. Instead, the bolt stump must be displaced in the opposite direction to disengage the detector latch (see Fig. 5.72). The correct key is required to achieve this outcome, as we explain next.

With the lock in the locked position and the detector mechanism triggered, the correct key is inserted and turned counterclockwise. The bolt-step of the key now contacts the right-hand talon of the bolt, pushing it to the left. Since the key has the correct bittings, it raises all the levers to the correct heights. This action aligns the gates and also the resetting notches on levers 1–6 (the detector lever has an enlarged outer pocket that accommodates the stump even in the overraised position). Turning the key further displaces the bolt stump to the left to engage the resetting notches. As the bolt is moved incrementally to the left, a step in its top-edge contacts the detector latch and moves it out of engagement with the nose of the detector lever. The key is then released, allowing the detector lever to move back under spring action to its rest position. The lock is now reset and can be operated in the normal manner by the correct key.

We described in the introduction to the chapter how in 1851 Alfred C. Hobbs picked a Chubb detector lock without triggering the detector lever. In practice, this requires a high degree of skill. For the Tann 7-lever lock covered in this section the task is classed as very difficult but not impossible since there are no false gates or serrations. Directly picking the lock open is difficult; however, it is possible to exploit the resetting mechanism to decode the lock, as we now explain.

The detector lever will only operate if it or the levers in positions 2–6 are raised fractionally higher than they would be by the shallowest cut on the key. Furthermore, even when the detector lever is engaged, it is relatively easy to disengage it by picking the lock in the reverse direction (counterclockwise). This is facilitated by the fact that the edges of the outer pockets on the levers are sloped. Once picked in the reverse direction, with the levers held in the correct alignment by the bolt stump, the lever bellies could in principle be read or decoded to produce a working key.

Chubb Butter's System

(UK) 5-lever (4)

A Chubb "Castle" 3G110 five-lever mortice deadlock employing the Butter's system is pictured in Figs. 5.73–5.76. The lock case is larger than a conventional 3G114 5-lever mortice lock. However, the increased space requirement is more

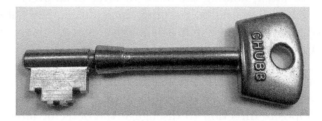

Figure 5.73: Symmetric single-bitted key for Chubb 5-detainer double-entry lock.

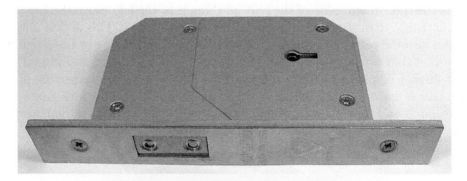

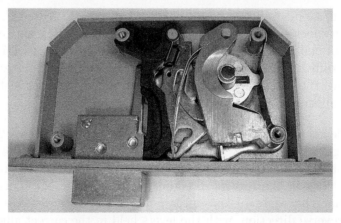

Figure 5.74: Chubb 3G110 mortice lock employing the Butter's system with cover on/off.

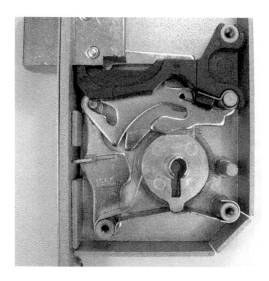
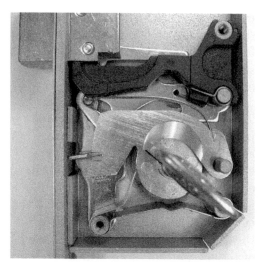

Figure 5.75: (Left) Chubb 3G110 with levers removed to show bolt and compound linkage. (Right) Stage 1 of opening sequence: key lifts lever to align gate with stump.

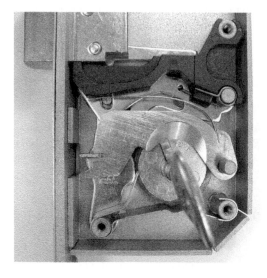

Figure 5.76: Stages 2 and 3 of opening: (left) key step causes stump to engage lever gate; (right) bolt retracted and stump disengaged.

than compensated by the lock's ability to be master-keyed while retaining a high degree of security against manipulation. The unit can also be supplied with a micro-switch for connection to an alarm system. In the Butter's system, all levers or detainers, as they are referred to in the Union and Chubb literature, have the same overall dimensions, including the same belly size. The Butter's system can be implemented with either pivoting or vertical-lift levers. In the case of the 3G110, the levers are pivoting and differ from conventional Chubb levers by having a much larger end opposite the pivot point. The gate has no pockets, and the usual bolt stump is replaced by a flange mounted on one member of a two-part linkage or crank.

The linkage consists of two arms, one horizontal and one vertical, each of which is pivoted at one end on a stud in the casing (see Fig. 5.75). The arms meet roughly at right angles at a sliding coupling. (In older models of the lock, the linkage is comprised of a single right-angled elbow pivoting at its corner with a protrusion on one arm forming the stump [21].) The horizontal arm that runs parallel to the bolt is shaped roughly like a standard lever and is raised at its belly by a cam on the curtain (in line with the keyhole) when the key turns between 11 and 1 o'clock. In this range of rotation, the cam also engages the talon of a slider, located at the rear of the mechanism, upon which the bolt is mounted.

As the cam raises the first part of the linkage, the movement is transferred through the angled slot of the sliding coupling, causing the vertical part of the linkage to present the flange to the active edge of the lever pack (see Fig. 5.76). Only the correct key will align all five of the lever gates at the height of the flange, permitting the key to complete its circle and withdraw the bolt. As the key turns past 1 o'clock in the clockwise direction, the linkage is returned to its rest position, the flange is disengaged from the lever gates, and the levers return under spring tension to rest against the curtain. The deadlocking of the bolt is achieved by a second stump on the horizontal linkage that travels between two pockets in the bolt stump. The flange on the vertical linkage is only presented to the levers during locking or unlocking when the horizontal linkage is raised.

The lock is designed for two-sided operation, taking a key with nine cut positions, as in Fig. 5.73. The bittings on the key must be of the form ABCDEDCBA, where each of A, B, C, D and E is one of the nine possible cut depths. Each cut depth corresponds to one of the gate positions along the edge of each lever. The gate width is about 3.5 mm with a 50 percent overlap between adjacent gates. Since there are only five independent key bittings out of the nine cut positions, the theoretical number of differs is $9^5 = 59,049$. The quoted number of usable differs is 25,000, which takes into account practical keying constraints such as the elimination of repeated adjacent cuts. A higher security version of the 3G110 is also available, called the 3G135, that has three different key blank and curtain profiles.

Since each of the five levers can contain multiple gates, the 3G110 lock provides a considerable amount of flexibility for master-keying compared with a conventional Chubb 3G114 lever lock, while retaining a good measure of security. Security is further enhanced by endowing the lever edges with indentations or serrations. The positioning of these false gates on either side of the true gate can be used to advantage in manipulating the lock, although this is by no means a trivial task.

Fichet-Bauche Sans Souci

(FR) 7-lever (4)

In a Chubb or conventional lever lock, the key bittings must be symmetric if the key is to operate the lock from both sides of the door. As explained in the Chubb lock section, the symmetry constraint drastically reduces the number of available differs. The French company Fichet-Bauche has circumvented this problem by designing a key with two separate bittings, that is, a twin-bitted key (see Fig. 5.77). The bittings are mirror images, so that the key can still be used from either side of the lock. The advantage of this construction is that the bittings on each of the two key bits are totally independent, each giving the full number of 7-lever differs (several million assuming nine cut depths). On the other hand, the key is more than 12 cm long, which is not a practical size for most people's pockets.

The Fichet-Bauche Sans Souci lock (Figs. 5.77 and 5.78) is rim-mounted onto the back of the door and has a two-turn (double-locking) operation with linkages for top and bottom bolts. The case of the lock is 22 cm long and 10 cm wide. The mechanism (Fig. 5.79) includes a single 7-lever lock operated by either one of the

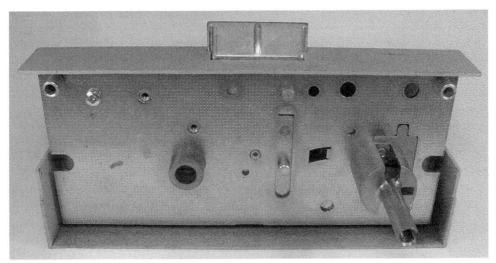

Figure 5.77: Fichet-Bauche Sans Souci 7-lever double-locking deadlatch and twin-bitted key.

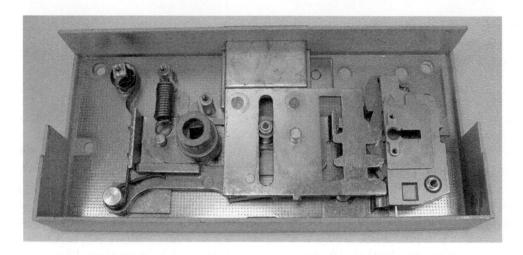

Figure 5.78: Fichet-Bauche Sans Souci lock with cover plate removed.

Figure 5.79: Fichet-Bauche Sans Souci mechanism. (Left) Key inserted in key guide. (Right): Key raises detainers to allow stump to enter gates.

two key bits, depending from which side of the door the key is inserted. For insertion from the outside, the bit nearest the tip of the key operates the lock. Conversely, it is the bit nearest the shoulder of the key that operates the lock when inserted from the inside. In both cases, the redundant bit of the key rotates in a dummy chamber mounted on the inside face of the lock. Each key bit has nine cut positions, the outer two of which are redundant.

The levers, which closely resemble the detainers in a Chubb Butter's system lock, are gated on the edge furthest from their pivot, and false gates are also included to make decoding and picking difficult. The bolt stump is mounted on a hinge that is raised by the key and drives a flat bar that enters the lever gates when all seven are aligned (see Fig. 5.79). The bolt is normally deadlocked by the stump and is driven by the bolt-step of the operating key bit as the stump is raised. Given the heavy-duty construction and level of security provided by this lock, it is no wonder that Fichet has chosen to call it Sans Souci, which is French for "without worry."

5.3   Double-Throw

Mottura

(IT) 6–12 lever (3–4)

Lever locks by Mottura, Multifort, Ezcurra, Elzett, and other European and South American companies utilize the system of Italian levers, usually containing between 4 and 12 of this type of tumbler. A 6-lever quadruple-throw Mottura lock (model 52571 DM) appears in Figs. 5.80–5.82, while a 6-lever double-throw lock from Nova is shown in Figs. 5.83 and 5.84. The levers are slideably mounted on pegs or stumps fixed in the lock case. Each lever contains a system of gates and three pockets similar to those in a 2-lever Barron lock. Double-locking (quadruple-throw) versions have four gates and five pockets. The bolt has two or more talons as well as a cut-out with a single active surface or belly. The levers are spring-biased toward the keyway. In a lock oriented so that its bolt is thrown to the left, the levers are sprung from below if the keyway is situated above the gates and from above if the keyway is below the gates.

Figure 5.80: Mottura 6-lever quadruple-throw lock and double-bitted key. Actual case dimensions: 206 mm × 136 mm.

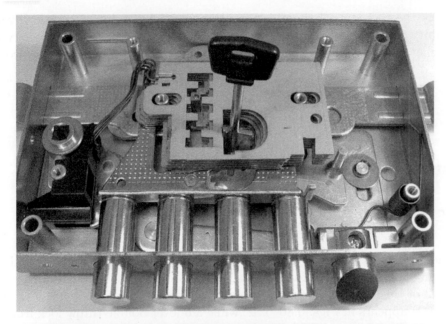

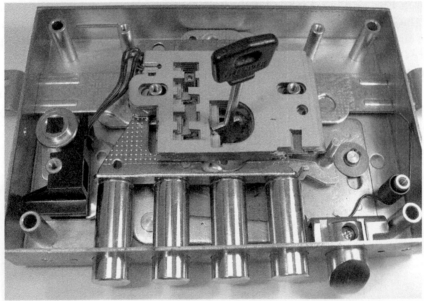

Figure 5.81: (Top) Operation of Mottura 6-lever lock as stump passes through third gate. (Bottom) Stump passing through fourth gate.

Imagine a vertically oriented lever with the bolt thrown to the left in the locked position, such as in Fig. 5.84. The stump of the bolt in this position will be in the top left pocket of the upward-acting levers. The asymmetric, double-bitted key is inserted in the correct orientation and turned clockwise in order to unlock the lock. As the key is turned through its first half-turn, only one-half of each bit is in contact with the belly of the lever. The movement is adjusted so that the bolt-step of the

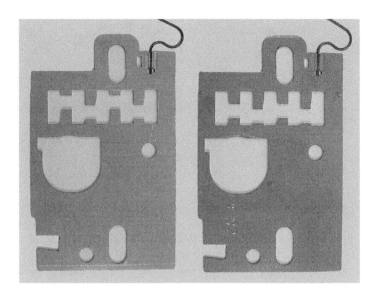

Figure 5.82: Two different types of Mottura levers: (left) lever with complementary gates; (right) lever with arbitrary odd and even gates.

Figure 5.83: Nova Acytra 6-lever double-throw lock and key.

key engages the right talon of the bolt at the same time that the lever has been raised sufficiently to align the stump with the left-hand gate. As the key completes its first half-turn, the bolt stump moves through the left gate and into the middle pocket.

Figure 5.84: Nova Acytra in locked position (left) and halfway to unlocked position (right).

The key's work is not yet done and the keyway is designed so that the key cannot be withdrawn until a full turn has been completed. The second half-turn brings the other half of the same key bitting into contact with the lever belly or cut-out. The other half-bitting now raises the lever sufficiently to allow passage of the bolt stump into the right pocket as the left talon of the bolt is engaged. As the key completes its full turn, the lever springs down under tension to lock the stump in the right pocket. The same description applies to quadruple-throw locks like the Mottura with the understanding that the key must complete two full turns (see Fig. 5.81). Naturally, all other bittings on the key must be of the correct depths to operate the other levers in synchrony.

In mortice locks designed to be operated from either side of the door, the center step of the key bit operates the bolt, and the sequence of bittings is reflected on either side of the key. For example, in a 6-lever lock like the Nova Acytra in Fig. 5.84, having three levers on either side of the bolt, if the left-hand side bittings from bow to tip were (2 3 5 0 1 4 4) (with the "0" bitting operating the bolt), the right-side bittings would be (4 4 1 0 5 3 2).

The odd-numbered gates are usually complementary to the even-numbered ones, as shown on the left side of Fig. 5.82. Thus, if the first gate is high, the second is low, and so on. If this is the case for all the levers in the lock, then the key has constant-width bittings, except for the bolt-step. This is not a necessary keying constraint, however, and it is common for manufacturers to include levers with arbitrary odd and even gate positions. In this case, illustrated on the right side of Fig. 5.82, the cut depths of the corresponding key steps are independent from one side of the bit to the other.

The system provides a reasonably good level of security while being easy to mass produce since the components can be made from stamped steel and brass. From a manipulation perspective, a standard 2-in-1 pick is not effective when the bolt talon, to which tension must be applied, is set in between the two lever packs. Instead, a pick with three coaxial sections (or 3-in-1 pick) with tension applied through the middle section, would be needed. Furthermore, the lock must be picked once per set of gates (i.e., twice in a single-turn lock or four times in a double-turn lock). The lever bellies are all cut to the same size, and the wear patterns on the lever bellies result from the passage of two cuts of different depths. These last two facts make it impossible to decode the lock by visual inspection.

Locks of this type are frequently used in safes (many different Italian safe lock brands have this mechanism [111]). There is plenty of room in such a lever lock for security features such as drill-resistant plate, notched levers, and bolt stumps. Microswitches can also be included to signal operation or to detect overlifting. For example, an Italian lever lock design from 1970 (UK 1,374,288) features a detector lever that operates in a similar manner to a Chubb detector lock. A 1998 patent by CISA (EP 0,903,455) incorporates Italian levers with antipicking notches. A key-changeable version of an Italian lever lock is presented in a 2003 patent by Mottura (EP 1,375,790).

5.4   Axial

Miller

(US) 6-lever (3–4)

The Miller Lock Company was founded in 1870 in Philadelphia and continued until the Great Depression of 1930 [50]. Miller's round-bodied or "pancake" push-key padlock, shown in Figs. 5.85 and 5.86, incorporated a compact 6-lever mechanism.

Figure 5.85: Miller 6-lever "Champion" padlock and end-bitted push-key.

Figure 5.86: Miller padlock with cover removed in locked and unlocked positions.

The design for the Miller lock was registered in England by B. Hunt in 1873 (UK patent 2793). Numerous embodiments were cited, including one with twin sets of opposing levers operated by a key with two rows of cuts on its tip. A number of patents were filed by other U.S. inventors for push-key padlocks similar to the Miller lock, in particular by F. Egge in 1878 (US 207,407), J. Loch in 1879 (US 228,656), and M. Jackson in 1886 (US 340,319). Pancake 6-lever padlocks were also produced by Walsall in England around 1916 and later by Union, continuing until the early 1960s [58].

The principle of the Miller lock is similar to but predates the Chubb Butter's system by about 80 years. It was also a much more secure mechanism than the later 6–8 "lever" flat-bodied padlocks made in the early to mid-1900s by Corbin, Fraim, Union, and other companies, in which the shackle was retained only by a set of opposing sprung hooks operated by a double-bitted pipe key (see Fig. 5.87). The Padlock Collector catalogue [1] lists scores of padlocks similar to the Miller, all with the same mechanism but having different brands and company names embossed onto the front face. Although locks of this kind are now more likely to be found at an antique market than in actual use, they provide a convenient example of the "end-bitted key" lever lock principle. The 1897 edition of the Sears, Roebuck & Company catalogue from Chicago lists the Champion 6-lever padlock at 50 cents, so it was clearly a household item.

The lock body is made of cast bronze with brass components inside. The key is flat with cuts in the end of the blade. The lock houses between six and eight pivoting levers of equal size and triangular shape. Their flat bellies rest directly on the end of the keyway slot. Each lever has a gate in its periphery, the position of which varies according to the combination assigned to the lock. The levers are biased toward the keyway by springs. One of the levers, which we refer to as the *actuator*, has a

Figure 5.87: British-made "6-lever" padlock: shackle retained by three hooked levers on each side.

slightly longer spring. The springs of all the levers except the actuator rest against a post at 3 o'clock in the case.

What is interesting about this mechanism is that the key never touches the bolt or the bolt stump. Instead, there is an L-shaped linkage or crank (see Fig. 5.86) that pivots on a post at a distance from the top left edge of the levers. One arm of the crank lies above the top edge of the lever pack. The other arm of the crank is hooked to provide a stump to match the lever gates. In the locked position the stump is pushed against the periphery of the levers by the spring of the actuator lever. The crank is at such an angle that its edge impinges on an indentation in the shackle, deadlocking it.

A correctly bitted key is required to align the gates of the levers as it pushes the entire pack into the lock cavity against the action of the lever springs. At a certain depth of insertion, the hooked portion of the crank enters the aligned gates of the levers. Once this happens, the crank no longer blocks the shackle and it moves to the open position under spring action. The shackle is stopped when a hook on its bottom edge engages a reciprocal hook on the left side of the case.

The clever aspect of this lock is that the actuator lever is the only one through which force can be applied to the stump of the crank. It is not known beforehand which of the levers is the actuator: it could be any one of them. Furthermore, the force on the crank is caused by only one lever spring, which is considerably less than the force required to maintain the other five levers at the correct depths to align their gates. Thus, all levers must be moved to the correct depths at once, confounding a would-be lockpicker.

Resistance to picking may not be of primary relevance in a 50 cent brass padlock, but it becomes crucial when the mechanism is located behind several inches of steel on a safe door. The principle is so effective that European safe manufacturers such as Fichet in France used it for many years.

NS Fichet

(FR) 7-lever (4)

The NS Fichet "pompe" lock (Figs. 5.88–5.90) is a push-key operated safe lock made by the Fichet Company in France from around 1920. It can be thought of as a type of side-bar cylinder lock with links to the Miller lock, except that the levers, or rockers as we shall refer to them here, are isolated from the key by a set of push-rods. The lock is mentioned in a 1928 patent (FR 654,495) in connection with an antitamper feature that allowed an incorrect key to turn the mechanism freely without deploying the clutch. The original design dates from 1884 [7, 41]. The reader is referred to US patent 4,187,705 (1978) for a modern embodiment of this lock (called the NS2i).

The end-bitted key (Fig. 5.88) is butt-joined to the stem, but slightly off-center so that it can only be inserted one way round. The key has bittings up to a maximum

Figure 5.88: Two views of end-bitted NS Fichet key.

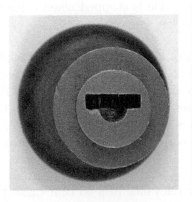

Figure 5.89: Front and rear views of NS Fichet safe lock.

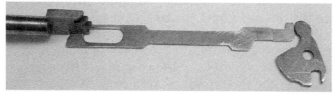

Figure 5.90: (Top) NS Fichet core and barrel assembly with crank engaging rocker gates. (Bottom) Operation of a single push-rod and rocker.

depth of about 5 mm that alternate in angle. The use of forward- and reverse-angled cuts is a copy-protection feature. The lock mechanism is mounted inside a domed housing made of 5 mm thick steel with a long barrel that extends to the front of the safe to form a fixed protective cover (see Fig. 5.89). The core of the lock is inserted from the back of the housing, and is fastened with a rear plate. The tail-piece that drives the locking cam is normally in a retracted position and can only be extended when the core starts to turn. A crescent-shaped hooked lever in the base of the housing limits rotation to one full turn in either direction.

The core, shown in Fig. 5.90, comprises a round barrel seated in a cast bronze foot. A set of parallel rods extends down the barrel, each rod is constrained to move in a channel cut into the rectangular keyway. The top ends of the rods are angled to match the alternating slopes on the key bit. The bottom end of each rod rests on the arm of a rocker, whose other end is a circular arc containing a gate. The rockers are pivoted in the foot of the core and are maintained with their gates in a downward position (toward the rear of the lock) by a strong spring-biased plate acting on their arms. Thin steel laminations separate the rockers and provide a degree of friction to damp their motion. Depressing one of the rods causes its bottom end to push down on the corresponding rocker arm, which pivots the gate edge of the rocker upwards.

The final ingredient is the bronze tail-piece, which is of unusual construction. The tail-piece is pivoted in the foot of the core and has an anvil-shaped appendage as well as the tail-piece proper. The appendage consists of a crank, as in the Miller lock, and two further projections or arms, one at the height of the crank and one lower down. The upper arm normally engages a radiused groove in the housing of the lock while the lower arm rests on a steel stud. The function of the stud is to protect the core from being forced inward by punching. This is backed up by a hardened steel pin, set in the barrel of the housing, that traverses the keyway via a set of holes in the rods.

The key bit depresses the rods so as to bring all the rocker gates into alignment with the crank of the tail-piece. With the key bit fully inserted, turning force can be transmitted to the core. Inward radial pressure is now exerted on the tail-piece's upper arm, causing the crank to engage the gates in the rockers. Once the core has turned through a small angle, the lower arm of the tail-piece is no longer blocked by the stud in the housing. At the same time, the upper arm rides out of its radiused groove, deploying the tail-piece from the base of the lock.

It is possible to apply inward tension to the crank by tensioning the core through the fixed keyway. However, the front of the core is rather deeply recessed, and the spring plate supporting the rocker arms is very stiff. A lot of force would therefore be needed to maintain the rockers in a picked position. The rockers also incorporate a shallow false gate that tends to cause slippage during a manipulation attempt. In addition, the spring-biasing of the rockers is only in one direction. This means that if a rocker is overraised when other rockers are still in contact with the spring plate, the overraised rocker would be set too high and could not be lowered by further pushing on its rod. This contrasts with the Bramah lock, in which the key acts directly on the sliders and, because they remain accessible in the keyway, can be moved either up or down to adjust the heights of their gates.

Muel

(FR) 6–10 lever (4)

Muel is a French lock manufacturer founded in 1896 by J. J. Muel. The Muel lock, described in French patent 303,376, has a rectangular keyway concealing a row of 6, 8, or 10 push-type levers, placed like teeth across the width of the keyway. Pictures of the 6- and 10-lever versions of the Muel lock appear in Figs. 5.91–5.95. Each lever is flat at the rear with a deep gate to accommodate the bolt stump. The levers are pivoted in a see-saw style across the square plug or "control box," with their smoothly curved, active end facing forward and flat springs on their undersides.

The distinctive end-bitted key (Fig. 5.93) has the word "incrochetable," French for "unpickable," stamped across the bow. The key blade is not reduced at the neck, unlike the end-bitted key used in the NS Fichet lock. The key cuts are at a 45-degree angle to the blade to pick up the curved ends of the levers. The cuts can be on either

Figure 5.91: Front and back views of Muel 10 push-lever cylinder.

Figure 5.92: Muel 10-lever key.

Figure 5.93: (Left) Rear of Muel 6-lever cylinder. (Right) Keys for 6-lever Muel locks.

side to suit the orientation of the levers, which are reversible. In addition, a pitted cut in the key (Fig. 5.92) accepts a scoop-type lever that would be overraised by a normal angled bitting. The cut for the scoop lever requires specialized machining, which acts as a form of copy protection. Blank keys are supplied by the dealer with between 0 and 3 scoop cuts in varying positions.

Figure 5.94: (Left) Plug and main spring from Muel cylinder. (Right) Selection of levers with scoop-type lever on right.

Figure 5.95: (Left to right) Muel 10-lever plug with incorrect and correct keys applied.

The end bittings of the key raise the levers as the key is inserted. A stop lug on the key limits the insertion depth, at which point the levers should be in the desired configuration, as shown in Fig. 5.95. Further force pushes the plug assembly (Fig. 5.94) toward the stump of the tail-piece (not shown) against a very strong spring. The stop lug in the key keeps pressure on the plug while it is turned. Small jags on either side of the lever gates (like those in Chubb levers) engage a mating V-notch in the bolt stump if the levers are not correctly aligned. The amount of force required to keep the plug at the depth where it meets the bolt stump, together with the antipicking notches, make picking this lock very hard. The French claim the Muel lock is unpickable, but they also make a manipulation tool for it.

Vak/Muel have also modified their push-lever key to incorporate a spring-biased pin in the place of one of the key bittings. This version is called the Vak Mobile, and it is designed to prevent unauthorized key duplication. There is also an electronic version of the Muel lock called the VAK "Genius." This is a 10 push-lever Muel cylinder with a microchip in the key stem. A set of pads on the key make an electrical connection when the key is inserted; the information carried on the memory chip is then interrogated by the lock (like the smart cards used in public telephones).

5.5 Radial

Cotterill

(UK) 8–13 lever (3–4)

The oddly named "climax-detector" lock is a radial lever lock with close mechanical ties to both Bramah and Bauche locks. The lock was patented by E. Cotterill & Co. in Birmingham in 1846 (UK 11,152) and was commonly made with 6 or 7 levers or sliders, although versions were produced with as many as 13 sliders. The lock pictured in Figs. 5.96–5.98, which we now describe, is an 8-lever Cotterill lock not featuring a detector.

Figure 5.96: Cotterill key and 8-lever radial lock.

Figure 5.97: (Left) Underside of the Cotterill lock. (Right) Core with retaining cap off.

Figure 5.98: Key displaces levers to align gates with channel.

The keyway is round with a drill-pin and a rectangular recess for locating the pipe key. The key itself has a round, hollow stem with eight angled cuts to varying depths in its end. One of the cuts is directly in line with the locating bit.

The Cotterill lock (Fig. 5.96) is seen to consist of a wide, shallow brass cap mounted on a flat backing plate. Only the smaller diameter knurled part of the cap would be visible from the front of the door or drawer on which the lock was mounted. Inside there is a round core holding the levers (see Fig. 5.97), at the back of which there is a cam that moves in a cut-out in the bolt. The core must be turned clockwise to withdraw the bolt.

The core is slotted radially in eight places to accommodate the miniature sliding levers. The levers are held in their tracks by a screw-on cover at the front of the core. Levers are made from flat steel and are irregular in profile, having a leading-edge bevel of roughly 18 degrees to the vertical. There is a cut-out at the base to leave room for a small coil spring and the top surface contains a narrow gate.

The outer edge of the core resembles a castle turret with an outer wall formed by the rim of the core and an inner circular channel. The underside of the lock cap, which is of cast and turned brass, contains an 8-toothed crown around the keyway. The teeth of this locking crown fit into the circular channel in the core and straddle the lever slots (similar to the monopole lock shown in Fig. 5.102, right).

The lock operates as follows: on insertion of the correct key, the angled bittings contact the beveled edges of the levers, causing them to slide radially outward. As the tip of the key contacts the bottom of the keyway, the lever gates align with the channel in the core as shown in Fig. 5.98, ensuring that its rotation is unhindered by the crown in the cap.

As the key rotates the core, it is held captive in the keyway by its bit. A spring arm on the underside of the cap clicks into a notch in the core to indicate the point where the keyway is aligned. The dimensions of the key stem only allow a maximum travel of about 3 mm on each lever; therefore, assuming a depth increment of 20 thousandths of an inch, there is room for only six depths of cut including a "0" cut corresponding to a dead-lift lever (one whose gate is already aligned with the channel in its rest position). An 8-lever lock would then have nominally $6^8 = 1.6$ million differs and a 10-lever lock around 60 million differs.

In more luxurious versions of the Cotterill lock, the lever ends are notched with their section (seen from above) resembling the letter "I" so as to catch on the teeth of the locking crown during a manipulation attempt. Like all cylinder locks with fixed front caps (e.g., Bramah), it is rather difficult to apply tension to the core without obstructing the sliders. As a bonus, the core relocks every one-eighth of a turn unless the sliders are maintained in their correct positions.

The detector part of the lock consists of a crescent-shaped pivoting arm mounted on the underside of the cap. When tension is applied to the lock without correctly setting all the levers, one of the arms of the detector crescent engages a notch in the outer edge of the core. It is necessary to return the core to its rest position in order to disengage the detector. Further details on the Cotterill mechanism are contained in [6] and [30].

In 1854, three years after the Great Exhibition, a 12-lever climax detector lock, similar to the one pictured in Fig. 5.99, was made available to A. C. Hobbs.

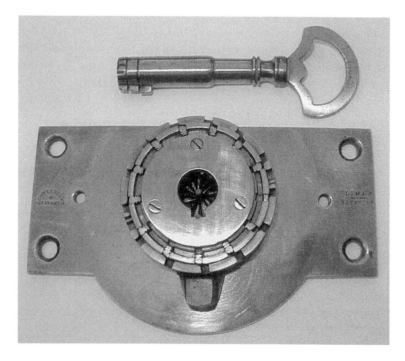

Figure 5.99: Cotterill's patent climax detector lock resisted Hobbs's picking attempts. (Courtsey J. M. Fincher).

Unknown to Hobbs, this particular embodiment of the Cotterill lock had a secondary slotted ring around the core that turned slightly if contacted by the beveled edges of the levers when tension was applied to the core. Unless all 12 levers were correctly positioned, the detector would block rotation. Cotterill offered a reward of £50 and gave Hobbs one day to take external measurements and a second day in which to pick the lock. This version of the Cotterill lock proved to be more than a match for Hobbs, who graciously admitted defeat after a 24-hour picking attempt, saying "The lock's yours; I give it up." The details of this gentlemanly contest were reported in the *Manchester Guardian* at the time and were also reprinted in [99].

Fichet-Bauche Monopole

(FR) 7-lever (4–5)

The Fichet-Bauche Monopole shown in Figs. 5.100–5.103 is a robust 7- or 8-lever lock designed for safes. Prior to 1967, the lock was produced by the Bauche Company [8]. The design is summarized in a 1928 French patent by Bauche (FR 659,113), although the lock was originally made in 1889 [41]. Essentially the same idea was also employed in Aubin's "vibrating guard" lock from 1850, covered in Price's book [99]. Our description is based on a modern version of the Monopole lock with seven levers.

The key (Fig. 5.100) has a round stem with a reduced-shank section leading up to the bit. Eight radial fins are arranged around the bit, one fin protruding more than the others to locate the key. The fins have round bittings to various depths.

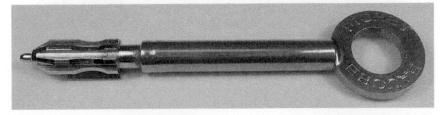

Figure 5.100: (Top to bottom) Bauche Monopole and Fichet-Bauche Monopole keys.

Figure 5.101: Front and rear views of Fichet-Bauche Monopole cylinder.

Figure 5.102: Fichet-Bauche Monopole core (left) and back plate (right).

Figure 5.103: Operation of Monopole core: key aligns lever gates with gap at rear.

Bittings may also be at one of two heights along the fins. The design is reminiscent of the Cotterill lock in respect of its radial symmetry.

The cylinder (Fig. 5.101) has a hard outer case that is fastened by screws to a flat mounting surface. The casing has a cast back plate with a turret (Fig. 5.102), similar to the one used in the Cotterill lock, with two concentric raised rings containing eight cut-outs. Inside the casing is a round core that is slotted radially in eight places to accept the seven levers. The eighth place is reserved for the linkage pin, which is normally retracted.

A round steel ring is recessed into the top of the core and serves as a hinge point for the seven levers, which point toward the rear of the lock. The levers are sprung radially toward the center by springs arranged around the core that are held in by the front cap.

Each lever has a flat end with a rectangular gate. With the rear cap fitted to the core, an annular gap is created that fits snugly in between the raised rings of the back plate. As the key is inserted, the lever ends move radially outward, and, at a certain point, their gates align with the annular gap, thus also clearing the inner ring (see Fig. 5.103). If a lever is overraised, however, it can protrude past the edge of the rear cap and impinge on the notch in the outer ring of the back plate. It is thus necessary for the fins on the key to align all seven gates simultaneously to create a free channel in the annular gap such that none of the notches in the base plate ring is obstructed and the core is free to turn. The locating fin of the key also depresses the linkage pin that connects through to the shaft of the lock to drive the boltwork of the safe.

Manipulation of the lock is difficult since the case is fixed and tension must be applied through the star-shaped keyway. The lock will also relock every one-eighth of a turn. In practice, the lock is usually mounted behind several inches of steel so that the moving parts are much harder to access.

Fichet M2B

A twin-cut version of the Monopole design was presented in US patent 4,196,606 (1978) by F. Guiraud of the Fichet-Bauche Company (see Fig. 5.104). In this embodiment, called the Fichet M2B, the levers are doubly-gated and can move both axially and radially. Pictures of a Fichet M2B lock appear in Figs. 5.105–5.107. The radial gating is in the form of a turret on the underside of the front of the housing. The axial gating involves a prong on the outward edge of each lever. The levers move in a set of longitudinal channels in the base of the housing. There is a circumferential groove in the housing that provides clearance for the prongs once they are depressed to the required depth. The key must simultaneously depress and shift the levers outward to correctly align both the radial lever gates and the prongs. The lock thus provides two independent degrees of freedom: axial depth

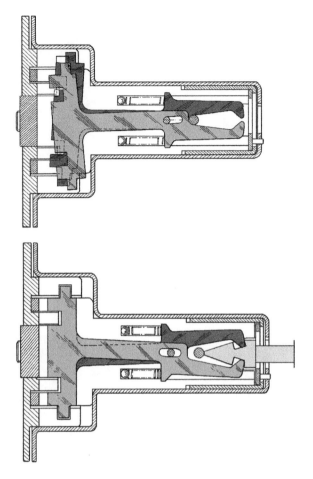

Figure 5.104: Twin-cut version of the Fichet-Bauche Monopole (US patent 4,196,606 by F. Guiraud).

Figure 5.105: Twin-cut key for Fichet-Bauche M2B lock.

and radial displacement. In addition to providing a huge number of key variations, the mechanism is highly manipulation resistant. Variants called the Fichet M2i and M3B are also available that have a movable element in the form of a ball bearing embedded in the key stem.

Figure 5.106: (Left) Fichet-Bauche M2B 8-lever lock. (Right) Underside of housing showing circumferential channel for axial gating and turret for radial gating of levers.

Figure 5.107: (Left) Fichet-Bauche M2B core with levers in rest positions. (Right) Core with key inserted.

5.6   Cylindrical

Chubb AVA

(UK) 10-slider (3–4)

The Chubb AVA lock, illustrated in Figs. 5.108–5.110, is a compact 10-slider lock for applications such as secure storage cabinets and safes. It is also made as a rim cylinder for door locks and as a cam lock. We use the term *slider* as opposed to lever, since this is the one used in Chubb's 1963 patent (UK 1,030,921). The sliders

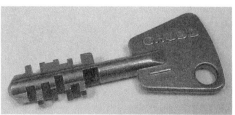

Figure 5.108: Chubb AVA 10-lever rim cylinder and key.

Figure 5.109: Chubb AVA core and sliders.

could equally well be referred to as wafers since the motion is one of sliding rather than pivoting.

The construction of the lock is similar to an Abloy lock, while its operating principles are very similar to those of pre-1931 versions of the Kromer lock in the next section. The design is also reminiscent of a plate-wafer lock patented in 1926 in France by R. G. M. Homolle of Bournisien Beau and Companie (Fichet's holding company) and registered the following year in the United States as patent 1,784,444. The sliding tumblers in Homolle's lock were operated by a flat double-bitted key with constant-width bittings.

The original AVA lock was proposed in a 1953 U.K. patent due to A. A. Saarento of Finland (UK 737,547), which described a 7-slider mechanism with a flat key. Chubb's patent for a 10-slider lock added antitamper features such as notched edges and serrations to the sliders. Undercut edges were also included to thwart decoding attempts. The description in this section is based on the Chubb AVA padlock mechanism, although it applies equally well to the AVA cylinder lock.

Figure 5.110: (Left) Chubb AVA core with key inserted. (Right) Key turned to align sliders with edge of shell.

The AVA lock (Fig. 5.109) comprises a cylindrical shell that is longitudinally slotted on two sides and housed in the cylinder body. On each side of the shell, in line with the slots, the lock housing contains a longitudinal chamber (as in a wafer lock). The sliders are oblong with an enlarged midportion and two radiused arms. The overall length of each slider (between the ends of the arms) is equal to the outside diameter of the shell. There is a central cut-out or aperture in each slider forming the keyway. This cut-out, which is the same shape for each slider, is wide enough to allow insertion of the key and also contains a contoured surface that contacts the key bittings.

The sliders, which have no separators and no spring biasing, are stacked 10 high in the shell and can slide transversely in either direction. The sliding motion is guided by the contact of the arms of the sliders on the sides of the two opposing slots in the shell. A profile plate may also be included at the front of the shell. The midportion of the slider does not completely fill the cavity in the shell, but allows a moderate amount of displacement from one side to the other stopped by the shoulders of the slider contacting the inner walls of the shell. The slider pack is sprung from below, pressing the sliders up against the front of the lock. In an AVA padlock, the mechanism is held in place by a retaining cover with a concealed C-clip.

In the locked position, the arms of the sliders protrude to varying degrees into the chambers on either side of the lock housing. As in a wafer lock, a slider can be either overlifted or underlifted, depending on which of its arms impinges on the chambers. This construction requires action in two opposing directions to bring all sliders in the pack to the shear line of the shell.

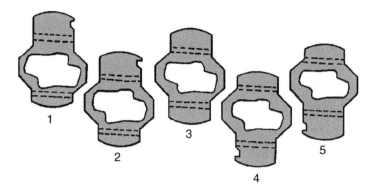

Figure 5.111: Five different slider sizes for Chubb AVA lock (UK patent 1,030,921 by W. H. Stanton and C. L. Brett).

There are three different slider types, varying in the relative positioning of the cut-out. It is useful to denote these as 1, 2, and 3 corresponding to the three sliders on the left in Fig. 5.111. The cut-out in the type 3 slider is centrally located, whereas type 1 and 2 sliders have one arm longer than the other. Since the cut-out is rotationally symmetric, the three slider types can be turned around. Clearly, this makes no difference for a type 3 slider. However, for type 1 and 2 sliders this makes their longer arm protrude into the opposing chamber. We refer to the rotated type 1 and 2 sliders as types 5 and 4, respectively. Thus the three different slider types lead to five different slider sizes, each with a different offset. These five offsets require differing key cuts.

The key blank has a narrow double bit that extends most of the length of the stem. The bit is mounted slightly off-center on the stem so that the key can only be inserted in one orientation. (AVA key blanks are also made with flat blades and a central ridge on one side.) Constant-width cuts, subject to the sectional width of the key, are made at 10 positions along the key blade, as shown in Fig. 5.112. The cuts are angled at 45 degrees to the plane of the key bit to match the contact surfaces on the slider cut-outs. The design allows a maximum angle of rotation of the key of about 45 degrees before the shoulders of the sliders contact the inner surface of the shell.

The respective bittings displace the sliders either to the left or right as the key is turned clockwise, as shown in Fig. 5.110. The bittings must be of the required offset to displace all 10 sliders so that they are fully retracted into the shell. Once this condition is met, the shell is free to rotate, which releases the locking balls (in the case of a padlock) or drives a cam (in a deadlock or latch). The key bittings retain the key in the lock when it is in the open position. When the key is turned counterclockwise back to the locked position, the discs are displaced to the right or left and their cut-outs are moved back into alignment. The key can then be withdrawn.

The design of the AVA lock is very versatile in terms of the keying combinations it provides. Recall that there are 10 sliders, each with five possible offsets. The MACS

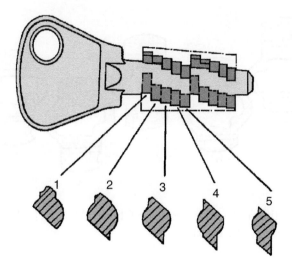

Figure 5.112: Five different cut types for Chubb AVA key (UK patent 1,030,921).

is unrestricted, and so the number of available keying combinations is very close to its theoretical maximum value of $5^{10} = 9,765,625$.

In regards to picking of the AVA lock: the sliders have a notch at either end that can bind against the sides of the slots in the shell if either overlifted or underlifted. Serrations or undercut edges can also be added to increase resistance to manipulation and decoding. The keyway cover is fixed, retaining the bit of the key as it is turned; this makes it difficult to apply tension to the sliders. Manipulation is also hampered by the absence of springs and separators, resulting in the movement of one slider being transmitted by friction to the adjacent sliders. The AVA principle has been employed in some European and U.S. models of Ford cars (see Chapter 7) as well as in the U.S. Miracle lock, which includes a side-bar and antipick notches on one side of the sliders.

Kromer Protector

(DE) 10–22 lever (5)

The Kromer Protector is somewhat of a masterpiece of lock engineering with a long development history, some of which was covered earlier. The original Protector lock was patented by Theodor Kromer of Freiburg in the Black Forest part of West Germany around 1870. The Theodor Kromer Company also took out many other patents for high-security double-bitted lever and combination locks such as Brisgovia, Certus, Reling, Novum, Central, and Integral [117]. A number of these were discussed in the chapter introduction. The Central and Integral (see Fig. 5.33) share similarities with U.S. key-operated combination locks like S&G and La Gard [44]. Kromer's combination lock, described in a 1921 German patent (DE 361,110) and also featured in [45], had features that were quite distinct from S&G-type wheel-pack locks (Fig. 5.36).

Theodor Kromer, who had worked in the U.S. and U.K. lock industries, was in regular contact with a number of U.K. lock manufacturers such as Chubb, Milner, and Chatwood, with a view to marketing his locks outside of Germany [46]. Kromer's double-bitted lever lock designs had a considerable influence on U.K. manufacturers like Chatwood, whose locks had previously been of the single-bitted type [43, 98]. As early as 1883, the Chatwood firm placed an order with Kromer for 600 Protector locks. In the 1930s hundreds more locks, of the type shown in Figs. 5.113 and 5.114, were supplied to Chatwood's for the production of night safes for the Midland bank [44].

The original Kromer Protector lock, which appeared around 1877, had six one-part sliders with a double-bitted key. The one-part alternating lever design is summarized

Figure 5.113: Kromer Protector 10-lever lock of the type supplied to the Chatwood firm in the 1930s. (Left) Cover removed to show one-part sliders. (Right) Operation by key.

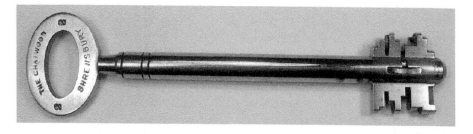

Figure 5.114: Key with cross-cut for Chatwood's Kromer Protector 10-lever lock.

in a Swiss patent from 1892 (CH 4935). It is possible that Kromer was motivated by the design of the flat-key disc-tumbler locks invented in the United States in the late 1860s. Kromer placed particular emphasis on protection against illicit key copying, as evidenced by the progressive addition of angled cuts and bevels to the key. A Protector lock with one-part levers could be made to operate under various configurations of the contact points on the key. An 11-lever lock required only 11 of the 22 possible contact points. Therefore, given a key with cuts in all possible positions, it was impossible to judge from inspection which 11 of the 22 cuts were actually required. This provided a safeguard against unauthorized copying since all 22 cuts had to be reproduced [44].

The split-lever Protector was mooted in an 1899 patent (DE 117,781), but it was not produced in commercial quantities until the 1930s. The main Kromer factory in Freiburg was destroyed in a 1944 bombing raid, eventually being rebuilt in 1951 [116]. In the remainder of this section we deal with the BP55K Protector, a VdS class 3 lock, which was produced until 1999 [79]. Pictures of this lock appear in Figs. 5.115–5.119. This is followed by a description of the picking of an earlier model Protector lock used in German government safes by a man named Bierhaus.

The BP55K Protector lock, which exists in both CW- and CCW-opening models, consists of a core and stator or body. The front cover of the lock is secured to the body by three blind bolts that are inserted from the front and tightened from the

 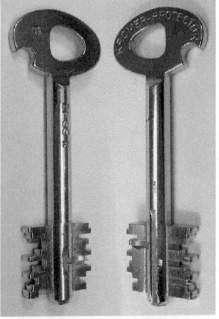

Figure 5.115: (Left) Kromer Protector BP55K safe lock. (Right) Front and back views of double-bitted key with millings and angled cuts.

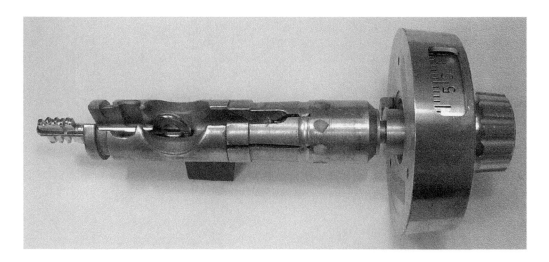

Figure 5.116: Lafette and dial mechanism with key loaded into breech.

Figure 5.117: Kromer BP55K lock with front cover off: (left) curtain plate removed to show lever pack; (right) operation by correct key.

rear via the driver slot in their ends. Prior to shipment from the factory, one of the screw holes is plugged with a lead seal.

The lock may be hand-operated by a double-bitted key, or, in larger safes, it may be recessed behind the safe door and operated by a lafette like the one shown in Fig. 5.116. In such cases, it is often teamed with a conventional combination lock. The lafette consists of a long sleeve connected to a dial and plate, which are mounted on the front of the safe door. The sleeve length is adjusted to match the thickness of the door (the lafette in the figure has a sleeve length of six inches). The dial is turned to a set position (e.g., 50) and pulled out from the door until a stop is reached. This reveals a breech into which the key is placed with the notched side of

Figure 5.118: Kromer BP55K core: (left) two split-lever pairs inserted; (right) action of key on pair of split-levers.

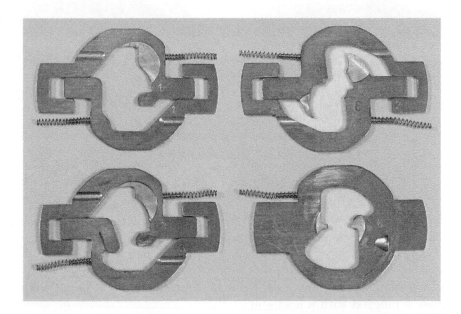

Figure 5.119: Kromer levers for clockwise opening: (left) two left-handed lever pairs; bottom lever with tab on left for angled key cut; (right) right-handed lever pair and single lever requiring pitted cut.

the bow pointing downward and the bit protruding toward the rear. As the dial is returned to the front plate, the lafette transports the key such that its bit finishes in the keyhole of the lock. The lock is then operated by turning the dial about a quarter of a turn.

The core supports a pack of 10 brass levers or sliders arranged around a central drill-pin (see Fig. 5.117). Both one-part and split-levers may be present in the pack, which also includes two steel curtain plates. The function of the plates is to provide profile control for the key and to restrict access to the keyway by coupling to the discs adjacent to the plate. The coupling is in the form of a small tab that presses into a recess in the adjacent lever, preventing it from moving until the curtain plate is turned. The keyway profile plate ensures that the double-bitted pipe key can only be inserted in the correct orientation.

Each member in a lever pair is independently spring-biased by a small coil spring that rests in a channel milled into the top surface of the lever beneath it (refer to Fig. 5.118). There are multiple types as well as sizes of lever: some require a normal cut to the key while others require radiused (pitted) cuts or angled cuts, as described in German patents DE 595,834 and DE 1,061,226. Versions of the lock exist with extended fingers that require cross-cut grooves in the key stem. There are also two handednesses, which we refer to as left (L) and right (R). In a CW-opening lock, viewed from above, a left-handed lever pair has its overlapping ends on the top left and bottom right (see Fig. 5.119). A right-handed lever pair is the reverse of this. Both handednesses of lever have contact points with the key bit on their top right and bottom left inner surfaces. A right-handed one-part lever has its driver spring on the right and conversely for a left-handed one. Referring to one-part levers as 1, lever pairs as 2, and curtain plates as P, the typical assembly sequence for a 10-lever pack, from front to back, is: P, 1R, 2R, 2L, 1L, P, 1L, 2L, 1R, 2L, 2R, 2L (a total of 12 layers).

Unlike a conventional wafer lock in which the wafers are depressed as the key is inserted, the lever pairs in a Kromer lock are actuated by the bittings on either side of the key as it is turned. For each lever pair, the split-levers slide in opposite directions against each other. In the locked position, the levers are biased by spring action into diametrically opposed chambers in the stator, preventing rotation of the core. Since the operation of the one-part levers is the same as in the Chubb AVA lock in the previous section (which is based on the Kromer lock), we restrict the remaining discussion to the action of the split-levers.

Whereas the sprung end of each lever is half as wide as the slot in the core, the other end is full width and overlaps the half-width end of the opposing lever in the pair. The overall length of a split-lever plus the width of the overlapping end of the opposing lever is equal to the diameter of the core. Thus, when driven by a bitting of the appropriate dimension on the key, the outer edges of the overlapping ends of the levers will be flush with the rim of the core and the half-width ends will also be in contact with the inner edge of the overlapping end as the key reaches 45 degrees rotation. It follows that if a key bitting is too long, the corresponding pair of levers will block against each other before the key reaches 45 degrees, leaving the other levers insufficiently displaced. Conversely, if a key bitting is too short, one of the overlapping lever ends will protrude into the stator chamber and block rotation in much the same way as in a conventional wafer-tumbler lock.

The book by Eras [30] describes the painstaking task of impressioning a key for a Protector lock. Eras, who knew Theodor Kromer well, claimed that no one ever picked a Kromer Protector lock. This may be true of the BP55K Kromer Protector, but the earlier RP37E Protector lock was successfully picked by an Austrian called Josef Bierhaus [27]. The following description of events draws on material from [118, 119].

Sellin's 1931 design of the Protector lock (DE 560,425), utilizing 11 split-levers and a key with 22 active contact points, was considered by the Kromer Company to be unpickable. It was approved by the German Post Office, the *de facto* standards organization for high-security locks at that time, hence the model name in which RP stands for Reich Poste. In 1950, reports reached the company that Bierhaus had picked a Protector lock in Vienna made by Steinbach and Vollman (STUV), similar to the one shown in Figs. 5.120 and 5.121. After four years of prompting by Kromer GmbH, Bierhaus finally agreed to travel to Freiburg to demonstrate the picking skills he had acquired over a 15-year period. He was initially supplied with a Protector lock taking a key with cross-cut grooves. Bierhaus refused to pick this lock on the grounds that cross-cut grooves were not part of the RP standard. He was then provided with a RP37E standard lock, which he duly defeated in three hours using his own manipulation tools. Unlike Hobbs, Bierhaus did not receive a reward, although Kromer did agree to pay his travel expenses and compensate him for his lost time. The upshot of this demonstration was that Kromer redesigned the lock to add interlocking split-levers as well as the two curtain plates coupled to the sliders to restrict the keyway. Aspects of the new designs are summarized in German patent 1,027,552 by O. Sellin and H. Münzer. Together with subsequent improvements, this led to the BP55 model described previously.

Figure 5.120: Asymmetric double-bitted key from STUV RP37EB Protector lock.

Figure 5.121: (Left) Steinbach and Vollman RP37EB Protector lock. (Right) Cover removed to reveal 22-slider mechanism.

CAWI

(DE) 22-lever (4–5)

The German company CAWI, established in 1857 and now trading as Carl Wittkopp GmbH, produces a range of VdS-approved high-security locks for safes and safe deposit boxes. The CAWI cylindrical lever lock, which we consider here, employs a different system to the Kromer Protector lock. The system is in fact a circular version of the Parsons 1832 balance lever lock, adapted for a double-bitted key. The CAWI lock also bears more than a passing resemblance to Hermann Kühne's Panzer safe lock from 1899.[6] Kühne's lock, shown in Figs. 5.122 and 5.123, was a 20-lever lock with an asymmetric double-bitted key in which the horseshoe-shaped levers formed a system of interlocking cams around the key. The core of the lock was prevented from turning by projections on the outer arms of the levers. The detailed operation of the lock can be gleaned from the description of the CAWI lock, to which we now turn our attention. Pictures of the CAWI lock appear in Figs. 5.124–5.126, while drawings from the patent are shown in Fig. 5.127.

The CAWI lock was patented by the Bode-Panzer Safe Company of Hannover in 1951, which ceased lock production following World War II. The Kromer Company, at the time a major supplier of locks to Bode-Panzer, immediately set out to find a weakness in CAWI's design. Kromer's expert, Ulrich Maurold, developed a rubber

---

[6]Kühne's Panzer lock should not be confused with the unrelated 1907 Panzer lock mentioned in the chapter introduction.

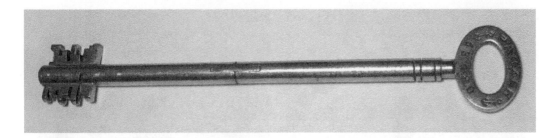

Figure 5.122: Double-bitted folding key for Kühne's 20-lever Panzer lock.

Figure 5.123: Two views of Kühne's 20-lever Panzer lock showing interlocking levers and action of key.

vibratory pick for the CAWI system that enabled the company to pressure CAWI into replacing their mechanism with Kromer cores and keys—an offer accepted by CAWI [80]. Our description of a modern CAWI lock assumes clockwise opening. The end of the lever further advanced around the clockface is referred to as the leading edge, with the other end being the trailing edge.

The CAWI cylindrical lever lock contains 20 crescent-shaped balance levers mounted in two stacks of 10 opposing pairs on a rotatable core (see Fig. 5.125). Each balance lever can pivot about its midpoint and is also equipped with two hooked ends that impinge on cavities in the lock housing, normally blocking its rotation. Some levers have extended arms that surround the drill-pin. As well as the springless balance levers, there are two further spring-biased levers that provide a blocking function. Each of these blocking levers has a Y-shaped inside edge that impinges on the keyway around the central drill-pin. The blocking levers have two degrees of freedom: they can shift slightly as well as pivot. The inside edge of each blocking lever possesses a ridge.

Figure 5.124: CAWI safe lock and double-bitted key. (Courtesy O. Diederichsen).

Figure 5.125: (Left) CAWI lock with cover removed. (Right) Keyway plate and top levers removed.

In addition to the levers, the core comprises an inner and outer rotor (visible in Fig. 5.126). Both of these parts are independently rotatable and are equipped with a pair of posts. The inner rotor is mounted on short stumps on the outer rotor and is limited in its travel to 45 degrees. The lever pairs pivot on the posts of the outer rotor. The function of the posts on the inner rotor is twofold. First, they maintain the balance levers so that in the locked state their trailing edges engage the cavity in the housing. Second, the posts, in conjunction with the ridges of the blocking levers, provide a deadlocking function for the inner rotor since the leading edges of

Figure 5.126: (Left) Insertion of key shifts blocking levers. (Right) Key turns to 45 degrees to bring levers flush with core.

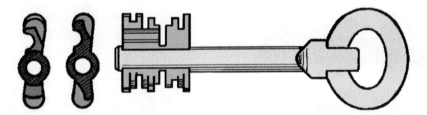

Figure 5.127: Design of CAWI key and various cross-sections from Bode-Panzer's 1951 patent (DE 911,220 by K. Langrehr).

the blocking levers normally impinge on cavities in the housing. The system of inner rotor and blocking levers acts as a release assembly for the rest of the mechanism.

The key is double-bitted and, like Kromer Protector keys, has nonsymmetric bittings as well as radiused or pitted cuts at some locations (see Fig. 5.127). The end of the key is hollow to accept the drill-pin, with a bevel at the tip that operates the release assembly.

The rotor assembly is loaded first, then the blocking levers followed by the first seven pairs of balance levers. Next there is a keyway plate, fixed to the inner rotor. The last three lever pairs are mounted on top of the keyway plate.

Underlifting a lever causes the hooked end of its trailing edge to engage a cavity in the lock housing. Overlifting can have two effects depending on the length of the extended arm of the lever. In the first instance, the hooked end of the leading edge of the lever engages a cavity in the housing. In the second case, the extended arm of the lever contacts the opposite bit of the key and prevents it from turning any further. This provides a means of checking for the presence of correct depth cuts on the sides of the key bit. Thus a key with the correct double bittings will not operate the lock unless it also has the correct side-millings.

When a correctly bitted key is inserted, the bevel at the tip of the key contacts the Y-edge of the blocking levers, shifting them so that the posts of the inner rotor clear the ridges on the inside edge of the levers. With the inner rotor now freed, the key can be turned to bring its bittings into contact with the protrusions on the leading edges of the balance levers. At the same time the bolt step on the key turns the inner rotor. The correct key pivots all 22 levers such that their hooked ends are flush with the rim of the core once the inner rotor has turned to 45 degrees, aligning its posts with those of the outer rotor. At this point, all the extended arms are gracefully resting against the radiused cuts in the sides of the key bits. Further turning force is then transmitted from the inner rotor to the outer rotor, which is now free to turn.

Unlike the original version of the lock, which was springless, the modern version of the CAWI provides high resistance to manipulation via notches in the leading edges of the levers and stator and the inclusion of the release assembly at the base of the rotor. The complexity of the key, featuring pitted and angled cuts, renders it impractical to reproduce without highly specialized equipment.

5.7   Geared

Fichet-Bauche 787

(FR) 10-lever + side-bar (5)

The Fichet-Bauche 787, illustrated in Figs. 5.128–5.132, is the successor to the Fichet-Bauche 484 and may be found adorning the doors of expensive Paris apartments. The lock has an extremely intricate mechanism comprised of over 50 components. It is intended for use in multipoint and bar-locking systems for steel-sheeted doors. The design patent for the lock was lodged in the United States in 1984 by L. Doinel of Fichet-Bauche.

The key has a round stem with a centrally fixed bit of rectangular section and a locating bevel on one edge (see Fig. 5.128). There are 10 cuts on each side, although on closer inspection only 5 cuts on each side are used and these alternate in sequence. The key bit has a reduced-width section at its top end. Correspondingly, at the front of the cylinder, which is fixed, there are two spring-loaded gates (see Fig. 5.131). The gates are normally closed, and clamp the key bit across its reduced-width section

Figure 5.128: Fichet 787 10-lever cylinder and key.

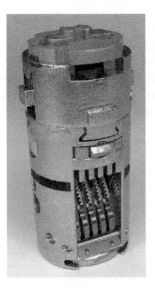

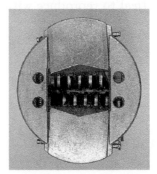

Figure 5.129: (Left) Fichet 787 core. (Middle) Core with rear section removed to show gears. (Right) View through gates at front of keyway.

Figure 5.130: Rear section of Fichet 787 core: (left) side view showing cross-piece and fence; (middle) tail-piece retracted; (right) tail-piece extended.

Figure 5.131: Two views of Fichet 787 core with key bit engaging spring-loaded gates.

Figure 5.132: (Top) Fichet 787 core with key partially inserted, gears scrambled. Inset shows contact of key bit with levers. (Bottom) Key fully inserted to align gates of gears. Inset shows blocking bar.

on insertion. The cylinder geometry is such that if the gates are opened any further than required, rotation of the plug will be blocked. This means that there is very little space left for the insertion of manipulation tools. A similar idea was mooted in a 1903 patent by W. H. Taylor (US 758,026), which proposed a guard or curtain for a pin-tumbler lock.

The cylinder has a compound core consisting of a front and rear section, as in Fig. 5.129. The front section houses a number of components including 10 levers or rocker arms and 10 toothed wheels or gears. The rear section, shown in Fig. 5.130, which engages the front section via a dovetail on one side, houses an interlock for the tail-piece that is normally retracted. The interlock is supported by a cross-piece that extends into a longitudinal cavity on either side of the cylinder. When the appropriate conditions are met, the core may be displaced longitudinally and subsequently rotated.

Before describing the function of the front section of the core, it is best to understand how the rear section works. The interlock mechanism, which is spring-loaded, encapsulates the tail-piece. The tail-piece is lightly sprung and will not extend past the rear of the interlock, which is heavily sprung, unless the core advances relative to the cross-piece. The cross-piece has an integral fence, centrally located at the front of the rear section, which controls the interlock mechanism. Any pressure exerted on the fence will block the advancement of the core and prevent extension of the tail-piece. In addition, the core cannot be rotated unless it has advanced sufficiently for the lugs on the rear section to attain a circular channel in the cylinder at the same depth as the cross-piece.

The advancement of the core is driven by the key being inserted into the front section, as shown in Fig. 5.133. There are three axles in the front part of the core: two at the front and one at the rear. The 10 levers are mounted on the two front axles while the 10 gears are mounted between fixed separators on the rear axle. The forward end of each lever is smooth with a belly that rides in a crescent-shaped pocket in the key bit. The rear of each lever is toothed, forming a rack that meshes with the teeth in the corresponding gear. Each gear is gated in one place along its circumference. The levers are arranged in an alternating sequence so that if pushing one lever causes a gear to rotate clockwise, then pushing the next lever along will cause a counterclockwise rotation of its gear. The levers are sprung from the side by a comb spring so that their bellies tend to cluster in the middle of the keyway (Fig. 5.129, right).

When a key is inserted, its bittings displace the rockers, causing the gears to rotate to various angles. Only the correct key will rotate all 10 gears to the position where their gates are in alignment and facing the rear of the cylinder, as shown in Fig. 5.132. With the gears properly aligned and pressure being exerted on the key, the core begins to advance. Since there is now a channel formed by the gates in the gears, the fence of the cross-piece can enter this space and the interlock allows the core to advance further without blocking. If any one of the gears is not properly aligned, the fence will contact the edge of gear and this will trigger the interlock, preventing further advancement of the core.

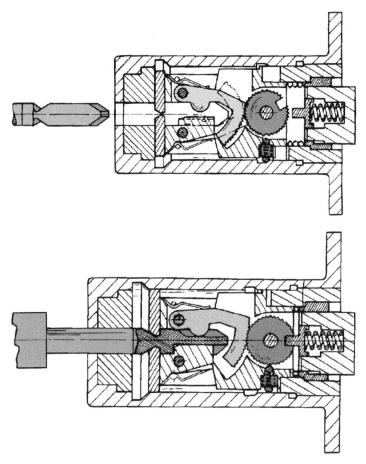

Figure 5.133: Detailed design of the Fichet-Bauche 787 from 1984 US patent 4,601,184 by G. H. D. Doinel.

As the core advances to the full extent allowed by the longitudinal cavity in the cylinder, the tail-piece now extends past the end of the interlock to engage the cam that drives the boltwork of the lock. The cross-piece of the interlock simultaneously attains the circular channel that allows the core to rotate, thus operating the lock.

A further security feature, traceable to the 1933 wafer lock restraint of J. W. Fitz Gerald (US 1,965,889), is included in the form of a spring-loaded blocking bar mounted parallel to the rear axle of the core (visible in the lower inset of Fig. 5.132 and also in the patent drawing in Fig. 5.133). The blocking bar is normally disengaged, but the advancement of the core causes the bar to impinge on the gears from the side. This prevents any further rotation of the gears and effectively freezes the combination that is presented to the fence of the cross-piece.

There are 10 possible depths of cut, and hence with 10 gears the theoretical number of combinations is 10^{10} or 10 billion. For the 8-gear (787 S) version, the theoretical number of combinations is 10^8 or 100 million, as stated in the manufacturer's brochure.

Fichet asserts that this lock is unpickable, and one is inclined to agree. However, it is claimed that a special tool developed by Falle Safe Securities can be used to decode and hence pick the 787 S lock in a matter of only minutes [122]. The tool allows the levers to be independently manipulated and set at the required positions to align the gears. It is not clear that the technique would work on 787 locks equipped with the aforementioned blocking bar, since the core cannot be advanced without freezing the gear positions. For further details on this ingenious lock, the reader is invited to consult the text of 1986 US Patent 4,601,184.

5.8   Trap-Door

Dény

(FR) warded 3-lever (2)

The Dény 3-lever lock in Figs. 5.134 and 5.135 is very unusual—a rare example of a plugless cylinder lock. The lock on which it is based was invented by George Davis in 1799 (UK patent 2,306) for use on cabinet dispatch boxes for the government of King George III [22]. Davis's patent lock was operated by turning the quite ordinary looking warded key 90 degrees, pushing it into the rear chamber of the lock, and then turning it a further 180 degrees to release the bolt. The Dény lock operates in a similar turn-push-turn manner. The lock comprises an elaborate system of wards and levers that operate a trap-door at the rear of the cylinder.

The core of the lock cylinder is fixed, being constructed with a stack of pressed steel laminations (see Fig. 5.136). The laminations are cut away in places to leave room for three levers mounted on rods. The stack of laminations form the intricate internal warding of the cylinder. The key is symmetric and double-bitted, with as

Figure 5.134: Front and rear of Dény warded 3-lever lock cylinder.

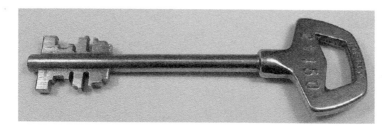

Figure 5.135: Dény double-bitted key: many bittings are redundant.

Figure 5.136: (Left) Dény core. (Right) Blocking latches displaced to allow through-access for key.

many as a dozen cuts in each side. The only active parts of the key are those that contact a lever inside the core. These bits must be sufficiently long to actuate the levers, which occurs in a particular sequence due to the angular positioning of the rods around the core. The rods pass through the back part of the core to a chamber whose entry is blocked by a movable cover. The cover is held fast by two interlocking latches fixed to the rods, with the last rod arranged to rotate the cover once the latches have been displaced by turning the key through 90 degrees.

With the cover rotated by a first 90-degree turn of the key, the key is in a position to slide through the gate and into the rear chamber. The stem of the key must then be long enough to enable the bit to engage a locking cam at the rear of the cylinder, which is slotted so that the key will turn it once inserted.

The difficulty in picking open a lock such as this is in locating the active parts, that is, the three levers, and operating them in sequence to access the gate. A fair portion of the warding is visible in the keyway, and the lock may be easily defeated by impressioning as only a skeleton key is required to operate it. Despite the relatively low level of security provided by the Dény lock, it is in widespread use in France by the railways and electricity industry due to its highly weather-resistant design.

5.9   Twin

Ross 700

(AU) 9-lever (4–5)

The Ross 700, appearing in Figs. 5.137–5.140, is the most recent addition to the range of Ross lever locks. It is closely related in its operating principles to the much earlier Reling lock produced in the late 19th century by Theodor Kromer (Fig. 5.22).

Whereas a conventional twin-lever lock such as the Chubb-Lips 6K207 (Fig. 5.26) has two sets of independently operating levers and two bolt stumps, the Ross 700 has an alternating stack of levers and a single bolt stump that projects on both sides of the bolt. Although it is not a true twin lever lock since there is only one lever in each position, independent bittings on either side of the key are required to operate the lock. In addition, the construction allows the double-bitted key to be reversible.

According to the assembly instructions, the levers are numbered from the rear casing of the lock. However, since the levers are mounted onto the front plate of the lock (containing the keyhole), we number them the opposite way here. This is in keeping with the previous convention in this chapter where the lever closest to the keyhole is referred to as number 1. As suggested in Fig. 5.137, our description assumes that the lock is viewed with the rear of the lock at the front, the bolt on the left, and the keyhole at the back. In this orientation, lever number 9 is frontmost.

The Ross 700 lock has a cast zinc alloy housing with a steel backing plate that is folded to provide a guide slot for the sintered stainless steel bolt. The bolt is attached to a slideable plate referred to here as the bolt tail. A folded steel cage (the "enclosure") encases the vertical-lift lever mechanism. This part of the lock is modeled on the preceding Ross 600 model with notable differences. First, the levers are mounted with the spring either up or down, providing action in both directions. Second, the bolt tail, sandwiched between the fourth and fifth levers, has a drive mechanism to throw the bolt. The design of the levers is also different from a Ross 600 lock.

The levers, shown in Fig. 5.140, are of stamped brass construction with an integral leaf spring (as in the Ross 102). Levers have a single pocket with a narrow gate and are open on the left side. All levers have the same overall shape: broadly rectangular with identical D-shaped cut-outs for the keyway. The curved part of the D accommodates the arc of the key bit as it turns through 180 degrees during operation of the lock. The flatter side of the D contains the lever belly that is contacted by the active faces of the key bit. The sliding motion of the levers is constrained by the vertical sides of the enclosure, with the right-hand edge (furthest from the bolt) protruding past the side of the enclosure to provide a stop.

Figure 5.137: (Top to bottom) Ross 700 series 9-lever safe lock; symmetric double-bitted key; case removed showing lever pack.

An upward-acting lever (↑) is mounted with its spring at the top of the enclosure and is reversed for a downward-acting lever (↓). The pocket width supports five gate offsets corresponding to the five depths of cut on the key. There is approximately a two-thirds overlap between gates in adjacent lever sizes. Although the levers are limited in the lift dimension, there is enough room for one false gate in each of the lever sizes. Above and below the bolt, the levers are generally loaded in an alternating sequence with either an upward- or downward-acting lever in position 1, for instance: ↑↓↑↓ (bolt) ↑↓↑↓↑.

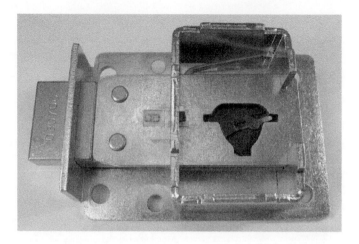

Figure 5.138: Ross 700 with levers removed to show bolt tail and drive cam.

Figure 5.139: Ross 700 with key operating the lock.

Figure 5.140: Ross 700 lever pack: (left) levers in locked position; (right) levers with gates aligned. Note: levers are identical except for gate position.

The bolt drive or cam mechanism, visible in Fig. 5.138, occupies roughly three cut positions in terms of its thickness. The drive consists of a rectangular plate affixed to slots in the enclosure in which a drive wheel is mounted. The wheel, made of sintered steel and containing a cut-out to match the keyhole, is equipped with a cam on one side that communicates with a cut-out in the bolt tail. The cut-out is trophy-shaped, with curved sides and a square base or talon. A spring-biased lever arm locates the neutral point on the drive wheel. The cut-out in the bolt tail functions in much the same way as a conventional Chubb bolt: the curved sides of the cut-out allow the key to turn a certain amount before engaging the talon; further turning then displaces the bolt.

The key, made from nickel-plated brass, is symmetric and double-bitted. The key cuts are radiused in order to reduce wear on the key and levers. This minimizes marking of the lever bellies—a precaution against reading and impressioning. Numbering the spaces along the key from 1 to 12, we note that spaces 5 and 6 are reserved for the bolt drive and may be left uncut. Space 7 is cut deeply enough to avoid contact with the bolt tail. Spaces 1 through 4 address the first four levers, and spaces 8–12 are for the remaining five levers. Denoting the sides of the key as left (L) and right (R) for a given key orientation (A or B), Table 5.4 summarizes the active bittings, assuming an alternating sequence of levers. The dashes in the table correspond to the bolt-step positions, which are not used to combinate the lock.

Since there are five depths of cut numbered from 1 (zero cut) to 5 (deepest cut), the theoretical number of key combinations is $5^9 = 1,953,125$ if all nine lever cuts (positions 1–4 and 8–12) are assumed independent. The MACS is 4 adjacent to the bolt-step positions, although some combinations may be excluded if they involve excessive repetition of cut depths or if they result in a very "spiky" key profile, for instance (1 5 1 5 1 5 1 5 1). The key pictured in Fig. 5.137 has code (2 1 3 4 5 4 2 1 1).

In some cases a pairing constraint is applied. For instance, all keys may use the same cut depths for positions 11 and 12. For each pair of cut depths that are constrained in this way, the theoretical number of combinations is reduced by a factor of five. For example, $5^8 = 390,625$ theoretical codes are available if there is one pairing constraint. Note that, although many different loading sequences for the nine levers are possible (the lock still functions with all levers inserted the same way round), the same key applies to each one because of its symmetric bittings and the central positioning of the bolt stump. Thus, different loading sequences do not lead to an increase in the overall number of combinations.

Cut position	1	2	3	4	5	6	7	8	9	10	11	12
Orientation A	L	R	L	R	-	-	-	L	R	L	R	L
Orientation B	R	L	R	L	-	-	-	R	L	R	L	R

Table 5.4: Active key bittings depending on key orientation.

The lock operates in response to the insertion of a correctly bitted key (see Fig. 5.139). Seen from the front (opposite to the view presented in Fig. 5.137), the key must be turned clockwise through half a turn to open the lock. As the key is turned, the alternate steps on the bit contact the bellies of the levers, displacing them upward or downward depending on their orientation. The four levers in front of the bolt tail as well as the five levers behind it must all be brought to the correct heights to align their gates. When the key has been rotated to about 70 degrees, the drive cam contacts the talon of the bolt tail and begins to displace both sections of the bolt stump toward the pockets in the levers. The bolt is drawn back until a point is reached where the drive cam falls out of contact with the talon. As the key turns to 180 degrees, the levers are returned to their rest positions, and the key may be withdrawn though the slot in the bolt tail. A captive key function is also available that stops the drive cam just before half a turn, retaining the key when the lock is unlocked.

Obviously, there is not a great likelihood of picking a lock such as this with conventional tools, given the difficulties presented by the small keyway and the need to manipulate levers in both upward and downward directions on both sides of the bolt tail. Recall that the bolt cannot be tensioned in the usual manner due to the bolt drive mechanism and that when this is turned to tension the bolt, the access to the five rear levers is severely restricted. Furthermore, the lock would normally be mounted behind the door of a safe with the keyway only accessible through a 60 mm toughened key guide. Nonetheless, a decoder-pick is available from Prescott's site [97] for around $3,000 Australian dollars that allows a safe engineer to open the lock nondestructively in 30 to 90 minutes. The European Certification Board [32] qualifies this lock as Class A under the ENV 1300 standard [19]; it is also approved to Class 1 under the VdS standard.

5.10   Variable

Sargent & Greenleaf 6804

(US) 7-lever (4–5)

The company Sargent & Greenleaf, founded in 1865 by James Sargent and now based in Kentucky, was a major force in the development of the keyless combination lock and the time lock for safes and bank vaults. As explained in the chapter introduction, recombinatable locks, which may be keyless or key-operated, allow the user to effect a rapid change of the lock's combination whether of necessity or as a routine security precaution. The time lock in particular had a major impact in reducing theft from financial institutions by precluding the lock from being opened, even by the bank manager, outside of office hours [57]. A picture of a S&G 6720 3-wheel keyless combination lock appears in Fig. 5.36.

Although S&G, along with many other high-security lock manufacturers, has moved toward electronic (digital) safe locks, it still produces a number of keyed safe and safe deposit locks. These must also allow the user to rekey the lock without the need to dismantle the mechanism. In this section we present the S&G model 6804 key-changeable safe lock in which recombination is effected by a variable bolt-stump array of similar construction to a Hobbs changeable lever lock (Fig. 5.39). Two versions of the 6804 lock are available, one of which is pictured in Figs. 5.141 and 5.142. Unlike the Hobbs lock, whose combination was set automatically to its operating key, the recombination of the 6804 requires the insertion of a change key, an Allen key in this case, via the rear of the lock. The relevant patent is US 4,462,230, which covers the application of the variable bolt-stump mechanism in a dual-control safe deposit lock (see Fig. 5.48).

The lock housing and the bolt are die-cast. There is a keyhole in the front cover of the lock that allows the key to be withdrawn at half a turn for the change operation. The key does not directly bear on the bolt; instead, the bolt-step of the key engages a slot in a drive disc seated behind the bolt. The drive disc has a cam that contacts the talon of the bolt at about 90 degrees of rotation.

The only other components in this simple but highly effective lock are the seven levers and their stumps. The levers are of stamped brass with an integral leaf spring. All levers are of identical outline with a peripheral gate flanked by a number of antipicking notches (see Fig. 5.142). The gate offsets and bellies of all levers are identical. The belly shape differs markedly from that of a Chubb lever in having a curve at the base of the lever that fully encloses the belly. The curved base of each lever rests against a molded stop on the bolt. The enclosed belly construction provides rigidity and limits access to the interior of the lock.

Since the levers are in all respects identical, differing is provided through the positioning of the stumps, which are also made from stamped brass stock and have a slotted rectangular body with a narrow perpendicular arm or fence. The seven stumps are stacked in an array with their rectangular portions slideably mounted on a pair of guide posts. A screw with a captive nut midway between the two posts secures the stump array in a given configuration.

The height of each stump may be set within a fixed interval. Different stump heights require differing degrees of lift of the levers. The gate in each lever is just wide enough to admit the fence of one stump. The triangular pocket in the lever is required since the angle of entry of the fence is different depending on the amount of lift of the lever.

In theory there is a continuum of different stump offset heights. As will be explained, however, the stump heights are adjusted to suit the key that is presented to the lock during the change operation. In practice, the number of depths of cut on the key is subject to the mechanical tolerances of the lock for effective differing and the overall height of the key bit. The 6804 lock pictured uses six depths of cut (including the zero cut) with an increment of 20 thousandths of an inch. A second version of

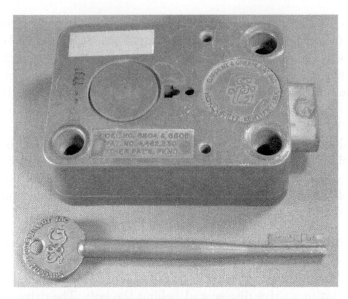

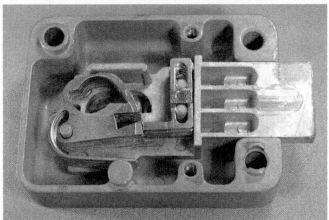

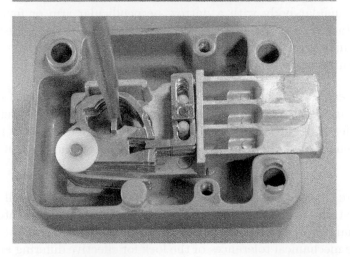

Figure 5.141: (Top) Model 6804 Sargent & Greenleaf 7-lever key-changeable safe lock and key. (Middle) Cover removed, showing lever pack and variable bolt-stump array. (Bottom) Operation by correct key.

Figure 5.142: Sargent & Greenleaf 6804 levers have identical bellies and gate positions.

the lock exists with eight depths of cut, providing more than two million keying combinations.

The brass key has a long stem and a surprisingly wide bit: there is enough space along the bit for around 14 cut positions. The reader may be wondering why there are so many cuts when only seven are needed for the levers. In fact, only positions 3–9 are used for the lever cuts, with positions 13 and 14 being for the bolt-step that operates the drive disc. The remaining positions 1, 2, and 10–12 are not required for operating the lock—in other words, they are redundant. Cuts 10–12 coincide with the cam on the drive disc, while cuts 1 and 2 turn in the void above the stump array. This space is taken up by a teflon washer on the pivot post above lever number one. A number of other locks in this book also have redundant or ornamental cuts on the key, viz., Bricard SuperSûreté and Dény.

We describe the operation of the lock by reference to the change operation [107]. The lock is assumed to be mounted with the bolt on the right. The correct key is first inserted with its bit to the right and turned 135 degrees clockwise to unlock the lock. During this stage the active key bittings contact the lever bellies, raising all seven levers to align the gates at the heights required to register with their respective stumps in the array. The key rotates the drive disc, displacing the bolt to the right so that the stumps fully enter the gate recesses. At this point an Allen key is inserted through a small hole in the back of the lock case to loosen the fence clamp screw. The key is then rotated to 180 degrees to the change slot position at 9 o'clock and withdrawn. All stumps in the array are then pushed back to their lowest height under the action of the lever springs.

A new key is inserted with its bit to the left and rotated counterclockwise about 60 degrees. This raises both the levers and the fences according to the bittings of the new key. The key is held in position, supporting the stumps at the required heights, while the fence clamp screw is retightened. The new key is then turned counterclockwise to 3 o'clock and withdrawn. This completes the key-change operation.

Naturally, a lock such as this offers a high degree of resistance to manipulation and impressioning due to the high precision of the mechanism and the presence of false gates on the seven levers. The small keyway size and the large back-set of the lock when installed in a safe add to the security level. The levers are truly identical, even down to the gate offsets and belly profiles, which are designed to minimize marking due to wear.

Mauer Variator

(DE) 11-lever (5)

The Mauer Company was founded in 1864 in Germany and has a long history of producing safe locks, especially double-bitted key locks. Among its product range we find the following mechanical safe locks (with model numbers in parentheses):

1. President A (71111) 8-lever.

2. Variator A (70091) 8-lever key-changeable.

3. Praetor B (70079) 11-lever.

4. Variator B (70076) 11-lever key-changeable.

5. Primus C (70011) 14-lever.

The A, B, and C ratings refer to the security level according to the ENV 1300 standard, as certified by the European Certification Board (Security Systems) [32]. The Mauer range is also certified by the German Loss Prevention Authority VdS and the French CNPP (A2P). The Primus is the top-of-the-line model, offering 13 billion (theoretical) combinations. We present in this section a detailed description of the Variator B 11-lever, key-changeable safe lock, illustrated in Fig. 5.143. For comparison, some pictures of the much simpler 8-fixed-lever model are shown in Fig. 5.144. Throughout we assume that the lock is viewed with the keyhole to the front and horizontal, and the bolt on the left, as in Figs. 5.145 and 5.146.

The lock is produced with a standard footprint for safes and has a die-cast case and bolt. Most of the internal components, including the levers, are made of zinc-plated steel. As in the S&G 6804, there is a variable bolt-stump array that allows the lock's combination to be changed. The twist with the Mauer is that the stumps form part of the levers and the array contains the gates. A similar construction is also found in Mosler key-changeable safe deposit locks [83]. Keys are supplied in either nickel-plated zinc alloy or nickel-plated high-strength brass. The keys are also available with a detachable bit or in a folding format, which is fortunate as the brass keys can be as long as 36 cm.

There are a number of novel aspects compared with a conventional lever lock. We start by listing the major internal components: bolt and release plate assembly,

Figure 5.143: Mauer Variator 11-lever key-changeable lock (top). Double-bitted keys for Mauer 8-lever (middle) and 11-lever (bottom) locks.

drive lever, blocking lever, set of regular levers, and corresponding gate laminations. The bolt has an extended tail portion, an enclosure for the gate laminations, a rack to accommodate the drive lever, and a cut-out that forms part of the keyway. The release plate consists of a subassembly that fits underneath the bolt, two stubs (one above and one below the keyway drill-pin), and a release bar for the gate array.

There is also a plastic disc with a toggle that is accessible through a slot in the rear of the case. This part of the mechanism, which engages with the release plate, is required for the change operation. The order of assembly with respect to the rear of the case is: release plate assembly, bolt, drive lever, blocking lever, regular levers, and gate laminations.

There are 10 regular levers of identical profile. The levers are pivoted at the right-hand end on a post fixed to the rear of the case. The levers are of German type, with a fixed-width belly equal to the bitting width of the key. The levers have an arm projecting to the left that replaces the usual bolt stump. There are 10 gate laminations forming the gate array stacked inside an enclosure on the bolt. The gate laminations have a rectangular recess on the right and a rack on the left. There are six possible vertical offsets for each gate corresponding to the teeth of the rack. A spring clip is inserted from the right that maintains the lever pack in a neutral downward position when the lock is locked.

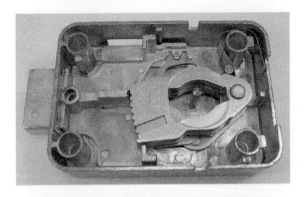

Figure 5.144: Mauer 8-lever lock with cover off (top) and operated by key (bottom).

The other two levers are irregularly shaped (refer to Fig. 5.147). We describe the blocking lever first: this is pivoted on a fixed post on the right-hand side of the case. The blocking lever has a cut-out, a post close to its pivot hole, and a protruding arm. The function of the post is to limit the angular motion of the regular levers, which have the same pivot point as the blocking lever. The blocking lever is spring-biased into a downward position where it normally prevents the rightward motion of the bolt and its subassembly. The eleventh bitting on the key operates this lever, raising the blocking arm to engage a recess in the bolt. The same action releases the regular levers so that they may be lifted to address the gate array.

The drive lever is a blend of German and Chubb levers. It pivots on the lower stub of the release plate subassembly; its upper pocket is constrained in its travel by the upper stub. The lower pocket is fashioned like a conventional bolt talon to engage the bolt-step of the key. Tangential motion is conveyed to the bolt by means of two teeth on the outside edge of the drive lever. The teeth mesh with a rack in the upper part of the bolt tail.

The key, which is double bitted, has a pipe end to match the drill-pin in the keyway. The bow of the key has a notched edge on the same side as the bolt-step to help the user identify the correct orientation. There is also a shoulder stop above the bit formed by the increased-diameter stem. The stop limits the insertion depth of the key.

Figure 5.145: Mauer Variator lock with cover off (top). Key aligns lever ends with variable gates (bottom).

One can discern 14 bitting positions along the bit of the key. The last three spaces are reserved for a compound bolt-step, which is on one side of the blade only. Spaces 1 through 11 on the key are constant-width, double-bitted cuts. The first 10 of these are used for combinating the key. Bittings 11–14 operate at the levels of, respectively, the blocking lever, the drive lever, the cut-out in the bolt, and the release plate.

In normal operation, the key is inserted with its bolt-step to the left and turned clockwise to open the lock. The bolt-step raises the blocking lever to engage the bolt recess. At the same time, the bittings in the key contact the bellies of the 10 regular levers. At about 90 degrees rotation, the lever stumps are fully raised and aligned with their respective recesses in the gate array, as shown in Fig. 5.145 (right). At this point the bolt-step on the key contacts the talon on the drive lever, the stumps of the levers engage the gate array, and the bolt slides to the right. The key stops

Figure 5.146: Mauer lock during key-change operation: key withdrawn with lock in open position and stump disengaged (top); new key inserted and turned counter-clockwise, reengaging stump and setting new combination (bottom).

before 180 degrees when its bolt-step contacts the inner shoulder of the now-raised blocking lever. It cannot be withdrawn since the bottom part of the bolt-step is retained by the release subassembly.

During the change operation, illustrated in Fig. 5.146, the toggle on the rear of the lock is moved toward the outer edge. Alternately, a change key, with a step one space lower than the end of the operating key, is inserted from the front of the lock and turned 90 degrees counterclockwise, then back to the right and removed.

The operating key is inserted in the usual manner and turned clockwise to operate the lock. The difference is that the release assembly now finishes just slightly to the left of its normal open position and the key completes half a turn. This permits the operating key to be withdrawn while the bolt is retracted. Moreover, it causes the release bar to disengage from the rack on the gate array. This frees the gate laminations, which are still meshed with the lever stumps, to return under spring action to a central position. (Note that the blocking lever continues to engage the bolt recess.)

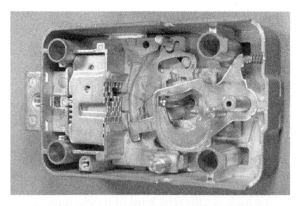

Figure 5.147: Mauer lock with lever pack removed, showing blocking and drive levers in locked position (top) and with blocking lever engaging bolt recess (bottom).

A new operating key can now be inserted with its bolt-step to the right. As the new key is turned counterclockwise through 180 degrees, it raises the regular levers and their gate laminations according to its bittings and returns the release subassembly to its usual position with respect to the bolt. The shifting back of the release bar locks the gate array into position, effectively memorizing the code of the new operating key.

There are six depths of cut. Bearing in mind that the cut depths on each side of the key must add up to the same fixed number for each cut position, the theoretical number of combinations is $6^{10} = 60,466,176$, or about 60 million. Naturally, not all of these are usable, depending on the bitting constraints applied in practice.

In addition, Mauer produces single- and dual-control safe deposit locks. It also produces a number of key-operated safe locks with electronic control interfaces to implement time delay and other functionality. Mauer, now trading as Kaba Mauer GmbH, is part of the Kaba Group of companies.

5.11   Dual-Control

(UK) **Chubb** twin 7-lever single-stump (5)
(UK) **Chubb** twin 8-lever double-stump (5)
(US) **Mosler 5700** (7 + 7)-lever single-stump (5)

Up to this point, with one exception, we have only considered locks with a single keyway. The exception was the Winfield bicentric hotel lock in Chapter 3, configured to accept two keys but operating on an "either/or" basis. In applications demanding supervised operation of a lock, a different kind of mechanism is required. In this mode of operation, called the "logical and," two different keys are required to unlock the lock. The class of mechanical locks satisfying this requirement is known as dual-control safe deposit locks and the primary application giving rise to the requirement is the banking sector. As we mentioned earlier, there are several possible implementations of the dual-control mechanism, the major variants being single- and double-keyway locks. Dual control can be achieved in a single-keyway lock by using multiple gates or by splitting the key.

An interesting example of a split-key dual-control lock is provided by G. A. Long's 1929 patent (US 1,863,525). This proposed a single-keyway 3-pin + 6-lever lock in which the key was split into two parts along its blade axis: the lever half of the key for the renter and the pin-half of the key for the custodian. The lock also allowed key-change operation via a variable stump array of the same type employed in the S&G 6804. The DOM iX-10 lock featured in Chapter 2 is also made in a split-blade configuration for safe deposit boxes.

In terms of twin-keyway dual-control locks, the S&G 4500 key-changeable lock was mentioned in the chapter introduction. The Chubb AVA lock is also used in this context. We present in this section two further locks of this type, both produced by Chubb. We will refer to these two types as double-stump and single-stump. The first type, shown in Fig. 5.148, is a twin 8-lever lock with two bolt stumps. This type of lock is of a similar design to the Sargent & Greenleaf 4440 series 11-lever lock from 1922 [84] and also the Diebold 175 series 14-lever lock from 1948 [85], both of which are widely used in the United States. The second type, pictured in Figs. 5.149 and 5.150, is a twin 7-lever lock with a single bolt stump. This lock is equivalent to the Mosler 5700 series 14-lever safe deposit lock from 1925 [86], now called the Kaba Ilco 5700. These locks all utilize flat lever keys with between six and eight depths of cut, with more differs allocated to the renter's key than the guard key.

Dealing first with the common aspects, we note that both types of lock are of rectangular construction with two barrels containing a rotatable key guide for a flat key. Since the two keyway barrels protrude from the front of the lock, the lock is sometimes referred to as double-nosed. Both locks require a renter's key to be inserted in the left-hand keyway (with the bolt on the left) and a guard key to be inserted in the right-hand keyway. The guard key is inserted first and turned counterclockwise until it stops. Next, the renter's key is inserted and turned clockwise to unlock the lock. In the double-stump lock the guard key must be rotated 180 degrees, whereas the single-stump lock only needs a 90 degree rotation during unlocking. Similarly, the renter's key must be turned through three-quarters of a turn in the double-stump and one-half a turn in the single-stump case.

Figure 5.148: (Top) Twin 8-lever double-stump dual-control safe deposit box lock; (bottom) with cover removed.

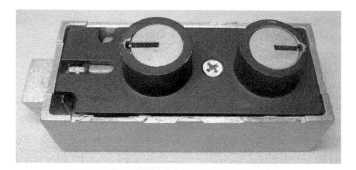

Figure 5.149: Single-stump dual-control safe deposit box lock and renter's key.

Figure 5.150: Single-stump dual-control safe deposit lock with cover removed.

The double-stump dual-control lock in Fig. 5.148 has an all-brass construction. The two slotted key guides pivot in holes in the back plate of the lock. The guide on the right has a ferrule that ensures that the guard key stops just above the bolt. The bolt has two lever stumps and two pivots, each supporting a pack of eight levers. The bolt stops short of the guard-lever pack on the right. Only the left-hand part of the bolt (the renter's section) has a talon. In each pack, the levers are end-gated with a slot to accommodate the bolt stump. Pocketed levers are not required since the lock is not designed to be left in the unlocked state.

All eight levers in the renter's pack are able to be used to combinate the renter's key. Since fewer guard key differs are generally required, only some of the eight levers in the guard pack may be active. Levers in the other positions, for instance, 7 and 8, may be inactive levers, the gates of which are already aligned with the stump. Inactive levers have a hooked belly that stops the rotation of guard key at 180 degrees to maintain the active levers at the position required to align their gates with the right-hand stump. Once this condition is met, the renter's key is inserted and turned clockwise to align the eight levers in the left-hand pack. The renter's key also makes contact with the talon of the bolt. If the levers in both packs are in proper alignment, the movement of the bolt is enabled and the lock can be operated by turning the renter's key past 180 degrees.

The single-stump dual-control lock in Fig. 5.149 functions in a slightly different way from that of its counterpart. The case is of cast alloy construction with a nickel-plated cast brass bolt. The lock is provided with two key guides: the left-hand one for the renter's key and the right for the guard key. The bolt, which stops short of the guard key section, has a cut-out for the talon and a guide slot. The bolt moves on a centrally located, threaded post onto which the front cover is fastened. The guide for the renter's key also has a cam that bears on the bolt talon (like the curtain in a Chubb 3G114 lock). As we mentioned previously, there is only one bolt stump.

There is a single pack of 14 thin brass levers that pivot on the central post (see Fig. 5.150). All levers have an open, slotted gate. The levers are of two distinct

Figure 5.151: Lever pack from single-stump dual-control safe deposit lock with gates aligned.

types and are interleaved in an alternating sequence, as shown in Fig. 5.151. Odd-numbered levers are sprung from the top edge of the case, and their identical bellies are centered above the guide for the renter's key. Even-numbered levers are sprung from the bottom edge of the case and pass underneath the renter's key guide to a tail section containing a belly located above the guard key guide. The bellies of the guard levers are also identical. Each set of levers acts as a separator for the other set.

The guard key operates the guard levers as it is turned counterclockwise through 90 degrees. This causes the gate ends of the levers to pivot downward, aligning their gate channels with the bolt stump. The renter's key is subsequently inserted and turned clockwise; the bittings act on the renter's levers, displacing them upward and aligning their respective gates with the bolt stump. At one quarter of a turn, the cam on the renter's key guide contacts the bolt recess to displace the stump toward the lever pack. Correct lifting of all 14 levers then ensures that the stump enters the gate channels, allowing the bolt to be thrown to the right.

For both types of lock, each half of the lock could be picked rather easily on its own, although reading/impressioning would be more difficult since all active levers have identical belly profiles. However, the interdependence of the two sets of levers, due either to sharing the same bolt or stump, precludes ordinary methods of manipulation since there are two separate keyways. In any event, these considerations are quite academic since the lock would normally be used inside a bank vault.

Chapter 6

Magnetic Locks

> Security is a relative term, and while no key lock is absolutely "pickproof," some of them are so difficult to pick as to defy attack except by an expert aided by all favorable circumstances. *A. A. Hopkins, The Lure of the Lock (1928)*[1]

6.1 Introduction

It had long been recognized that the blocking elements of a lock could be moved either by mechanical or magnetic force, but the development of magnetic locks was stymied by a lack of appropriate materials until the second half of the twentieth century, when powerful, lightweight permanent magnets became available in commercial quantities.

The motivation for magnetic locks is easy to understand: magnetic force can be transmitted through nonmagnetic metallic materials (e.g., brass). It is therefore possible to devise a lock that does not even possess a keyway! This is quite a twist on conventional lever-tumbler and pin-tumbler locks. The implication is that the blocking elements of a magnetic lock could not be reached or observed without destroying the lock. In practice, it is rather difficult to achieve this goal, and a number of magnetic locks have been made that had fairly serious security flaws, for instance, a susceptibility to being rapped open with a hammer.

Replacing mechanical pins with magnetic ones has other potential drawbacks. Magnetic materials are very brittle and break easily if shear force is applied to them. Magnetic pins are inherently unsuited to master-keying because they cannot be stacked (although there are ways to alleviate this deficiency). The number of

---

[1]See [56].

magnets needed to obtain a respectable number of system codes is quite large since a magnet only has two polarities: north (N) and south (S). In order to achieve a high level of security, for instance, against drilling, the lock mechanism should be provided with a thick wall separating it from the key; but this in turn requires strong magnets to work at the increased distance. Traditionally, the strength of a magnet was very much proportional to its size, and no one would want to carry a huge horseshoe magnet in his or her pocket! A more recent drawback, and one that has caused mechanical magnetic locks to be unpopular, is that a key containing permanent magnets is capable of erasing data stored on magnetic media such as magnetic stripe or swipe cards and floppy disks.

Magnetism and Magnetic Materials

The subject of magnetism and magnetic materials will be unfamiliar to many readers of a book such as this, and for this reason we first provide a brief overview of these matters. This is followed by a historical perspective on the development of magnetic locks together with short descriptions of the various operating principles. This leads naturally to the more detailed descriptions of a number of commercially produced magnetic locks. The scope of the presentation is limited to mechanical locks utilizing permanent magnets; thus magnetic stripe cards and electromagnetic (solenoid) door fasteners are excluded, as are electronic locks based on magnetically operated reed switches or other sensors. Our initial discussion of permanent magnets draws on material from [2].

Since ancient times, the Greeks and Chinese knew of and used the substance magnetite, a naturally occurring iron ore. The principal application of magnetite was in navigation where it was used to determine magnetic north, hence the name "lodestone," which means "stone that leads." The phenomenon of magnetism was studied scientifically in Europe from the 1600s and was linked to electricity in 1820 when Hans Oersted discovered that a magnetic needle could be deflected by an electric current. Around 1830, the inverse effect of electromagnetic induction was discovered independently by Joseph Henry and Michael Faraday. This discovery led to ways to generate electricity through rotating magnetic fields. The idea that materials exhibiting permanent magnetism were comprised of tiny magnetic domains was put forward by Pierre Weiss in 1907. The modern (quantum mechanical) viewpoint is that domains are formed by groups of atoms with parallel magnetic moments.

In iron, the magnetic domains are not necessarily aligned, but this can be achieved under the influence of a strong external magnetic field. This can be arranged by placing the sample of iron in the air gap of a C-shaped former with a winding to carry an electric current (Fig. 6.1). The current produces a magnetizing force or magnetization (symbol H, Fig. 6.2). As the current in the coil is increased, the strength of the magnetic field through the sample, as measured by the magnetic

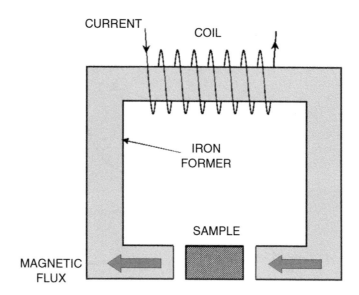

Figure 6.1: Electromagnetic circuit for testing the response of a magnetic material.

induction (symbol B), reaches a saturation point B_s on the curve in Fig. 6.2. If the current is then turned off, so that the applied field diminishes to zero, the magnetic induction relaxes to a value B_r called the remanence of the sample.

Now consider the effect of applying the magnetization in the opposite direction, so that the north and south poles of the electromagnet are reversed. The magnetic induction of the sample changes from $+B_r$ to $-B_r$ as its magnetic domains align the opposite way. The point at which the curve passes through the horizontal axis can be thought of as the external field required to demagnetize the sample, that is, reduce its magnetic induction to zero. This value of negative magnetization is called coercivity H_c [93]. If the cycle of magnetization and demagnetization is repeated, the graph of B versus H traces out what is known as a hysteresis loop; for clarity only the top half of this has been drawn in Fig. 6.2. A simple way to compare the magnetization properties of different materials is to form the magnetic energy product $H_c B_r$, which is a measure of the performance of the magnetic material in terms of the strength of its magnetic field together with how hard it is to demagnetize. This quantity is usually expressed in units of kJ/m^3 (kilo-Joules per cubic meter).

Since the 1920s, considerable effort has been expended in the development of magnetic materials with high performance figures that are also relatively inexpensive to produce, easy to work, resistant to corrosion, and stable against changes in temperature. Today many different types of magnetic materials are available including Alnico, ferrite (ceramic), and rare earth.

The Alnico magnets, made from an alloy of aluminium, nickel, and cobalt, were available prior to World War II and had a performance figure of around 10. This was improved to 55 with the introduction of Alnico V in 1940, which was used for

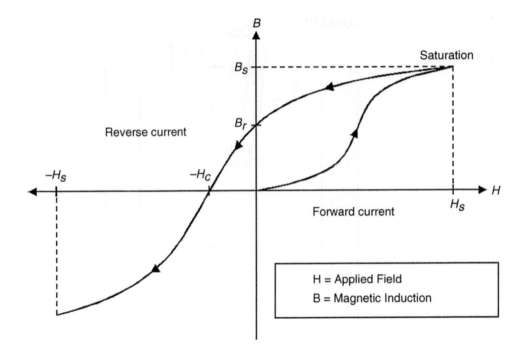

Figure 6.2: Magnetization curve for a typical magnetic material.

moving-coil loudspeakers in radio and television sets and other equipment. Alnico magnets are mechanically strong and can be cast into various shapes and sizes, such as rectangular and round bars, cylinders, discs, rings, bridges, and U-shapes.

In 1951 the first ferrite ceramic magnets were produced by Philips Corporation. These nonmetallic magnets are inexpensive to produce but are fragile and difficult to machine. They are only available in simple shapes like rectangular and round bars, discs, and rings. Later, plasto-ferrite magnets, in which the magnetic material is embedded in a flexible plastic substrate, were developed. (These are now found in the ubiquitous "fridge magnet.")

Rare earth permanent magnets based on samarium-cobalt compounds appeared in 1969. With a performance figure of 160, these magnets represented a vast improvement on the previous technology. A second generation of rare earth magnets[2] was pioneered in 1983 called neodymium-iron-boron, having a performance figure of 260 or five times that of Alnico magnets. This type of material produces a very intense magnetic field compared with other materials of similar size. Despite their high strength-to-weight ratio, the rare earth magnets are very brittle and quite expensive relative to ceramic and Alnico magnets.

---

[2]The rare earth elements or lanthanides have atomic weights 58–72 in the periodic table. They include neodymium and samarium.

Development of Magnetic Locks

The development history of locks utilizing permanent magnets has been heavily dependent on the technology of magnetic materials. Although there have been scores of patents for magnetically operated mechanical locks since the start of the 1900s, only a small number of operating principles are in evidence. If we restrict the focus to purely magnetic locks, that is, those that do not use auxiliary pin-tumblers, the principal categories are as identified below (with examples in parentheses).

1. Magnetic pin-tumblers formed by pairs of rod-shaped permanent magnets, one in the body of the lock and the other in the key (Miwa EC, Ankerslot).

2. Slideable ball bearings driven by magnets in the key (Eagle).

3. Blocking elements that slide transversely, containing one or a pair of magnets (Miwa 3800, Anker).

4. Tilting pins with a magnetic head, oriented by the position of a magnet in the key (MagLok, Genii, Sima).

5. Rotating magnetic tumblers controlling the radial motion of a side-bar (Parnis, Elzett).

6. Rotating magnetic tumblers controlling the longitudinal motion of a side-bar (EVVA MCS, IKON System M).

7. Card-in-slot locks with a matrix of magnetic pin-tumblers controlling a knob-set or handle clutch (Schlage CorKey, Schulte-Schlagbaum).

8. Code-changeable magnetic locks.

In terms of combinations, it is useful to think of the type 1 and 7 locks as having binary magnetic tumblers, since there are only two possible orientations for each one. Type 3 locks can be described as possessing *multipole* tumblers, since, by omission of a magnet on the key, more than one combination of magnets can operate the blocking element in the lock. Magnetic locks with multipole tumblers are inherently better suited to master-keying than locks with binary tumblers, a point we return to later in this section. Some of these designs have been enhanced by the addition of conventional pin-tumblers, in which case we can refer to them as *hybrid* magnetic plus mechanical locks. The Miracle Magnetic lock produced by Liquidonics Industries around 1970 fits this profile [121]. Magnetic locks with additional pin-tumblers or profile pins provide far more differs and are more amenable to master-keying.

It is instructive to view these lock mechanisms in a historical perspective since many of them are based on earlier ideas that could not be commercially realized until the appropriate materials and manufacturing processes came into existence. Some of the mechanisms, particularly the magnetic card-in-slot types, are very complicated, and many years elapsed between the prototype and the final form of the product.

Of course, like other types of locks, many of the ideas proposed in the patents did not receive commercial consideration and faded into oblivion.

Although we are mainly concerned with locks that fit the above categories, we mention for continuity of ideas some of the more obscure locks. As before, in our references to patents from the United States and other countries, we cite the filing date of the patent rather than its publication date, as this more accurately reflects when the mechanism was first developed. Appendix D contains a list of magnetic lock patents.

Locks Based on Magnetic Attraction

One of the very first magnetic locks, proposed in an 1890 U.S. patent by William Fenner, was of the tilting-pin type (see Fig. 6.3). The specification called for a lock with two pivoting pins, resembling vertically mounted compass needles, actuated by a large horseshoe magnet. In their rest positions the pins, which were top heavy, leaned over under gravity so that their pointed ends were out of kilter with holes in the back of the lock case. A magnet was required to align the two pins and push them down to engage a gear that drove the bolt. The lock did not provide much in the way of different permutations, with any suitably sized magnet sufficing to align the pins.

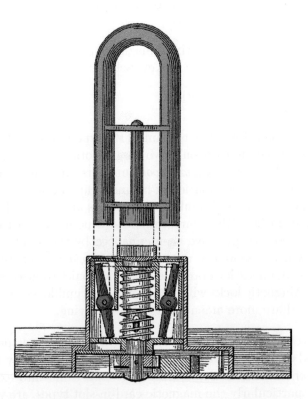

Figure 6.3: W. W. Fenner's 1890 pivoting-pin magnetic lock (US 428,247).

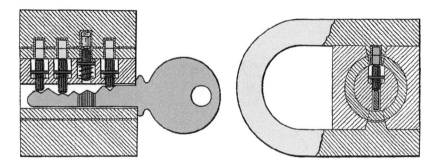

Figure 6.4: P. W. Fuller's magnetically levitating pin lock (US 1,172,203).

Another early invention, by P. W. Fuller in 1914 (Fig. 6.4), is best described as an "antigravity" lock. In this contraption the lock cylinder was constructed between the poles of a large magnet. A system of doubly sprung pin-tumblers of ferromagnetic material provided the locking function. In the locked position, the upward-acting springs forced the bottom pins *up* past the shear line. When a ferromagnetic (iron) key was inserted, reducing the air gap between the poles of the magnet, a strong magnetic field was created between the blade of the key and the bottom pins, attracting the pins *down* onto the key bittings. If the bittings were at the correct heights, the pins would be set at the shear line.

A. E. Anakin's patent of 1925 (US 1,669,115) put the magnet back into the key, which was formed by folding a piece of magnetized steel into a double-bladed key with a separating piece to ensure an air gap between the poles of the magnet. In addition to pin-tumblers, the lock had a number of passive profile balls that were attracted up into V-cuts in the bottom of the blade when the key was inserted. With no springs to tension the profile balls, the blocking mechanism would not have been difficult to circumvent.

The magnetically driven sliding ball idea was articulated in a 1938 U.S. patent by H. H. Raymond of the Eagle Lock Company (see Fig. 6.5). The lock in question was a cam lock with two steel balls in longitudinal grooves between the plug and barrel. A magnet in the key brought the balls into line with a circumferential channel, releasing the plug.

Sliding Magnetic Tumblers

We have so far only covered magnetic locks containing magnets in either the body or the key, but not both, the working principle being that a magnet attracts a ferromagnetic object like a steel pin or ball. This principle fails to capitalize on the obvious advantages of having two magnets that can generate either an attractive or a repulsive force according to their polarity and/or orientation.

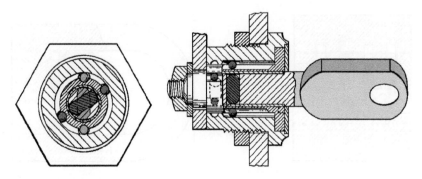

Figure 6.5: H. H. Raymond's magnetically driven sliding ball lock (US patent 2,177,996).

The earliest examples of locks with magnets embedded in both the body and the key are from the United States and Germany around 1937. A U.S. patent from that time by E. M. Ractliffe describes a hybrid lock with both pin-tumblers and blocking magnets in the barrel. A correctly bitted key carrying magnets arranged to oppose the polarity of the magnets in the barrel was required to operate the lock. This type of lock came shortly after the development of a number of new magnetic alloys including Alnico, Cunife (copper-nickel-iron), and platinum-cobalt.

A further and more significant phase in the development of magnetic locks had to wait until after World War II. By the early 1960s, both Alnico V and ceramic ferrite magnets were commercially available. The previous ideas of tilting magnetic pins, rolling balls, and various arrangements of magnetic pin-tumblers were pursued with renewed vigor. C. V. Allander's 1959 patent (US 3,056,276) laid the foundation for the tilting magnetic-tumbler lock, with the magnets located at 60-degree increments around a circle. This idea was subsequently taken up in the design of padlocks and other locking devices marketed under the brand names MagLok, Sima, Genii, and others during the early 1970s (see Fig. 6.6).

In the mid-to-late 1960s, we also see the beginnings of the matrix magnetic-tumbler lock in patents by E. L. Schlage (1964) and by B. S. Sedley (1966). These prototypes laid the groundwork for the CorKey and magnetic card-in-slot locks developed in the United States and Germany from the 1970s and improved upon in the 1980s in patents by Sedley and A. Eisermann (of the firm Schulte-Schlagbaum). Diagrams from some of the relevant patents are shown in Figs. 6.7 and 6.8. The design of these locks was motivated by the hotel sector and other large organizations where correspondingly large numbers of system codes and frequent key changes are necessary. The card-in-slot lock evolved from a purely mechanical/magnetic form through electromechanical embodiments utilizing reed switches to the now popular swipe card or magnetic stripe card microprocessor-based access systems.

The hybrid magnetic plus pin-tumbler lock called the Miracle Magnetic (Fig. 6.9) is traceable to a 1968 patent by M. M. Check, G. Mauro, and S. R. Valentinetti of the company Liquidonics Industries. The presence of four magnetic tumblers in between

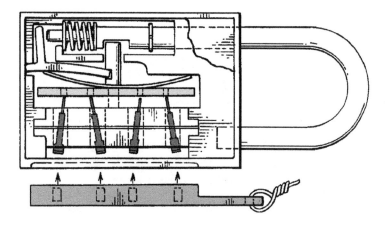

Figure 6.6: Genii/Sima 4-pin magnetic padlock from a patent describing a magnetic lock-pick (US 4,073,166 by W. H. Clark).

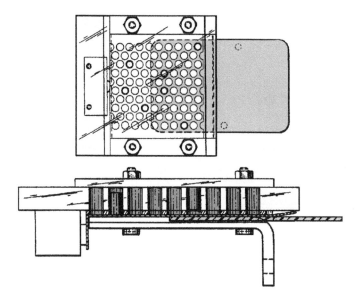

Figure 6.7: Early form of the magnetic pin matrix by B. S. Sedley (US patent 3,444,711).

the five conventional pin-tumblers increased the number of locking combinations by 2^4, or a factor of 16.

Purely magnetic tumbler locks, as illustrated by the Miwa EC and Ankerslot 14-magnet locks, were invented in Japan by K. Wake in 1967 (see Fig. 6.10). Such a large number of tumblers is needed because only two polarities are available, so that the number of combinations is theoretically 2^N when there are N tumblers (absent magnets do not add to this total). A 14-magnet lock therefore provides 2^{14} or 16,384 different combinations. The extension of the magnetic tumbler principle to blocking elements carrying more than a single magnet is also due to Wake and dates from the

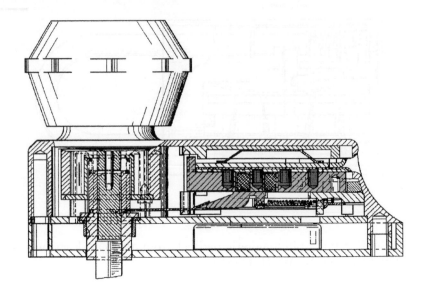

Figure 6.8: A. Eisermann's design of a magnetic card-in-slot lock (US patent 4,932,228).

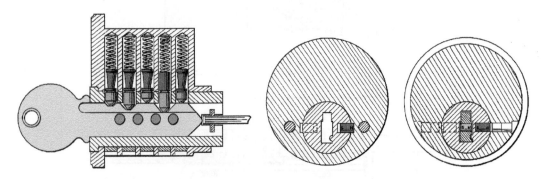

Figure 6.9: Hybrid magnet plus pin-tumbler lock (US patent 3,512,382).

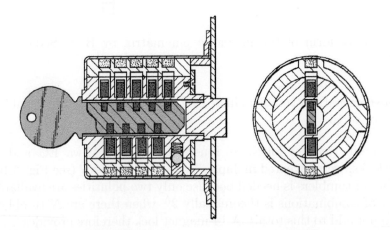

Figure 6.10: Design of the Miwa magnetic lock by K. Wake (US patent 3,518,855).

early 1980s. An embodiment of this idea is the Miwa 4-tumbler lock containing up to eight magnets, each with one of four orientations: north/south in the longitudinal or perpendicular axis. The Miwa 3800 8-magnet cylinder, introduced in 1983 [66], operates in a similar manner and includes an additional four pin-tumblers for added differing and security. We describe the Miwa EC and Miwa 3800 locks in detail in Sections 6.2 and 6.3.

Rotating Magnetic Tumblers

In contrast to the preceding locks that employ magnets to effect a sliding or translational motion in their tumblers or blocking elements, it is feasible instead to base the locking principle on rotational motion. One of the first proposals for a lock with rotating magnets was from G. Heimann in a 1966 German patent. The Heimann lock incorporated a number of rotating magnetic tumblers disposed in one or more lines or in a radial formation. The rotation was fixed at increments of 30° so that each tumbler could take any one of 12 angular positions to align its slotted end with a matching protrusion in the cylinder. The lock required complicated milling of internal grooves in the cylinder bore, which would have made it costly to produce. A number of other lock patents with rotating tumblers were filed in the late 1960s and through the 1970s. The more successful mechanisms used one of two basic principles.

The first mechanism, evidenced by Hallman's 1973 patent for MRT Magnet-Regeltechnik in Germany and also by Böving's 1974 patent, employed a set of rotating magnetic tumblers, as shown in Fig. 6.11. The tumblers were indented on the outside edge and controlled the radial motion of a blocking side-bar or pin. This idea was later commercialized by Parnis in the United States and Elzett in 1988 in Hungary, both of whom produced a six magnetic-rotor lock with twin side-bars taking a key with three magnets on each side of the blade. The design of this lock is depicted in Fig. 6.12. Further details may be found in [66]. More recently, Chubb produced a 4-magnetic rotor lock of much the same design called the 3G222. The Chubb magnetic lock, which is primarliy intended for high-security applications such as prisons, is described in UK patent 2,151,295 (1984) by J. Rogers and W. K. Robinson.

The second type of mechanism employs rotating magnetic discs, each containing a notch or gate at one or more points on their circumference, to control the lateral motion of a side-bar as illustrated in Fig. 6.13. The earliest example of this type of lock is by K. Prunbauer of the Austrian company EVVA-Werk, who in 1976 proposed an eight magnetic-rotor (four per side) lock with five conventional pin-tumblers. The commercial version of this lock was patented by A. Burger and A. Paar of EVVA-Werk in 1979. By using compact, high-strength samarium-cobalt magnets, which had been available since the 1970s, it was feasible to produce a flat key of average dimensions that incorporated magnets at each of the four

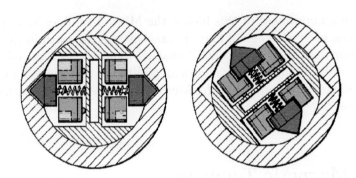

Figure 6.11: H. Hallman and B. Perkut's rotating-magnet side-bar lock (US patent 3,855,827).

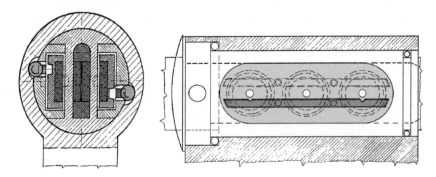

Figure 6.12: Design of the Elzett/Parnis 6-rotor magnetic lock from T. Kassza's 1989 patent (UK 2,214,226).

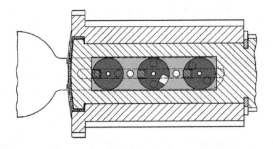

Figure 6.13: Magnetic control of lateral side-bar motion from E. Böving's patent (US 3,935,720).

locations along the key stem. This is the high-security 8-rotor, twin side-bar lock, marketed as the EVVA MCS or IKON System M, which we describe in detail later on. More than any of the preceding locks mentioned in this chapter, this particular one exemplifies what is achievable in the field of magnetic high-security locks. The most recent patents on magnetic locks are from the 1990s and are adaptations of hybrid locks to the horizontal keyway and Mul-T-Lock concentric pin formats.

Master-Keying of Magnetic Locks

Master-keying in a magnetic lock is addressed according to the type of tumbler mechanism. In terms of the list given earlier in this section, we distinguish three cases: "binary" magnetic tumblers (types 1 and 7), multipole tumblers (type 3), and rotating magnets that control a side-bar (type 5 and 6).

For a lock with binary magnetic tumblers (e.g., the Miwa EC/Ankerslot 14-magnet or the CorKey), master-keying can be accomplished by omitting magnets from some of the locks and keys in the group—a technique known as positional master-keying. For example, in a two-tier MK system some locks may be equipped with, say, 7 tumblers (level 1) and others with 10 (level 2). Keys with 7 magnets in the correct positions and polarities will operate level 1 locks but not level 2. A key with 10 magnets is required to operate a level 2 lock. One can see that by omitting some tumblers, the range of system codes is being sacrificed in order to implement master-keying. Of course, using less than the full number of tumblers is also detrimental to system security (as in any maison-keyed system).

Multipole magnetic locks are more like ordinary pin-tumbler locks in respect of master-keying. A multipole magnetic tumbler may be operated by more than one combination of permanent magnets in the key by virtue of the fact that only one correctly oriented magnet in the pair is required. For instance, a tumbler with two south magnets (SS) may be operated (attracted) by any one of NN, N−, or −N in the key, where "−" denotes an absent magnet. By reversing the argument, a two-pole tumbler with a single magnet −S is operated by magnetic key combinations NN, SN or − N. Thus, in this example, a two-pole magnetic tumbler can be operated by three different keys. The same argument also extends to tumblers containing more than two magnets or magnets polarized in different directions (as in the Anker/Miwa 3800). It is easy to see that master-keying for multipole tumblers can be accomplished by omitting magnets from the key or the tumbler (without completely removing it) to enable more than one key to operate the same lock.

Magnetic locks with rotating disc tumblers are easily master-keyed by allowing more than one angle of rotation on a particular tumbler to operate the side-bar. This is achieved by adding one or more gates to the periphery of the tumbler disc.

In addition to the methods described above, any hybrid magnetic lock, that is, one containing conventional pin-tumblers or passive profile pins, may also be master-keyed as described in Chapter 2. For such locks, the total number of system combinations equals the product of the combinations due to mechanical and magnetic elements (see Appendix A).

Chapter Organization

The remainder of the chapter presents detailed descriptions of the following magnetic locks:

1. Miwa EC 14-magnet.

2. Miwa 3800/Anker 8-magnet.

3. MagLok.

4. Miracle Magnetic.

5. EVVA MCS.

6. Schlage CorKey.

6.2   Miwa Magnetic

(JP) 14-magnet (3)

Miwa EC and Ankerslot inline magnetic locks have a rectangular keyway with no visible moving parts except for a key-retaining ball. Figs. 6.14–6.17 show a Miwa EC rim cylinder made of cast brass. The cylinder houses up to 14 magnetic tumblers in two rows of seven positioned at 3 and 9 o'clock, although typically only seven or fewer pins are installed. In the example shown in Fig. 6.15, the tumblers are housed in two plastic inserts that slot into the cylinder body from the rear. Each tumbler comprises a rod magnet encased in a nonmagnetic metal sleeve, as shown in Fig. 6.16. The sleeve is hourglass-shaped to act as a spooled tumbler. In the locked

Figure 6.14: Miwa EC magnetic cylinder and key.

Figure 6.15: (Left) Miwa EC cylinder from rear. (Right) Plug removed to show pin chambers.

Figure 6.16: (Left) Miwa EC plug. (Right) Magnetic tumblers and springs.

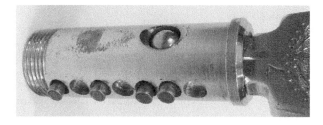

Figure 6.17: Miwa EC plug with key inserted (magnets reversed for photo).

position the tumblers, which are biased by very fine springs, engage blind chambers in the plug to block rotation.

The nonsymmetric key is made of nickel silver with a provision for two rows of seven magnets. In order to prevent incorrect insertion, the key is bulleted on the underside and there is matching warding in the keyway. The upward face of the key has a dimple close to the tip that registers with the retaining ball in the plug. Tiny rod-shaped magnets are embedded in holes along both edges of the key blade and crimped into place. The outward-facing end of each key magnet is the opposite polarity to the corresponding tumbler magnet. When the key is inserted, magnetic repulsion between the key and tumbler magnets causes the tumbler magnets to be forced back against their driver springs a sufficient distance to clear the bores in the plug. The locating ball captures the key while the plug is being turned. (Note that, for illustrative purposes, in Fig. 6.17 the magnets have been reversed since they would otherwise not stay in place with the correct key fully inserted.)

As mentioned previously, a 14-magnet Miwa lock provides a maximum of 2^{14} or 16,384 different codes. The omission of one or more magnets from the cylinder allows more than one key to operate the lock. Only as many magnets need be installed in the key as there are magnetic tumblers in the cylinder, with the remaining holes filled by dummy magnets. In a master-keyed system, different positions on the key can be used to actuate different locks. Extra profile balls may also be used to increase master-keying possibilities (see Chapter 2).

Drill protection for the pin chambers is provided in the form of a hardened steel washer at the front of the cylinder. The lock is susceptible to decoding with a suitably designed magnetic probe, as described in US patent 4,229,959 (1979) by T. E. Easley. This device had the capability to decode the combination and actuate the tumblers electromagnetically to open the lock. The Miwa EC has been superseded by a hybrid model with four additional pin-tumblers and more recently by a six pin-tumbler model. Further information on the Miwa magnetic lock is contained in [113].

6.3   Anker

(NL) 8-magnet + 4 pins (3–4)
(JP) **Miwa 3800** 8-magnet (3)

The Anker or Miwa 3800 lock, illustrated in Figs. 6.18–6.20, comprises four twin-magnet tumblers teamed with four conventional pin-tumblers. Aspects of the design are discussed in a 1980 patent by K. Wake (US 4,398,404). The pin-tumblers are arranged along a central groove at 12 o'clock inside the horizontal keyway and are actuated by bittings on the top side of the key. There are four lower pin sizes, ranging from 0.100 to 0.175″. As shown in Fig. 6.20, the bitting edge for the pin-tumblers is raised above the surface of the blade, providing increased potential for differing and at the same time making it harder to reproduce the key. (A similar principle is employed in other more recent horizontal-keyway locks like the DOM iX-5 HT.) Our description of the Miwa 3800 also covers the Miwa 8-magnet lock, shown in Figs. 6.21–6.23, which does not contain any pin-tumblers.[3]

The plug (Fig. 6.19) has four transverse bores in its lower half to receive the four twin-magnet tumblers. Each magnetic tumbler consists of one or two samarium-cobalt magnets mounted on a spring-biased slider with a hooked metal end

---

[3]Some of the information for this section was supplied in [124].

Figure 6.18: Anker/Miwa 3800 cylinder with top and bottom views of the key.

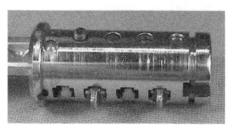 

Figure 6.19: Anker plug with key partially and fully inserted.

Figure 6.20: Positioning of magnetic tumblers (left) and pin-tumblers (right) by Anker key.

(see Fig. 6.23). Tumblers are arranged in an alternating sequence in the plug chambers in terms of their directions of action. A metal clip runs along the bottom edge of the plug that provides a back-stop for the very fine tumbler springs. The inner wall of the cylinder has two longitudinal channels, at 3 and 9 o'clock,

Figure 6.21: Miwa 8-magnet cam lock and key.

Figure 6.22: Barrel and plug from Miwa 8-magnet lock.

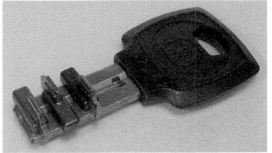

Figure 6.23: (Left) Miwa magnetic tumblers. (Right) Key with three magnetic tumblers in position.

into which are fitted slotted steel "ladders." The slots in the ladders provide anchor points on the left and right sides of the cylinder for the hooked ends of the magnetic tumblers; without the conventional pin-tumblers, the key could be withdrawn with the plug in the unlocked position since the magnetic tumblers disengage from their anchor points during operation. The Miwa 8-magnet lock in Fig. 6.22 has a key-retaining pin at 12 o'clock to fulfill this function.

The tumbler magnets measure roughly 60×80 thousandths of an inch in the horizontal plane. Since samarium-cobalt is a very high-permeability material, only one small-sized magnet is needed to compress the slider spring when a key magnet of opposite polarity is placed directly underneath it. This arrangement is achieved by positioning the top of the tumbler chambers just slightly above the bottom of the keyway, leaving only a thin wall of brass to isolate the chambers from the keyway.

System flexibility is provided by the fact that the rectangular tumblers, which support up to two magnets, may be magnetized in either longitudinal or normal axes. Thus, looking along one of the tumblers, the polarization axis can either be along the tumbler or in the up/down (perpendicular) direction. Denoting north by N and south by S, the longitudinal polarization is either NS or SN; the normal polarization is either N or S (depending on the pole of the magnet facing up). Combined with the fact that a tumbler magnet may also be absent (denoted by "−"), we obtain the $5 \times 5 - 1 = 24$ different tumbler types listed in Table 6.1; the exclusion is for the double blank tumbler.

The key has four pairs of magnets set into the bottom of the blade. Since only one real magnet in each pair is required to operate a tumbler, one of the magnets in each pair is usually a blank. In a non–master-keyed Miwa lock, only a single magnet is required per tumbler, each of which is actuated by a single magnet in the key. In a master-keyed system, one or more tumblers are equipped with two magnets (as described in the introductory section on master-keying): thus the lock may be made to operate with more than one key.

In terms of keying combinations, one is tempted to use the fact that there are 24 different magnetic tumbler types in the computation. This is incorrect, however, as many of these are operated by the same key (recall that only one magnet per key position is required to operate the tumblers). We can calculate the number of combinations requiring different keys as follows. Each position in a tumbler may carry one of four different magnetic orientations, totaling eight; each of these requires a different magnet in the key. We can now see that the total number of different magnetic key codes is 8^4. We then multiply this by the number of different pin-tumbler codes, which is 4^4, obtaining a final tally of $8^4 \times 4^4 = 1,048,576$.

NS	SN	N	S	NS	SN	N	S	NS	SN	N	S
NS	NS	NS	NS	SN	SN	SN	SN	N	N	N	N
NS	SN	N	S	NS	SN	N	S	-	-	-	-
S	S	S	S	-	-	-	-	NS	SN	N	S

Table 6.1: The 24 magnetic tumbler types for Miwa 3800 locks.

The system, while being immune from manipulation by conventional lockpicking tools, is susceptible to decoding with magnetic probes or other devices and materials that are capable of detecting magnetic polarity. Once the magnetic tumblers have been decoded, a "skeleton key" could be made up that contains the correct magnetic code. The skeleton key is then inserted, leaving a small space for the manual picking of the four pin-tumblers. Alternately, the loose tolerances of the magnetic tumblers could be exploited to allow the four pin-tumblers to be picked first, leaving the four magnetic tumblers to be actuated (note that these will relock after the plug is turned through 180 degrees). The light springs on the magnetic tumblers make the lock susceptible to rapping, which is a common problem with magnetic-tumbler locks, particularly if not all the tumbler positions have been filled. Fortunately, the dual-action (mechanical plus magnetic) nature of this lock makes it considerably more resistant to manipulation.

6.4   MagLok

(US) 4-magnet (3–4)

This unusual tilting-pin magnetic lock, now obsolete, was used primarily for padlocks and vending machines. It is similar to a 1975 design by W. F. Stackhouse of the American Locker Company (US 3,948,068). Although not constructed as a high-security lock, it demonstrates how magnetic tumblers can be made in such a way that it is next to impossible to pick or decode them manually without exhaustively trying combinations. Locks built on equivalent principles were also produced in Germany under the names of Sima and Genii [66].

The MagLok comprises a cylinder and tail shaft, and as can be appreciated from Fig. 6.24, is designed for flush mounting. It comes close to the ideal of not having a keyway. The key consists of a round flat disc, with a handle and two locating tabs, covered with a stainless steel cap. There are four magnets mounted inside the disc.

Figure 6.24: MagLok magnetic lock and key (top and bottom views).

At the front of the lock is a round recess into which the key fits snugly, located by the tabs, and which captures the key once it is turned. Inside the cylinder housing is a core divided into a front and rear section, as shown in Fig. 6.25. The front section is thicker, containing four axial chambers that house the magnetic tumblers.

A centrally located shaft is anchored in the front core, and this forms the tail-piece of the lock, to which a linkage or cam may be affixed. The rear section of the core is thinner than the front and has on its face four irregularly spaced blind holes. There are also two guide holes, in line with two similar holes in the front core, through which two stainless steel posts are fitted. The drive shaft passes straight through the rear core so that the two sections can slide longitudinally with respect to each other. The rear core has a milled groove that completely encircles its rim. The groove is set back from the face of the core except in the two regions away from the guide posts, where it is positioned further forward.

On each side of the cylinder housing there is a threaded pin whose end is located in the forward part of the groove when the core is in the locked position (Fig. 6.26, left). A spring-loaded ball bearing in the front part of the core provides positive location. The function of the groove is to control the relative displacement of the rear core with respect to the front as the core is turned. (Recall that the two guide posts do not permit the two parts of the core to rotate relative to each other.)

Figure 6.25: MagLok core consists of front and rear sections; front section contains four tilting magnetic tumblers.

Figure 6.26: MagLok mechanism with key inserted and pins aligned (left); front and rear sections of core engaged (right).

Thus, ignoring the action of the tumblers, the rear core will be displaced away from the front part in the locked position, leaving a small gap between the two. As the core is turned from the locked position, the action of the pin in the groove is to slide the rear core forward until the pin attains the back part of the groove; at this point the back part of the core is brought into contact with the front part.

We have so far omitted the magnetic tumblers from the equation. These are of a most unusual design: each one is in the form of a steel spike with a disc at its midpoint and a small toroidal magnet mounted at the top (see Fig. 6.25). The tumbler is a rigid structure but is mounted loosely in its chamber in the front core. A strong C-clip is inserted in a groove about halfway down the chamber. Since the chamber bore is smaller at the front than at the rear, the tumbler is captive in the chamber: its disc resting on the C-clip, and its upper surface caught on the lip of the thin part of the bore. There is still a small vertical gap in which the tumbler disc can move, so this allows it to tilt slightly off its axis. The direction of the tilt is unconstrained: it can be anywhere inside a circle. The extent of the tilt is set by the magnet contacting the side of the top bore. Any tilt in the magnetic tumbler causes its spiked end to lie off the central axis of the chamber.

It is now straightforward to see how the mechanism operates when the key is inserted in the front cap of the lock. Given the appropriate polarities, each magnet in the key attracts a magnetic tumbler, which moves as far as possible toward it. Each magnetic tumbler is tilted in a direction determined by the positioning of the corresponding magnet in the key. If the key has magnets in the correct places, the spiked end of each tumbler will be properly aligned with the blind holes in the rear of the core, as illustrated in Fig. 6.26. Now, as the key is turned, the threaded pin encounters the sloped part of the groove in the rear core, which is forced toward the front. The forward translation of the rear core relative to the front is possible since the tumbler spikes can enter the blind holes. The key then turns both parts of the core together to operate the drive shaft.

The number of key combinations in this system depends on the number of tilt positions for each tumbler. The polarity of the magnets is immaterial as long as they always attract the tumblers (repelling the magnets merely causes them to move to the opposite tilt). The mechanical precision of the lock is such that it can easily support eight positions per magnet. If we imagine a little dial drawn around the chamber of each magnet, then the number of codes is the number of settings on these dials, viz., $8 \times 8 \times 8 \times 8 = 4,096$. Note that the absence of a magnet is excluded from this calculation since one could not expect the tumbler to remain perfectly central in its chamber, especially if the lock is mounted on a vertical surface.

Now we come to the interesting part of the discussion: how could such a lock be picked? (Note that the lock is easily bypassed by forcibly removing the thin stainless steel front plate to expose the tumblers, which could then be manipulated by hand.) As far as nondestructive techniques are concerned, there is no point trying to probe the position of the magnets since they tilt. The polarity is relevant, but this would

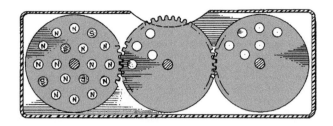

Figure 6.27: Magnetic lock-pick for tilting-pin padlocks (US patent 4,073,166 by W. H. Clark).

be easy to establish. The hard part is figuring out how to tilt each tumbler to align it with its hole in the rear core. A trial-and-error method, suitable for the Genii/Sima four inline magnet padlocks, was proposed in a 1977 patent by W. H. Clark (see Fig. 6.27). This tool, which generates an almost random magnetic field distribution, would not be very effective with the MagLok since applying torque to the lock prevents the tumbler spikes from moving due to the advancement of the rear core. One could design an electromagnetic pick to sequence exhaustively through all the codes, but this is stretching the bounds of practicality.

6.5   Miracle Magnetic

(US) 4-magnet + 6 pins (3)

Around 1970, Liquidonics Industries produced a hybrid lock featuring both mechanical and magnetic components. The so-called Miracle Magnetic lock, illustrated in Figs. 6.28 and 6.29, was a standard six pin-tumbler cylinder lock incorporating an additional set of four magnetic tumblers. The lock was offered in a variety of packages including rim and mortice cylinders and padlocks [121].

By combining both pin-tumbler and magnetic tumblers that function independently, the number of different keys for the pin-tumbler lock is multiplied by the number of permutations of the magnetic tumblers. Since there are four magnets, the number of permutations is $2^4 = 16$, so theoretically the system could support 16 times more codes. In reality, the number of available codes is somewhat less than this, since the positioning of the four magnets on the key blade prevents the use of large depths of cut in positions 1–5 as these would weaken the key excessively. The second and potentially more important aspect is that the presence of magnetic tumblers prevents the lock from being picked open by conventional means.

As seen in Fig. 6.29, the construction of the lock is almost identical to a standard pin-tumbler cylinder with a number of small modifications that could be retrofitted. The barrel has two neatly milled channels at roughly 4 and 8 o'clock in addition

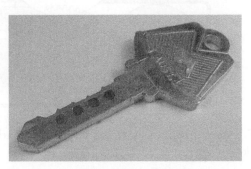

Figure 6.28: Miracle Magnetic hybrid magnetic lock and key.

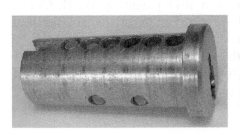
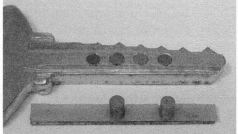

Figure 6.29: (Left) Miracle Magnetic plug showing location of two of the four magnets. (Right) Key, magnets, plastic and steel strips.

to the usual line of pin chambers at 12 o'clock. Each channel is fitted with a thin steel strip. In front of the steel strip is a plastic strip with two indentations for the magnetic pins.

The core is a standard 6-pin core except for two 0.1″ holes on each side at 4 and 8 o'clock to house the magnets. The holes are blind so that the magnets do not encroach on the keyway. The magnets on the left-hand side reside between pins 2 and 3, and pins 4 and 5. The right-hand side magnets are located between pins 1 and 2, and pins 3 and 4. Note that the plug is brass and therefore nonmagnetic. The keyway section is quite square, with a ward at the lower right and a thinner top portion. The key itself is made of an aluminium alloy. It has six top bittings and four rod magnets embedded in the blade.

In the locked position, the four magnets are attracted to the steel strips in the two channels. The ends of the magnets are captured by the indentations in the plastic strips and, although they are only about 15 thousandths of an inch proud of the rim of the plug, it is sufficient to block the rotation of the plug. This last remark still holds even if all six pin-tumblers are at the shear line (although it is not clear how much torque would be needed to make the magnets pop out of the channels). To unlock the lock, the key must simultaneously align all six pin-tumblers and have the right combination of magnets. There are 16 possible magnetic profile codes (NNSN, NSNS, etc., with N = North, S = South).

Certainly, it is not possible to pick the lock without taking care of the magnetic tumblers. It is doubtful that it could be rapped open since there are three lines of action. However, there is more tolerance in the magnetic tumblers than the pin-tumblers; therefore the pin-tumblers could probably be picked first and the plug turned forcibly. Although not a practical burglary proposition, it would be relatively easy to read the magnetic code using a Hall effect probe or magnetic (reed) switch. A key blank could then be prepared with the right magnets in order to impression the lock.

6.6   EVVA MCS

(AT) 8-magnet + 4 pins + 2 side-bar (5)
(DE) **IKON System M** (equivalent)
(IT) **Mottura MC** (equivalent)

The EVVA Magnetic Code System (MCS) and the equivalent IKON System M cylinder locks [66] feature eight magnetic rotors, twin side-bars, and a 12 profile-ball system for key control. Pictures of the EVVA MCS cylinder and key appear in Fig. 6.30, while Fig. 6.31 shows the IKON System M. The MCS system is subject to a number of patents held by the firm EVVA-Werk of Vienna, the first dating from around 1975, lodged in Austria. Early designs for the MCS lock by K. Prunbauer suggested a six magnetic rotor design, with three rotors per side as in Fig. 6.36, supplemented by five regular pin-tumblers. Like the later EVVA 3KS, the pin-tumblers were excluded from the production version of the lock, which appeared in 1979. A description follows of the operating principles of an EVVA MCS lock, with accompanying illustrations in Figs. 6.32–6.35.

The EVVA MCS is supplied in standard Europrofile, oval, rim, and mortice cylinder formats with a nickel silver key. Unlike conventional pin-tumbler locks, there is almost no friction during key insertion. The key, which contains four in-line magnets,

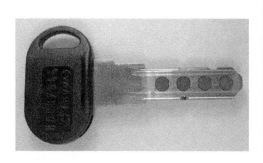

Figure 6.30: Key and cylinder from EVVA MCS equivalent by Mottura.

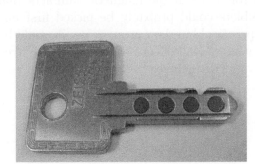

Figure 6.31: Key and cylinder from IKON system M lock.

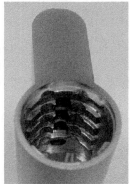

Figure 6.32: (Left) Top view of EVVA MCS plug showing profile-control balls and driver pins. (Right) Cylinder ribbing and driver pin chambers.

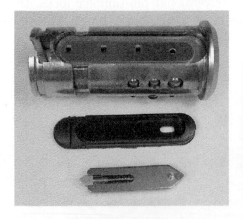
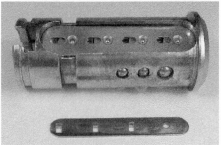

Figure 6.33: (Left) Side view of MCS plug with outer side-bar and membrane removed. (Right) Inner side-bar removed to show strip covering rotor chambers.

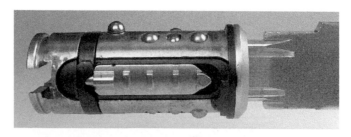

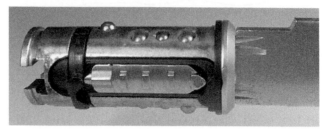

Figure 6.34: Operation of MCS Plug: side-bar slides forward on contact with plastic ring.

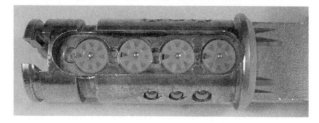

Figure 6.35: (Top) MCS plug with key partially inserted and rotors scrambled. (Bottom) Rotor gates aligned by key.

is not reversible. Each of the key magnets can be magnetised independently on its two faces. The locking plug comprises two racetrack-shaped recesses, at 3 and 9 o'clock, that house the eight magnetic rotors. The top and bottom sides of the plug house the profile-control balls. Only the profile balls communicate with the keyway via narrow channels at 6 and 12 o'clock.

Each magnetic rotor consists of a samarium-cobalt magnet that can be aligned by a magnet on one side of the key in one of eight possible orientations (similar to an 8-pole stepper motor). The rotors are arranged in two rows of four, one row on each

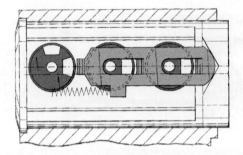

Figure 6.36: Initial design of the EVVA MCS from K. Prunbauer's 1976 patent (US 4,084,416).

side of the plug (see Fig. 6.35). Each rotor also contains a gate in its circumference. Magnets in the key cause the magnetic rotors in the plug to rotate to the correct orientations to align the gates at 9 o'clock (toward the rear of the plug). The rotors are damped so as to minimize oscillatory motion and ensure rapid and accurate positioning by the key. The magnetic tumblers alone account for $8^8 = 16,777,216$ keying combinations.

The side-bars (Fig. 6.33) have a two-part construction with an outer heavy-duty part and an inner light-duty part. The outer side-bar has a single stump that links through a plastic membrane to the inner slider equipped with three further stumps: the membrane serves to isolate the magnetic rotors. The outer side-bar is sprung from the rear of the plug by a spring-loaded pin that rides on a ring containing a wide dip. As the plug is turned by the key, the side-bar pin encounters a ridge in the ring that causes the spring to apply axial tension to the side-bar. This action forces the side-bar stump and the three stumps on the inner slider against the rotors. The rotors must all be correctly aligned so that the side-bar and slider stumps can enter the gates, allowing the side-bar to be displaced axially toward the front of the plug (see Fig. 6.34). The same logic also applies to the rotors and side-bar on the opposite side of the plug. If any of the rotors are incorrectly aligned, the inner slider will not be able to move forward and the ribs in the outer side-bar will not mesh with the corresponding crenellations milled into the inner face of the cylinder (Fig. 6.32, right), blocking rotation of the plug.

The profile-control system consists of four sets of three profile balls, each of one of two possible sizes (small and large). There are also three spring-loaded blocking pins in the pin chambers (Fig. 6.32, left). A fourth pin is ball-driven and functions to retain the key while the cylinder is in the unlocked position. The profile balls are passive, being activated by bittings in the four edges of the key. A large profile ball in one of the channels at 1, 5, 7, or 11 o'clock will block if the corresponding key bitting is not present. A small profile ball will not block the cylinder channel, but, if the key *is* bitted (so that the ball is *not* maintained at its maximum radial distance from the center of the plug), then a cavity will be created that will trap one of the three blocking pins as the plug starts to rotate.

The profile-control system multiplies the combinations due to the magnetic rotors. For illustrative purposes, we assume that each profile ball provides a binary operation corresponding to whether a small or large ball is installed. If we further assume that all 12 positions are used, then the number of possible profile options is $3^{12} = 531,441$ since a ball in a given position may also be absent. The number of theoretical combinations taking both the magnetic rotors and profile balls into account would therefore be $531,441 \times 16,777,216 = 8,916,100,448,256$ (roughly 8.9×10^{12}). This is a truly enormous number of keying combinations, though somewhat less than the quoted number of 2.99×10^{26} (299 quadrillion).

More recent versions of the EVVA MCS include a wavy milling on both the top and bottom edges of the blade that replaces the system of profile balls, although a key-retaining ball is still present. The millings, which are limited to the width of the key blade, operate a system of transerve sliders mounted in cavities in the plug at 6 and 12 o'clock. Viewing the cylinder as in Fig. 6.31, there are three sliders at 12 o'clock and 4 at 6 o'clock. The sliders are slotted so that the millings on the key displace them to the left or right as the key is inserted. A key with the correct top and bottom millings aligns the ends of all seven sliders at the rim of the plug.

In a similar vein to DOM and Winkhaus, versions of the EVVA cylinder are available that contain a chip in the key that transmits a digital code to the electronic part of the lock; this variant is called the EVVA ECS (Electronic Control System).

6.7 Schlage CorKey

(US) 24-magnet (3)

Having more than 20 parts just in the handle part of the lock and over 20 magnetic tumblers, the Schlage CorKey lock is without doubt a complicated device. Photographs of the lock and its component parts appear in Figs. 6.37–6.41. A 1989 article in the *Locksmith Ledger* stated: "locksmiths either love it or fear it." Perhaps the ones who fear it are those who have had to strip it down and reassemble it! The lock design evolved over a long period starting in the mid-1960s, traceable via a string of patents by Sedley (see Table D8 in Appendix D). At roughly the same time, a German magnetic card lock was also developed by Eisermann. The commercial version of the CorKey was introduced around 1974. CorKey locks have been installed on many large sites in the United States, including the San Francisco Mint and the FBI Building in Washington, DC.

CorKey Control Systems sells the CorKey lock as part of a "CorKit" replacement for key-in-knob sets, dead bolts, and rim locks [95]. The CorKit transforms a standard key-operated lock into a card-operated lock, which is a strong selling point for upgrading old key-operated MK systems. Supporting equipment and software are supplied to enable the user to erase and reprogram the card keys, which do not need to be replaced. This is important for large installations wishing to minimize maintenance costs since it allows the building supervisor to recode the keys without calling a locksmith. The lock is also mechanically reconfigurable via the so-called TriSec mechanism, and we will see how this is achieved shortly. In the subsequent description, it is important to distinguish between the nonmagnetic metal parts (copper, stainless steel, and alloys) and the magnetic steel parts.

We start by describing the key, pictured in Fig. 6.37, which is far simpler than the rest of the lock mechanism. The key measures approximately $2\frac{2}{3}''$ in length and $1\frac{1}{2}''$ in width, and is only 50 thousandths of an inch thick. It consists of a plasto-ferrite sheet magnet (like a fridge magnet) sandwiched between a stainless steel front plate and alloy backing plate crimped around the edges. The key is inserted in the direction indicated by the arrow into the slot in the knob with the

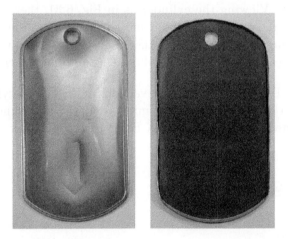

Figure 6.37: (Left) Schlage CorKey card key. (Right) Cover removed to expose flexible magnetic strip.

Figure 6.38: Schlage CorKey magnetic card-operated entrance set.

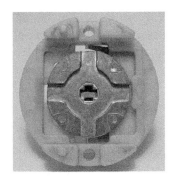

Figure 6.39: (Left) Front part of knob with armature removed. (Right) Underside of armature and coupling.

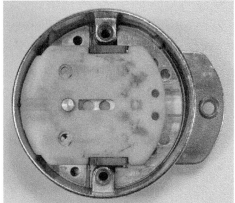

Figure 6.40: Inserting CorKey card to depress carriage.

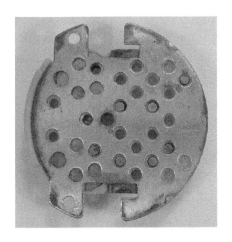
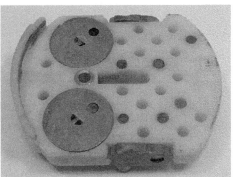

Figure 6.41: CorKey carriage: fixed copper base-plate (left); magnets and turntables for combination change (right).

alloy backing to the rear, as shown in Fig. 6.38. The nonmagnetic backing for the key allows the permanent magnets in the lock to respond to the magnetic pattern on the card. It is possible to repeatedly code and recode the rubber sheet with a new magnetization pattern. This can be done electromagnetically, like recording a signature on a magnetic stripe card, resulting in a different distribution of north and south poles on the rear surface of the card.

Next we consider the construction of the lock, focusing only on the parts of the knob that are relevant to the locking function. When locked, the knob turns without engaging the spindle of the lock. The action of inserting the correct card key and pressing it down until it clicks into place, connects the knob to the spindle and allows the lock to be operated. When the knob is returned to the upright position, the key may be removed.

The front of the knob is made of a cast, nonmagnetic alloy with a slot in the top for the card. Inside the back part is a teflon armature that is fixed by two screws to the front of the knob (see Fig. 6.39). A tail-piece connects through the collar of the knob to the bolt mechanism. In between the teflon armature and the knob there are six layers of components consisting of, from front to back:

1. Copper leaf spring.

2. Front keyway plate (steel).

3. Back keyway plate (alloy).

4. Perforated copper plate.

5. Teflon carriage.

6. Cast alloy coupling.

Other components, including the magnetic tumblers, are inserted into the teflon carriage and we deal with these presently. The first four of the above components are slotted onto the mounting posts of the knob. The front keyway plate can move forward a small distance against the leaf spring to allow the key to be inserted.

The carriage is mounted on the armature and can only slide up and down. The coupling, which is pinned to the tail-piece, can only rotate. There is a brass stud on the back of the carriage with a return pin in its base. The default position for the carriage is up, toward the key-slot. The coupling is shaped like a disc that is flat at the back where it connects to the tail-piece, but with several ridges at the front. The ridges are at the rim of the disc so that it is higher there than in the middle. The ridges are punctuated by two or more radial channels. The interplay between the coupling and the stud on the carriage is such that (i) when the carriage is up (the locked position), the stud is in the middle part of the disc and cannot drive the coupling; (ii) when the carriage is down, the stud slips into one of the channels between the ridges on the disc, engaging the coupling.

The teflon carriage (see Fig. 6.40), apart from the stud, is flat at the back and peppered with blind holes at the front. There are 22 holes in all as well as two curious discs, also having a hole, whose function we cover later. In addition, there is a flange on the front of the carriage that occupies the bottom of the keyway. Downward pressure is exerted on the carriage when a card key presses on the flange. A rear-facing swing hook is mounted on a pivot on each side of the carriage.

The magnetic tumblers are inserted into the matrix of holes in the carriage, giving a maximum total of 24 positions, counting the ones in the discs. The matrix is somewhat irregular (refer to Fig. 6.41 to understand the layout). The perforated copper plate is located in front of the carriage. In the locked position, the holes in the carriage match the perforations in the plate. Suppose now that the carriage is populated with magnetic tumblers. These are small but powerful magnets about 0.2″ in length and 0.09″ in diameter.

Consider now what happens to the tumblers in the locked position. The carriage is teflon, as is the armature, so there is nothing magnetic for quite some distance behind them. The perforated plate and back plate to the keyway are both nonmagnetic. However, the front keyway plate is steel, and, since the keyway back plate is quite thin, the magnets move toward the front plate and latch on. At the same time, the magnets have passed through the perforated copper plate, which is held in position on the mounting posts of the knob. Any downward pressure on the flange of the carriage results in the magnets binding in the perforations of the copper plate, which is thick enough to withstand considerable force.

Now suppose that a card key is inserted whose magnetic coding is precisely the same as the magnetic tumblers on the carriage. The key intervenes in the space between the two keyway plates, with its nonmagnetic (alloy) side to the rear of the lock. It is now the magnetic sheet in the card that is the closest magnetic object to the tumblers. The magnetic repulsion exerted on the tumblers causes them to be displaced into the blind holes in the carriage, freeing them from the perforations on the copper plate. Clearly, if one or more tumblers is not matched by a pole on the card, it will continue to protrude into the plate and the carriage will remain locked.

If all the tumblers match the coding of the card, the carriage can be depressed into the knob. As a result of this action, the two hooks on the carriage pivot back as they move out of a slot in the keyway plate. The hooks engage a copper clip in the armature that holds the carriage down. At the same time, the brass stud reaches the channel in the coupling, thereby engaging the tail-piece of the lock. The CorKey thus completes the linkage and the knob operates in the usual manner. Withdrawing the key releases the two swing hooks and the carriage springs back under the tension of the return pin, ejecting the card key.

One can appreciate that with 24 possible tumbler positions, the number of system codes is effectively limitless. There are more than 16.7 million (2^{24}) permutations since not all the tumblers need to be present. In practice, the lock is set up with seven or more tumblers.

But what of the two curious discs we alluded to earlier? These are actually turntables that enable the pin that they contain to be rotated to any one of five positions (2, 3, 4, 7, or 11 o'clock). Thus the total number of tumbler positions is in fact $24 + 8 = 32$, of which 24 may be used at the same time. The discs contain a spring-loaded bolt that locks them into one of the five positions. In the center of each disc is a semicircular hole. The lock is cleverly constructed so that these holes line up exactly with two holes in the front of the knob. When the correct key is used to depress the carriage and then withdrawn slightly, there is through access for a change key to be inserted to reorient the discs. With two such changeable discs there are 25 available changes that can be used when it is desired to recombinate the lock. The marvelous point is that it is not necessary to dismantle the lock to change the configuration of the magnets. Naturally, new card keys can be issued via the recoding unit.

Master-keying is accomplished by arranging for different locks to use different tumbler positions. A master key can be coded for all the tumbler positions, whereas the submaster and change keys use only a subset of these positions.

As for manipulation resistance, the key-slot is very narrow, and there is little space for a magnetic probe or decoder. In a high-security installation, the lock should be mounted with a protective collar to prevent it from being snapped or sawed off. Following the development of the original invention in the mid-to-late 1970s, further patents were filed for modifications to improve the lock's resistance to rapping.

Chapter 7

Car Locks

> It is a mistake to hold the fatalistic idea that thieves will get in if they are sufficiently determined, no matter what. *F. J. Butter, c. 1958*

7.1 Introduction

As the era of horse-drawn transport was winding down and the petroleum age was ramping up, people were seeking practical means of providing individual transportation that did not involve stoking the boiler of a steam engine. Two Germans, Gottlieb Daimler and Wilhelm Maybach, invented an internal combustion engine and were among the first to install it in a four-wheeled vehicle, achieving in 1886 the breakneck speed of 14 miles per hour. Although Daimler died in 1890, his business partner Maybach went on to develop the Mercedes car. Daimler's engines were among the most reliable of their time, but they were not aimed at providing an affordable car for a mass market.

The first recognizable motor vehicle companies in the United States started up in the late 1890s, leading in the early 1900s to the two giants we recognize today—the Ford Motor Company and General Motors Corporation. By 1910, the General Motors Company was offering closed car bodies as standard equipment. At that time, automobiles had latches on the doors but locks were not generally fitted. One of the earliest patents for an automotive lock is from 1909 by F. P. Pfleghar (US 914,669), who proposed a push-key operated cylinder lock for limousine doors having a sliding dust guard. The following year R. D. Markham patented a pin-tumbler cylinder for locking the starting crank (US 958,815). In 1911, Charles Kettering of General Motors' research department devised an electric self-starter motor, which was installed in a Cadillac. A number of steering wheel locks were also proposed around this time. The earliest of these was due to E. R. Creamer (US 915,416), whose patent was filed in 1908. A different design was patented by M. R. Rosen and R. E. Fischel in 1917 (US 1,266,161).

Through the early to mid-1920s, automotive locks were generally only available as an after-market item. By the late 1920s, with most vehicles having doors and self-starter ignition systems and with the continuing growth in sales, it became necessary to incorporate locks as a standard feature in motor vehicles. Key-operated locks were first used for the doors, steering wheel, and transmission [50]. Later, locks were also installed on the glove box and trunk (or boot) and on the filler cap for the fuel tank. In those days, ignition locks were of limited deterrent value, with many vehicles having the same key [63]. H. A. Kendall's lockable steering column clamp (US 1,345,014), invented in 1919, formed the basis for the combined steering wheel and ignition lock used in modern cars. A diagram from Kendall's patent is shown in Fig. 7.1.

The early automotive locks were generally of the pin-tumbler or single-throw wafer type. In the early 1930s, pin-tumbler locks were used by Ford while both pin-tumbler and wafer locks were in use by Chrysler [63]. Both types of locks were used by the major car companies until the 1970s and 1980s. During the post–World War II period, one of the few innovations in conventional car locks was the introduction in the 1960s of "convenience keys," which had the same cuts on both sides to enable them to be inserted either way.

As early as 1935, General Motors adopted the Briggs & Stratton six-wafer side-bar lock for their automobiles. This represented a significant departure from the other

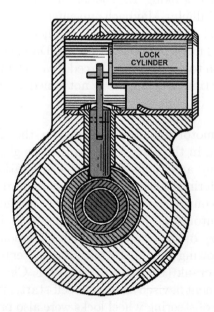

Figure 7.1: H. A. Kendall's 1919 steering column lock (US 1,345,014).

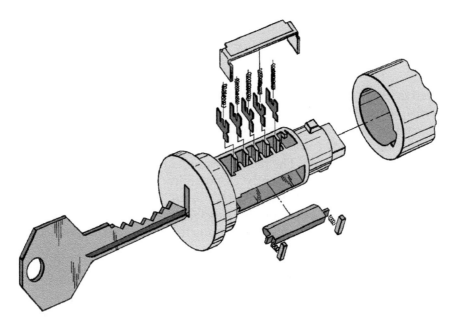

Figure 7.2: The General Motors' side-bar cylinder from a patent by W. B. Nail describing a decoding tool (US 4,185,482).

types that were in use at the time. The General Motors lock, illustrated in Fig. 7.2, uses a unique inwardly sprung side-bar design that we describe later in this chapter.

With the accelerating growth in sales of motor cars (Ford produced its 100 millionth car in the mid-1950s), the security deficit of single-sided wafer locks was starting to show. The issuing of new key profiles went part way to increasing the security of car locks, but there is only so much you can do with a one-sided wafer lock. The softer alloys used in their construction lessen the effectiveness of differences in plug broachings through wear. The net result is a car lock that can be opened by a number of different keys, a fact to which anyone who has opened someone else's car by mistake can attest. Gradually, manufacturers began to introduce real double-sided wafer locks with reversible keys. Combined with improvements in steering locks, this was a vast improvement over the single-sided wafer lock. The double-sided wafer lock is still a popular choice for motor vehicles today, although it is giving way to the two-track system, which we return to later on in this section.

The Ford five pin-tumbler system was replaced in 1984 by the "10-cut" system, having a 6-wafer side-bar lock for the ignition and a 6-wafer non–side-bar lock for the doors. The first six cuts (gauged from the tip of the key) operate the door lock, while cuts 5–10 operate the ignition. Five different depths of cut are used, with key codes assigned according to the following bitting rules:

1. The MACS is 2.

2. No code can have fewer than three different depths.

3. Cuts 1–3 cannot be the same.

4. Cuts 8–10 can be the same.

For a system like this, one cannot derive the ignition cuts from the door lock cuts or vice versa. However, it is possible to "progression" an ignition key from a door barrel by trying successive combinations of cuts that satisfy the bitting rules. Since cuts 5 and 6 are shared, only the remaining four cuts have to be established by trial and error. The manufacturer supplies progression charts so that this can be accomplished as efficiently as possible. The idea of progressioning had been around since 1967 for General Motors locks. Since it is easy to understand, we have provided a description of the process in Section 7.6. Extensive lists of codes and progressioning methods are contained in the Reed Code Books [60], for example.

The Chrysler Corporation was formed around 1930 from the Dodge and Plymouth companies. Chrysler originally used pin-tumbler locks on their cars, but moved to a new 7-wafer, 4-depth keying system in 1989. All seven cuts are needed to operate the ignition, while the door locks only use the first five cuts (from the shoulder of the key). Thus, knowing the door code, a locksmith could progression the sixth and seventh cuts to obtain a working ignition key. The system uses single-throw wafers with a double-sided key. A master key operates all the locks in the car including the trunk, while a valet key only operates the doors and ignition. The wafers also have serrated top and bottom edges to increase manipulation resistance.

Apart from the General Motors side-bar lock (Fig. 7.2), the ubiquity of inline wafer and pin-tumbler automotive locks with flat keys was largely unchallenged until about 1968 when BMW started using the 7-pin Pebra lock. This is a KESO-type lock with two rows of pins (three on one side and four on the other) operated by a 6-faceted KESO-type key. The key is reversible and has indentations or dimples in four of its faces. From the early 1970s, we see a steady stream of higher security locks typically found in luxury and sports cars. These superior quality locks have gradually percolated down to the lower end of the automotive market. For example, the Lancia Giobert lock, introduced in 1972, is a 10-pin lock with a specially cut key and two opposing rows of five pins.

In the late 1970s, a number of side-track wafer lock designs were adopted by automotive manufacturers. These designs are based on the Dudley split-wafer concept appearing in the 1934 patent of G. D. Full and J. Muntner of the Dudley Lock Corporation (US 2,030,836) and subsequent patents (see Fig. 7.3). Another influential patent was Neiman's 1971 German patent (DE 7,203,658), shown in Fig. 7.4. This introduced the idea of a reversible two-track key to operate a set of alternating wafers in a two-sided wafer lock. The wafers are supported on one side of their central cut-out, which slides along the milled track in the key. The requirement of reversibility, very important from a customer convenience perspective, imposes constraints on the bitting codes that can be implemented.

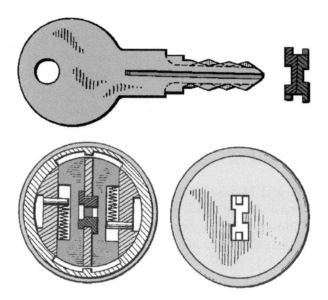

Figure 7.3: Depiction of a Dudley four-track wafer lock from a 1936 patent (US 2,279,592 by H. Machinist).

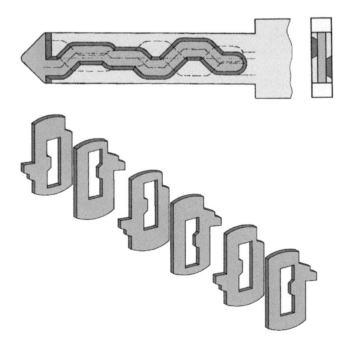

Figure 7.4: Reversible two-track key and wafer design from Neiman GmbH & Co.'s 1971 patent (DE 7,203,658).

This is because each track must operate the wafers on either side of the cylinder, depending on the way the key is inserted. The cut spacing is therefore halved and this imposes a more severe MACS constraint, which reduces the number of available codes.

Porsche started using its 10 split-wafer system in 1977. The Mercedes two-track system, also manufactured by Huf-Ymos, was launched 1979 with 7- and 8-wafer barrels and later 10-wafer models. An extended-barrel type with 13 wafers is also produced that operates an alarm switch on the ignition lock, similar to the system used by Porsche. The two-track system was later adopted by Vauxhall, Saab, and GM Holden (Australia) among others. In 1985, Ford began incorporating the Chubb AVA 5-disc lock mechanism into its U.S. and European cars. At roughly the same time, Renault in France was using the 6-pin Vachette system, having a round key profile operating three pins on top and bottom edges.

Around 1988, BMW and Mercedes introduced four-track systems—a variation on the Porsche split-wafer system made by Huf-Ymos (see Fig. 7.5). By 1989, the Japanese Toyota Company had put out a four-track system with a "sidewinder" or "wave-type" key. In the same year, Jaguar, followed by Daimler and then Ford, took on board the Abloy lock principle in the form of the Tibbe 6- and 8-disc locks. By this time, car manufacturers were beginning to look beyond mechanical high-security locks to systems with additional electronic security features.

One of the first combined electrical and mechanical car key system was the VATS/pass-key system, launched in 1986 by General Motors. While retaining the 6-cut wafer side-bar key, it integrated a precision resistor pellet into the upper part of the key blade. Two pads near the head of the key provided electrical contact to the lock, allowing the resistance to be checked. A correctly bitted key would operate the ignition lock but, unless the resistor was the correct one out of the 15 possible values, the ignition system would remain disabled.

During the early 1990s, a number of new mechanical locks for cars were ushered in. The Simplex 9 (= 4 + 5) pin lock for the XM model Citroën was released in 1990. Around the same time, Neiman produced a similar system with 11 (= 6 + 5) pins for use by Peugeot. Both of the preceding systems feature cruciform convenience keys working two rows of pins in the lock. Neiman also made a four-track, 8-wafer system in 1991 for the Volvo 760 series.

Figure 7.5: BMW E36 four-track key for a 10-wafer cylinder.

There have not been any significant developments in mechanical locks for cars since 1995. This can be understood in light of the introduction of key-top transponders in the same year by most of the major car manufacturers, including Alpha Romeo, BMW, Daewoo, Fiat, Ford, GM Holden, Honda, Lancia, Mazda, Nissan, Opel, Vauxhall, Renault, and Volkswagen. Transponders are electronic devices that are powered either by an internal battery or by connection to the car's power supply and transmit a radio frequency code to the car's computer. The car is effectively immobilized until the correct code is transmitted. Transponder keys can also be powered by electromagnetic induction in a coil that responds or is tuned to an RF transmitter in the lock, although these are not well suited to automotive applications since they only operate at very short ranges and require a continuous power supply.

In a transponder system, the key code is stored as a binary sequence in a small memory chip embedded in the key head or fob. A 32-bit code can take any one of $2^{32} = 4,294,967,296$ possible values. Without considering emergency start procedures, the only way to defeat something like this is to capture the code electronically using an RF scanner when it is transmitted by the original key. To foil this method of attack, cyclic cryptographic codes that change from one transmission to the next have been used since 1998 in key-top transponders. The reader is referred to [77, 112] for coverage of such topics, which are outside the scope of this book.

Returning to the subject of mechanical automotive locks, we can say that the main classes of car locks in use today are:

1. Single-sided wafer.

2. Two-sided wafer.

3. Single-sided wafer side-bar.

4. Multiple inline pin-tumbler.

5. Two-track wafer.

6. Four-track wafer.

7. Four-track split-wafer.

8. Chubb AVA.

9. Abloy (Tibbe).

The remainder of the chapter is devoted to detailed descriptions of locks that fit into each of these categories. Where appropriate, the reader may wish to refer to previous sections of the book covering locks with similar principles, particularly Chapter 3 on wafer locks. Master-keying is not generally applicable to automotive locks since, even in a car fleet, each vehicle should have a unique key; however, a limited type of master-keying can be applied to locks inside the same vehicle, whereby the same key operates locks for different functions that have been coded differently.

The locks covered in this section are not considered to be high-security locks, although some may have quite high manipulation resistance. In order to keep costs down, most car lock barrels are made of cast zinc alloys that are not very strong. On the other hand, no practical purpose is served by installing a very secure lock on a car door that can be defeated with a "slim jim" or on an ignition column that may be attacked with a slide-hammer. Furthermore, despite the use of 10 or even more wafers in modern car locks, the number of usable combinations for a given key blank is often limited to a few hundred thousand. The reasons for this stem primarily from the requirement of a reversible key, which also means a small MACS. The MACS may be as small as two depths of cut in some instances. From a servicing perspective, a small value of the MACS makes it easier to perform progressioning since fewer cuts and fewer key blanks are required.

7.2   Double-Sided

(US/EU/JP) 10-wafer (2–3)

The overwhelming majority of car locks are single- or double-sided wafer locks. We focus our attention on the double-sided variety in this section. A typical double-sided wafer lock from an Audi vehicle is pictured in Figs. 7.6 and 7.7. The lock cylinder has a composite plug with a spring-loaded inner core that slides inside an outer shell. When the key is inserted, a shutter at the front of the keyway causes the inner core to slide forward and engage the steering wheel locking mechanism.

The plug has 10 wafers that are alternately sprung and arranged in a sequence of singletons and pairs: 1-2-1-2-1-2-1, with wafers in a pair acting in opposite directions. In other double-sided car locks, wafers may be arranged in five pairs: 2-2-2-2-2. Wafers are of conventional construction, made from flat punched steel or brass, and contact the key at the bottom of the cut-out (the opposite end to the spring shoulder).

Figure 7.6: Audi 10-wafer core and key.

Figure 7.7: Audi core with key inserted.

The key is symmetrically bitted and therefore functions either way. On insertion, it activates a shutter that transfers the forward movement to the inner core, disengaging the steering lock. At the same time, the key blade picks up the wafers to position them so that their edges are flush with the outer shell, enabling the plug to be turned. There are 10 bitting points along the key blade on each side. In each of the two possible orientations, five bitting points on each side of the key set the heights of the wafers.

Clearly, a much larger number of differs is available due to the increased number of wafers with further differing provided by variations in the keyway profile. However, the number of combinations provided by a two-sided wafer lock with 10 wafers falls well short of that provided by a 10-pin cylinder lock; this is due to the requirement of a symmetric key: each side of the blade must carry all 10 cuts. Since the spacing of the wafers is generally quite small compared with a pin-tumbler lock, only a relatively small MACS (for example 2 or 3) can be supported. This results in a much smaller number of usable differs than the unconstrained theoretical number. Bitting rules are also applied that further reduce the number of available key codes. As an example, the Toyota 8-cut key has a MACS of 2 and the following bitting rules: (1) no more than three adjacent cuts may be the same; (2) no more than four of the cuts may be the same; and (3) at least three cuts must be different.

Two-sided wafer locks offer higher manipulation resistance than single-sided wafer locks. They require double-sided and ball-shaped picks that can set the wafers in opposite directions without the need to reverse the pick. However, these locks still suffer from susceptibility to so-called try-out or "computer keys." This method, described briefly in Chapter 3, employs a set of keys either with intermediate cuts or with a special shape optimized to approximate a large number of different keys when jiggled. A two-sided wafer key is easy to duplicate with standard key-cutting equipment, and this is a further impediment to security.

Some features that manufacturers have added to increase security include wafers with cut-away tops and bottoms to hamper picking. Other devices include staggered-edged wafers and notched wafers that tend to wedge in the plug or cylinder cavity if an attempt is made to open the lock by manipulation (see Fig. 3.9 in Chapter 3).

7.3 Holden Commodore

(AU) 7-wafer (3)

Double-sided wafer keys are not always cut along the top and bottom edges of the key blade. An example is the Holden VN-model Commodore lock (Figs. 7.8–7.10), as supplied in Australia and also found in European cars such as Opel, Mercedes, Volvo, Saab, and Vauxhall. This lock has seven alternating wafers of nonconventional construction. The system utilizes a type of symmetric two-track key of rectangular profile with "internal cuts" milled into each side of the blade. The contour is single-faced, unlike keys for Bell locks or side-winder keys having a milled channel.

Whereas a conventional wafer has a rectangular cut-out, Commodore wafers possess an inside shoulder (as well as the usual outside shoulder that contacts the spring). The inside shoulder contacts the contour on the key rather than the outer edge of the blade. Apart from this feature, the rest of the lock operates in the same way as a conventional two-sided wafer lock (see Fig. 7.10). There are nine different wafer

Figure 7.8: Holden Commodore 7-wafer barrel and two-track key.

Figure 7.9: Set of wafers from Holden Commodore lock.

Figure 7.10: Operation of Commodore double-sided wafer lock.

MACS	Codes Satisfying MACS
1	4,845
2	75,229
3	375,091
4	1,042,167
5	2,025,831
6	3,148,905

Table 7.1: Number of combinations for 7-wafer locks with nine sizes, subject to MACS constraint.

sizes, shown in Fig. 7.9. With seven wafer positions, the theoretical number of key codes is $9^7 = 4,782,969$, although this assumes an unrestricted MACS. Since the key is symmetric, the cut spacing is the same as the spacing of the wafers in the plug, which implies a relatively small value of MACS, (e.g., 3). The number of codes subject to a given MACS constraint for this type of system is listed in Table 7.1.

An advantage of this type of lock is that key duplication requires a special key-cutting machine with a high-precision milling tool, rather than the usual cutting wheel. Since the distance between the inner shoulder and the bottom of the cut-out does vary for different-sized wafers, the lock is susceptible to decoding in order to read off the combination and hence cut a key. However, this requires specialized tools and is considerably more difficult to perform than opening the lock by force.

7.4   Ford Tibbe

(DE) 6-disc + side-bar (3)

One car lock that does not fit the wafer-lock mold is the Ford Tibbe. It comes in two varieties: Chubb (AVA) and Abloy. In Australia the Tibbe is the Abloy type, which is fitted to EB and later model Ford vehicles. In the United Kingdom, both types are in use, with the Chubb type fitted to the Ford Transit and a number of other models. Both locks have a reversible key; however, with the Ford Chubb it is not immediately obvious how this is achieved. We describe both variants in detail in this section, starting with the Abloy type.

The Ford Tibbe (Abloy) lock is pictured in Figs. 7.11 and 7.12. Locks are supplied in both 6-disc and 8-disc versions. Although the vast majority of vehicles fitted with Tibbe locks use the 6-disc type, the 8-disc Tibbe has been used from around 1990 in some luxury vehicles like the Daimler Sovereign and Limousine and the Jaguar XJ6 and XJS [112].

Figure 7.11: Ford Tibbe 6-disc barrel and double-bitted key.

Figure 7.12: (Left) Ford Tibbe core. (Middle & right) Key aligns discs to retract side-bar.

The Tibbe (Abloy) key is a symmetric version of the ABUS Plus key covered in Chapter 4, having a cage or drum containing a stack of disc tumblers and a side-bar mechanism. Starting with a narrow double-bitted blank, the same bitting pattern is cut on both top and bottom edges of both key bits. The symmetric bittings allow the key to operate the lock, both clockwise and counterclockwise, regardless of which way it is inserted. The key has a reduced-width shank, which is necessary since the cylinder has a fixed front plate. This is an advantage, especially for a car lock, where a potential thief's first move is often to insert a heavy-gauge screwdriver into the keyway. Notwithstanding, this crude form of attack may succeed if enough force is applied to turn the entire lock cylinder in its mounting.

Since the key must operate in both directions to drive the linkage to the door lock, it would normally be necessary to endow each disc with a pair of gates. However, since the door lock can only be operated in one direction, it is an acceptable economy to allow the cylinder to turn freely in the locking direction until a stop is reached. This is achieved via the shouldering of the inner surface of the barrel. The arrangement is such that the side-bar only needs to be retracted for operation in the unlocking direction; hence only one set of gates is needed. When the discs are rotated to the required angles by the correct key, a channel is formed by the gates, permitting retraction of the side-bar (see Fig. 7.12).

The key symmetry requirement implies that only a 45-degree sector is available for the angled bittings in each quadrant of the key bit, which results in a relatively limited number of combinations. There are only four possible code discs requiring 0, 15, 30, and 45 degree cuts to the key (numbers 1 to 4, respectively). As discussed in the section on Abloy locks in Chapter 4, at least one zero cut must be included on the key to set the maximum angle of rotation of the discs. Applying the same reasoning as for the Abloy lock, it follows that the theoretical maximum number of combinations is 3,367 for the 6-disc Tibbe lock and 58,975 for the 8-disc one. These figures are adequate considering that the key may also be teamed with a transponder to immobilize the vehicle or activate the alarm system.

The lock cylinder has an inspection slot, visible in Fig. 7.11, that enables the combination to be read off the markings on the side of each disc. This obviates the need to dismantle the lock to replace a lost key, which is important from a servicing perspective.

If the location of a number 1 disc is known, then manipulation of the discs is possible by applying tension to the number 1 disc while rotating the other discs to engage the side-bar. In the absence of false gates, this method should produce results due to the lower manufacturing tolerances and presence of fewer discs compared with an Abloy lock. Naturally, the question of picking a car lock is academic when much faster bypass or destructive techniques would be used in practice.

7.5   Ford Chubb

(DE) 5-slider (3)

We have seen how the Abloy lock can be made to take a reversible key, as evidenced by the ABUS Plus mechanism and the Ford Tibbe lock. The trick is to move the keyway to the middle of the discs and make the broaching symmetric. The Chubb AVA mechanism has also been put to use by the Ford Motor Car Company since 1985 in its European and U.S. models (see Fig. 7.13). But how can the AVA mechanism be made to operate with a reversible key? The short answer is that it cannot, if we wish it to retain its full complement of sliders. A workaround that is used by Ford, and that is adequate for this purpose, is to halve the number of sliders. We will clarify this point further on after briefly revisiting the AVA mechanism. The reader may wish to refer to Chapter 5 for a full description of the operating principles. Our description assumes that the lock cylinder is oriented so that the keyhole is vertical with the key as pictured in Fig. 7.13. The automotive version of the AVA lock is specified in US patent 4,385,510 (1981).

The AVA lock has a constant-width variable-offset system. Thus, all sliders have the same-sized cut-outs, but the position of the cut-out with respect to the active ends of the slider is variable (see Fig. 7.14). There are five slider offsets, numbered 1 to 5.

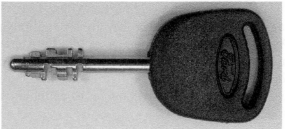

Figure 7.13: Ford Tibbe (Chubb-type) lock barrel and key.

Figure 7.14: (Left) Ford Chubb slider pack with separators. (Right) Sliders.

The overall length of each slider is equal to the diameter of the core. The offset of a cut-out can be zero, positive, or negative. By zero offset we understand that the ends of the slider are level with the edge of the core. This offset corresponds to a #3 slider, and the cuts on either bit of the key will be the same depth for this size of slider. The #3 slider is special in that it is a dead-lift lever in the terminology of lever locks: it must not be displaced either left (9 o'clock) or right (3 o'clock). The cuts for a #3 slider are therefore such that the left and right bits of the key just reach the contact points on the slider as the key reaches its maximum turn angle of about 35 degrees clockwise relative to the core.

Slider numbers 1 and 2 have negative offsets. In the locked position, their left ends will be proud of the core (at 9 o'clock) by 2 and 1 units, respectively; similarly, their right ends will be recessed at 3 o'clock by the same amounts. These sliders are shifted by the top side of the key contacting the slider cut-out on the upper right side. The corresponding bottom bitting for sizes 4 and 5 will be undercut on the key so as not to contact the lower left side of the cut-out until the key reaches 35 degrees.

Finally, slider numbers 4 and 5 have positive offset, protruding at 3 o'clock when in the locked position. These must be shifted by the bottom bitting of the key contacting the lower left side of the cut-out. Therefore, the top side of the key bit will be undercut so that it only reaches the upper right side of the cut-out as the arms of the slider arrive at the shear line of the core.

There is no essential difference between sliders with positive and negative offsets, except that they are operated by different sides of the key. Indeed, the two are interchangeable. A #1 slider working off the top bit of the key works equally well as a #5 slider operated by the bottom bit of the key if the slider is rotated 180 degrees. Similarly, a #2 slider can double as a #4 slider. Put another way: a size n slider will function as a size $6 - n$ slider when mounted the opposite way in the stack. This gives a convenient economy of production since only three components are actually required to make the five sliders.

So far, everything we have said about the Ford Chubb lock applies equally well to the Chubb AVA mechanism. The original AVA lock had 10 sliders. If we remove every second slider, that is, all the even-numbered ones, and replace them with a spacer of equal thickness, then we end up with a 5-slider AVA mechanism with gaps in between the sliders. Concentrating on this 5-slider mechanism, suppose that the odd-numbered sliders are sizes 1, 2, 4, 1, and 3. A 10-cut AVA key with these cuts in positions 1, 3, 5, 7, and 9 would then operate the lock.

Suppose further that we make complementary cuts (i.e., six minus the cut number) in all the even positions. Thus we make cuts 5, 4, 2, 5, 3 in positions 2, 4, 6, 8, and 10. If we could turn the key around and shift it up by one position (so that cut 2 was in position 1, cut 4 in position 3, etc.), then it would still operate the lock. This is true since the #5 cut in position 2 when turned 180 degrees is equivalent to a #1 cut and, when shifted to position 1, matches up with the #1 slider. A similar argument applies to the other even positions.

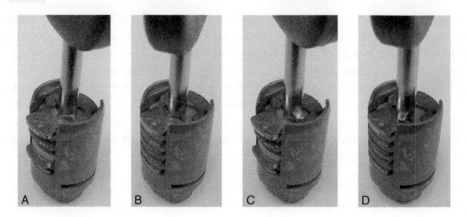

Figure 7.15: Operation of the Ford Chubb lock: key inserted right way (A & B); key inserted upside down (C & D).

In the Ford Chubb lock, the key blank has the same bit width on either side and a symmetric section; it can therefore be inserted either way. In addition, there is an eleventh bitting on one side of the blank. At the bottom of the keyway, the core has an uneven floor: it is raised on one side. Referring to Fig. 7.15, when the key is inserted the right way, cut position 1 is level with the first slider. Conversely, when the key is inserted the other way, the eleventh bitting rests on the raised part of the floor. This shifts the key stem up so that cut position 2 is level with slider position 1. Therefore, either way, the correct sized bittings are applied to the five sliders in order to operate the lock. The key is still a 10-cut key, in this case (1 5 2 4 4 2 1 5 3 3); however, there are far fewer system codes since we only can choose 5 of the 10 cuts independently. The resulting tally of $5^5 = 3,125$ codes is still adequate for an automotive lock. With only five sliders in the stack, the Ford AVA is easier to manipulate than its high-security 10-slider version, although this is not a practical option for a car thief.

7.6   General Motors

(US) 6-wafer (3)

For many years, General Motors (known as Vauxhall in the United Kingdom, Opel in Germany, and GM Holden in Australia) used a 6-wafer lock on their cars. The lock is illustrated in Figs. 7.16–7.18. The original design was patented by J. W. Fitz Gerald of the Briggs and Stratton Corporation in 1934 (US 1,965,336). The key looks the same as a conventional single-sided wafer lock key. The GM lock, however, is far from conventional, and, despite being made from a zinc alloy casting, the design is highly manipulation resistant. Our description assumes that the lock is viewed from the front with the side-bar at 3 o'clock.

Figure 7.16: General Motors 6-wafer side-bar lock and key.

Figure 7.17: Selection of wafers from a GM side-bar lock.

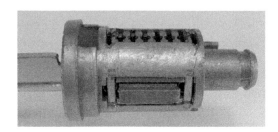

Figure 7.18: GM plug in locked position (left) and with side-bar retracted (right).

The six wafers (see Fig. 7.17) are in the shape of an inverted "h" and are spring-biased from the top. The springs that drive the wafers are rather strong. Each wafer has a single triangular notch on the outer edge of its longest side. If this were a conventional side-bar lock with a beveled side-bar riding in a channel in the cylinder, it would probably not be very hard to pick open. However, the outer edge of the side-bar is square in cross section, as indicated in Fig. 7.19. So instead of being urged toward the wafers as the plug is turned, the side-bar merely skews and blocks the mechanism if tension is applied. The side-bar is in fact sprung *inward*. The side-bar springs, which are quite fine compared with the wafer springs, supply a constant, light tension that keeps the side-bar edge in contact with the wafers.

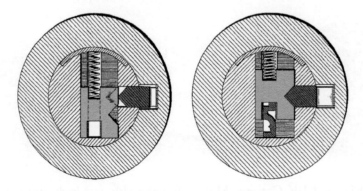

Figure 7.19: Section view of a GM side-bar lock from US patent 1,965,336 (1934) by J. W. Fitz Gerald.

The clever thing about biasing the side-bar radially inward, rather than outward (which is usually the case with side-bar locks), is that it denies the possibility of using torque on the plug in order to manipulate the wafers into position. The side-bar is effectively inaccessible from the front of the lock, which also denies any feedback to the would-be lockpicker. Without feedback of some kind, manipulation becomes pure guesswork, and since the wafer springs are much stronger than the side-bar springs, the wafers will not remain in position if raised. Thus one can appreciate the difficulty of picking or impressioning this mechanism.

There are five depths of cut for each of the six wafers. The theoretical number of combinations is therefore $5^6 = 15,625$. The bitting rules that constrain the sequence of key codes for the GM lock are: (i) the MACS is 2; (ii) the sum of the six cut depths is even; (iii) at most three cuts in a row can be the same. When we take the MACS into account, the number of system codes is reduced to 10,727. Taking the other two constraints into account further reduces this to 5,228. (The source code for the software used to generate the code lists is given in Appendix F.) While the presence of bitting rules has greatly lessened the number of available codes, it also makes practical the technique of progressioning a key for the GM lock. Progressioning is the narrowing down of the possible keys to be cut according to the constraints on the coding sequence, as summarized next.

Suppose we know the first four of the six cuts from inspection of the glove box key, for example. Our aim is to make a working 6-cut key for the ignition. Since there are five depths of cut, as described in [105], we could cover all 25 possibilities by progressively cutting five keys in positions 5 and 6 in the following sequence:

key 1: 11, 12, 13, 14, 15, 25, 35, 45, 55
key 2: 21, 22, 23, 24, 34, 44, 45
key 3: 31, 32, 33, 43, 53
key 4: 41, 42, 52
key 5: 51

Note how $9 + 7 + 5 = 21$ of the total 25, or 84 percent of the possible bitting pairs are covered with the first three keys. Note also that for each line the sequence of left-hand numbers and the sequence of right-hand numbers are both nonincreasing. This is because the depth of cut must increase as more of the key is cut away.

The process can be further streamlined by applying the bitting rules mentioned previously. Applying the MACS constraint eliminates six of the above combinations. The evenness constraint eliminates 50 percent of the possibilities, depending on whether the sum of the first three cuts is odd or even. For example, if the first three cuts were 4 5 3, then we would only need to test the 11 even combinations 11, 13, 35, 55, 22, 24, 44, 31, 33, 53, 42. This could be done with only three key blanks according to

key 1: 11, 13, 33, 53, 55
key 2: 22, 24, 44
key 3: 31, 42

Numerous patents have been filed for tools and methods to decode and pick GM side-bar locks. Among the contending schemes, we find Nail's 1978 shim-based decoder (US 4,185,482) and a Sputnik-type decoder by Dobbs in 1992 (US 5,224,365). A bypass tool for tensioning the side-bar via the front of the cylinder was described in Embry's 1993 patent (US 5,325,691). The tool was maneuvered into place to apply inward tension to the side-bar, making it possible to manipulate the six wafers to align their notches with the wedge of the side-bar. Once in alignment, the wafer sizes could be measured with a gauge and a working key cut. Robinson [105] describes a picking technique for this lock that involves a snake pick used without any tensioner to align the wafers. This method is based on the fact that there is a perceptible increase in tension on a wafer when its notch is aligned with the side-bar. Tobias [122] points out that GM wafer locks have a tolerance of half a depth of cut, which increases their susceptibility to try-out keys.

7.7 Mercedes Two-Track

(DE) 10–13 wafer (3–4)

Since 1979, Mercedes cars have used a 10-wafer double-sided cylinder with various key profiles. The lock is pictured in Figs. 7.20–7.23. The construction is similar to the Holden Commodore lock but with a larger complement of wafers. The lock has a cast zinc alloy body with a hardened steel cap to protect against the insertion of burglary tools. The plug is inserted from the back of the barrel and therefore cannot be forcibly extracted. The steering lock linkage is provided by a twist-bar running down one side of the barrel, driven by a nylon insert at the front of the plug.

Figure 7.20: Mercedes two-track key and ignition barrel.

Figure 7.21: Mercedes two-track 10-wafer plug.

Figure 7.22: Wafers from Mercedes two-track lock.

The plug contains 10 wafers loaded in an alternating sequence (see Fig. 7.21). A 13-wafer version of the lock is also in use. The wafers are equipped with an internal step of varying height that is depressed by the internal cut or track in the side of the key blade. Viewing the lock with the keyway vertical, the contact points on the key blade are at 2 o'clock and 8 o'clock. Each side of the key operates five wafers at once, and because it is reversible, all 10 cuts must be made on each side of the blade. The wafers are double-throw with antipicking notches on both top and bottom edges. Fig. 7.23 shows the lock being operated by the key.

There are five wafer sizes (see Fig. 7.22) so that the 10-wafer mechanism offers around $5^{10} \approx$ 9.7 million theoretical keying combinations. Naturally, the number of usable combinations is considerably less than this due to MACS and other bitting

Figure 7.23: Operation of Mercedes two-track plug by key.

rules. However, with several different key profiles in use, the chances of your key fitting someone else's Mercedes are very slight.

7.8   BMW Two-Track

(DE) 8-wafer (3–4)

Numerous car companies have adopted the Bell lock or "side-winder" principle, including Mitsubishi, Toyota, and Volkswagen. Instead of bar-wafers, most automotive locks taking side-track keys utilize wafers or split-wafers with a centrally located stub that is picked up and guided by a track in the key. Although in principle a Bell lock does not require wafers that are sprung, the wafers in side-track automotive locks are generally spring-biased to ensure positive locking and to enable the lock to function reliably in the presence of dirt and grit. We consider in this section a side-track automotive lock made for BMW by Huf-Ymos.

The BMW E46 two-track lock is illustrated by the ignition barrel in Fig. 7.24. This particular model, used from 1999 on 3-series and 5-series BMW cars, also includes a key-top transponder for remotely operating the door locks and trunk. The ignition lock incorporates a toughened steel cap at the front to protect the mechanism. The plug, shown in Fig. 7.25, which we use to illustrate the operating principle, is from a model E46 door lock.

The lock comprises a barrel and cast zinc alloy plug equipped with a tail-piece. Rear shouldering on the barrel of the lock limits the maximum angle of rotation of the

Figure 7.24: BMW E46 8-wafer ignition barrel and two-track key.

Figure 7.25: BMW 8-wafer plug.

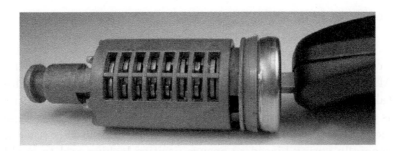

Figure 7.26: BMW plug and cage with correct key fully inserted.

plug to ±90 degrees with a strong return spring to center the plug for insertion and removal of the key. A slotted cage surrounds the plug (see Fig. 7.26). The inside surface of the barrel is fluted at the rear to accommodate the return spring assembly and a coupling that anchors the slotted cage.

The plug contains eight wafers arranged in an alternating sequence: ↑↓↑↓↑↓↑↓. All wafers are shouldered on the left side. Upward-acting wafers (↑) are loaded from the top in odd-numbered positions and have their stub on the left side. Conversely, downward-acting wafers are loaded from the bottom in even-numbered positions and

Figure 7.27: Selection of wafers from BMW two-track lock: left three top-loading; right two bottom-loading.

have their stub on the right. All wafers contain serrations on the top and bottom edges to enhance manipulation resistance.

The wafers have a constant-sized cut-out: they are made to differ by varying both the offset of the stub and the position of the shoulder. There are four differs for both upward (1—4) and downward (11—14) wafers. Three of the differs are illustrated in Fig. 7.27. The travel of the wafers is limited by the central rung of the cage that bisects the slots. In order to allow for the extra travel due to insertion of the key, wafer numbers ending in 1 and 2 are notched at the top end, while wafer numbers ending in 3 and 4 are notched at the bottom. Ignoring bitting constraints, there are nominally $4^8 = 65,536$ different combinations.

The key is a typical Bell lock type with the same constant-width track milled into both sides to allow it to be reversible. The ends of the tracks are fashioned with pick-up ramps on the top and bottom surfaces. The ramps guide the wafer stubs into the smooth contour of the track, positioning the wafers according to its height. Since the key is reversible, eight bitting points are required along the track on each side of the key, with only four bitting points per side being active for a given key orientation. When the key is inserted, bitting points 1, 3, 5, and 7 on the left track operate the upward-acting wafers, whereas points 2, 4, 6, and 8 on the right track operate the downward-acting wafers. All eight wafers must be brought to the shear line of the plug to operate the lock. Incorrect lifting of a wafer causes the serrations on its top or bottom edge to engage the beveled edge of the slotted cage.

7.9 Mercedes Four-Track

(DE) 3-wafer + 3 split-wafer (3–4)

A number of car manufacturers have upped the stakes on their double-sided wafer locks by splitting the wafers down the middle. Mercedes, BMW, and Porsche are all examples of this practice. This section concentrates on the Mercedes 9-wafer lock, illustrated in Figs. 7.28–7.30. It is in essence a modern version of the Dudley lock from the mid-1930s.

Figure 7.28: Mercedes four-track lock barrel and key.

Figure 7.29: Two views of Mercedes four-track plug.

Figure 7.30: Operation of Mercedes four-track lock.

The Mercedes four-track key has contoured tracks milled into the blank in four places: top and bottom, left and right, leaving a ridge in the middle of the key blade. The key has a reduced-width stem reminiscent of the Bricard SuperSûreté, which has a fixed front cap obscuring the keyway. The difference in this case is that the reduced section of the stem is required to accommodate a blocking wafer that disables the steering lock: the key, being symmetric, must have a notch on on both top and bottom surfaces.

The plug houses nine wafers arranged in six slots, as can be seen in Fig. 7.29. The first three are taken by solid wafers that contact either the left or right side of the key blade only, unlike normal flat cut-out wafers that contact the key across its entire width. The last three are reserved for six split-wafers. Each of these slots houses a left and a right half-wafer that function independently and are respectively driven by the left or right side of the key contour. In addition, the wafers alternate in their directions of action along the plug, with each split-wafer in a pair of wafers having the same vertical orientation.

As well as being hard to defeat by picking due to the presence of nine separate wafers, there are antipicking barbs on both the split and full wafers that engage corresponding grooves in the molded alloy housing. Access for implements is restricted somewhat by the broaching of the keyway. The multiplicity of wafers also gives substantially more key combinations than a standard six-wafer lock.

7.10   Mitsubishi

(JP) 4-wafer + 3 split-wafer (3–4)

The Bell lock principle was already mentioned in connection with the BMW two-track lock earlier on. Although automotive lock manufacturers do not use the same type of bar-wafer as the traditional Bell lock described in Chapter 3, many car locks employ side-track keys that resemble the keys used in Bell locks. In this section we examine the workings of the Mitsubishi car lock shown in Figs. 7.31–7.33.

Figure 7.31: Mitsubishi 10-wafer lock barrel and four-track key.

Figure 7.32: Two views of Mitsubishi plug.

Figure 7.33: Mitsubishi wafers: split and conventional.

A cursory look at the key reveals that this is not a standard Bell-type lock in that the width of the side-milling is not constant. Each side of the key has two milled contours or tracks, the same millings being replicated antisymmetrically on either side to ensure that the key may be used either way round. Whereas Bell locks like SEA and EVVA 3KS utilize driverless bar-wafers that are guided by the key tracks, the wafers in a Mitsubishi lock are spring-biased.

Mitsubishi wafers (Fig. 7.33) are different from standard wafers in two respects. First, there is a stub protruding about one-third of the way across the cut-out. The side of the stub opposite the shoulder of the wafer rides on one of the side-millings in the key. The second nonstandard feature is the use of split-wafers, also having a central stub. Naturally, the offset of the stub determines the depth of cut required on the key to align the ends of the wafer with the edges of the plug. A milling of incorrect depth will either insufficiently raise the wafer or overraise it, exposing antipicking barbs that catch in matching grooves in the lock housing. This greatly increases the manipulation resistance of the lock.

All told, there are 10 independently operating wafers in the lock arranged in 7 slots. Counting from the front, the first four slots contain full (i.e., nonsplit) wafers, whereas the last three slots accept two split-wafers each. The wafers operate in an alternating sequence with the first sprung upward on the right, the second downwards on the right, and so on. Full wafers that are sprung upward on the right have their stubs on the left and are therefore driven by the upper milling on the left-hand side of the key. Conversely, wafers that are sprung downward on the right are actuated by the lower milling on the right-hand side of the key. Each split-wafer pair is sprung on the left and right, and the pairs are arranged in an up, down, up configuration. For each pair, therefore, millings on both sides of the key blade contact the wafer stubs.

A sequence of 10 bitting points must be internally milled on both the upper and lower tracks in order for the key to function in both orientations. The theoretical number of key codes is enormous (of the order of 5^{10} or 9,765,625). However, this is subject to a maximum adjacent cut specification, which, together with the symmetry constraint, greatly reduces the number of possible track profiles. On the other hand, the mechanical key is only one of the parts required to start the car, the other part being a transponder code.

7.11 Porsche

(DE) 10 split-wafer (3–4)

Many car manufacturers base their locks on various high-security locks. We have already described car locks that use Bell- and Dudley-type wafers, side-bars, and the like. However, the requirement of low production cost tends to result in the use of lower grade materials, especially zinc alloy castings, which are typically found in low-end domestic locks. The Porsche 928 lock is one example where a higher emphasis has been placed on physical strength and security.

The Porsche 10-wafer lock in Figs. 7.34–7.36 is manufactured by Huf-Ymos. It differs significantly in a number of respects from conventional double-sided wafer locks. The lock housing and plug are of cast alloy, but a number of features are present that

 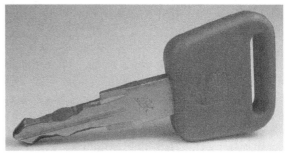

Figure 7.34: Porsche 928 four-track lock barrel and key.

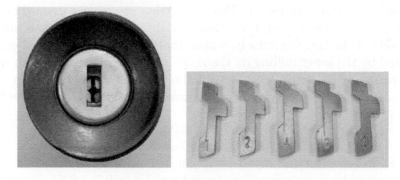

Figure 7.35: (Left) Front view of Porsche 928 lock. (Right) Split-wafers.

Figure 7.36: Operation of Porsche 928 plug by key.

add to its robustness: (i) the housing is bolted into the door handle assembly; (ii) a 4 mm steel plate is integrated into the front of the plug; and (iii) the plug is secured by two brass pins that ride in a groove around the front of the cylinder. The lock is therefore considerably more difficult to attack with a screwdriver or slide-hammer than cheaper wafer locks fastened by spring clips. In newer model cars, the rear part of the lock is extended to house an electronic alarm linkage, and additional wafers are provided.

The plug contains two rows of five chambers to house the 10 split-wafers, which are arranged in pairs as shown in Fig. 7.36 and driven in an alternating sequence: ↑↓↑↓↑. Wafers come in four sizes and have serrated top and bottom edges that are matched by serrations in the lock housing to increase manipulation resistance.

In contrast to the Bell-lock principle, where the wafers are driven by a stub protruding into the keyway, Porsche half-wafers have an active surface on a 45-degree angle. The presence of half-wafers sprung in both upward and downward directions

MACS	Codes satisfying MACS	Codes satisfying all Constraints
3	1,048,576	956,184
2	363,314	318,488
1	21,892	14,708

Table 7.2: Number of combinations for generic 10-wafer locks with four sizes, subject to MACS and differing constraints.

requires a key that is cut in four different planes. Thus, both edges of the top and bottom of the key blade must be cut at ± 45-degree angles at five bitting points. An additional five bitting points per track are required to allow the key to operate in either direction. The action of the wafers is similar to that of the split-wafers in a Mercedes lock, but with both active surfaces at a 45-degree angle rather than in a horizontal plane. As with other car locks employing split-wafers, picking difficulty is greatly enhanced over a standard two-sided wafer lock.

Since each wafer can be one of 4 possible sizes, there are nominally $4^{10} = 1,048,576$ differs. Table 7.2, which also applies to 10-wafer locks other than Porsche, shows how this is affected by the MACS constraint and additional keying constraints (at least three cuts different, up to three adjacent cuts the same). For instance, with a MACS of 1 there are only around 15,000 usable differs, with a code series ranging from (1 1 1 2 1 1 1 2 2 3) to (4 4 4 3 4 4 4 3 3 2).

7.12 Alpha Romeo

(IT) 10 pins in 2 rows (2–3)

The lock in this section is a pin-tumbler lock as opposed to a wafer lock. With the advent of the KESO dimple-key lock in the 1960s, a number of car manufacturers including BMW fitted Gemini-type locks to their vehicles. Although these have now been replaced with two- and four-track wafer locks, they provide an interesting case study.

As early as 1981, Alpha Romeo and Lancia were producing 10-pin car locks with dimple-type keys, like the one shown in Fig. 7.37. Alpha Romeo fitted such locks to the Giulietta and GTV models. In a similar vein to the Japanese X-key system, locks for these cars employ two opposing rows of five pins, each arranged to contact the key blade width-ways.

The pins are arranged in opposing pairs (see Fig. 7.38). Since the key blade is 3 mm thick, the maximum cut depth is limited to 1.5 mm, being half the blade-width. This limits the number of available pin lengths, but since there are 10 positions to

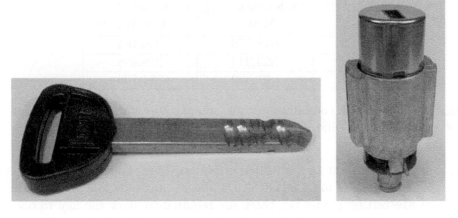

Figure 7.37: Lancia/Alpha-Romeo key and 10-pin lock barrel.

Figure 7.38: (Top) Lancia plug. (Bottom) Core with key inserted.

be filled, the total number of differs remains large. The lower pins have a reduced-diameter shank with a conical tip. The smaller pin size and mushroom-head of the bottom pins naturally hamper picking.

Blank keys are cut in an unusual way that requires a special high-precision machine for key duplication. Since the key is symmetric, two rows of five cuts are required on each side of the key. The cuts themselves are produced by a cutting wheel in a plane that, instead of being perpendicular to the key blade, is angled back toward the bow of the key. This allows a slightly deeper maximum cut without undermining the adjacent cut on the same side of the key. In newer models, a more standard dimple-type bore is used. The end of the key blade is beveled on the top and bottom to ensure pick-up of the pins on both sides of the barrel regardless of the key orientation.

7.13 Volvo

(DE) 8-wafer (3–4)

Yet another variation of the wafer-tumbler theme is the 8-wafer lock used by Volvo since 1991 on the 850- and 960-series cars. The lock and its components are illustrated in Figs. 7.39–7.41. As in Mercedes locks, the reinforced front cap and rear-anchored plug are intended to increase the resistance to attack by screwdrivers and slide-hammers. In addition to the usual turning action, the plug also slides forward on its axis to disengage the steering lock. A pair of roller bearings mounted in a cavity at the front of the plug trigger this action as the key is inserted.

The key, which is of basically rectangular section, is internally milled on both sides of the top and bottom surfaces of the blade to produce four tracks. A similar statement applies to the Porsche 928 key, but, in the case of the Volvo key, the cuts are perpendicular to the key blade rather than at a 45-degree angle.

Like the Dudley patent in Fig. 7.3, the plug has eight chambers arranged with four on each side of the plug. There are four different sizes of wafer (1–4) of a type

Figure 7.39: Volvo four-track key and 8-wafer ignition lock barrel with front cap removed.

Figure 7.40: Volvo four-track plug.

Figure 7.41: Volvo four-track wafers: right-handed on top row, left-handed on bottom row.

illustrated in Fig. 7.41. For the lock in Fig. 7.40, viewed from the front with the wafers on the right side of the figure facing up, the loading sequence is ↑↑↓↓↑↑↓↓. There are also two different wafer handednesses: left (L) and right (R), having a step on one side of the cut-out or the other. Taking handedness into account gives eight different types of wafer: 1L, 2L, 3L, 4L, 1R, 2R, 3R, 4R. With both top- and bottom-acting wafers as well as left- and right-handed wafers, there are four possible lines of action (one for each key track), which we denote by compass points: NW, NE, SW, SE.

An innovative feature of 2- and 4-track keys, such as the Volvo key, is that they permit a limited form of master-keying to be applied to locks of different functions within the same car, (e.g., the door and ignition locks). This is achieved by using bittings on either the left or right side of the key to operate differently coded locks. This makes it impossible to determine the ignition lock cuts by reading the wafers in the door lock barrel (without progressioning). It also means that special-function keys, like valet keys, can be issued that operate only a subset of the locks in the vehicle. The lock in the illustrations provides a convenient example of this concept.

The sequence of contact points for the ignition lock from front to back is SW, SW, NE, NW, SW, SW, NW, NE, although many other sequences are possible. The wafer sizes from front to back are 1L, 2R, 2R, 1R, 3L, 4R, 4L, 4L. Since the key is reversible, the eight wafers require $2 \times 8 = 16$ bittings points on the key. The bitting matrix for the four tracks in this case is as shown in Table 7.3. However, since each track on the key can support eight cuts and there are four tracks, we are left with up to 16 additional bitting points—the spaces in the table—that can be freely set (subject to a MACS constraint).

Suppose that the door lock had the following sequence of wafers loaded: 1L, 4L, 2R, 1R, 3L, 4L, 4R, 2R. Thus only the wafers in positions 1, 3, 4, and 5 are common to both locks. The combined bitting matrix for the master key that operates both the ignition and door locks is then given by Table 7.4.

NW	-	-	-	1	-	-	4	-
NE	1	2	2	-	3	4	-	4
SW	1	2	2	-	3	4	-	4
SE	-	-	-	1	-	-	4	-

Table 7.3: Example bitting matrix for Volvo ignition key.

NW	-	4	-	1	-	4	4	2
NE	1	2	2	-	3	4	4	4
SW	1	2	2	-	3	4	4	4
SE	-	4	-	1	-	4	4	2

Table 7.4: Combined bitting matrix for Volvo door and ignition key.

MACS	Codes Satisfying MACS	Codes Satisfying All Constraints
3	65,536	60,480
2	28,642	25,384
1	3,194	2,096

Table 7.5: Number of combinations for 8-wafer locks with four sizes, subject to MACS and differing constraints.

To gain an appreciation for the number of key codes that can be supported by such a system, we note that there are four tracks, only two of which are distinct. For each distinct track, the number of theoretical permutations with four sizes of wafer and eight positions is $4^8 = 65,536$. Imposition of MACS and typical keying constraints reduces this substantially, as indicated in Table 7.5. However, the overall number of codes taking both independent tracks into account is the square of the corresponding entry in the table.

7.14 Citroën Simplex

(FR) 9 pins in 2 rows (4)

The Citroën Simplex lock, employed since the early 1990s in the XM, ZX, Xantia, Xsara, and Berlingo models, is in a class on its own. Any key with the right profile will turn the plug; however, only the correct key can operate the lock. What is the mechanism that achieves this? The key idea is that both a translation (sliding) and a rotation of the core are required. Almost all cylinder locks operate by turning the plug of the lock, but a small number also require a translation. The idea is traceable to a 1933 patent by the Yale & Towne Manufacturing Company (US 2,049,742), in which a sliding-core wafer lock was proposed. Another lock in this category is the Fichet-Bauche 787, which we dealt with in Chapter 5. In the early 1990s, Vachette patented a doubly-sleeved wafer lock concept for car locks that closely resembles the Simplex design (FR 2,650,021 and FR 2,657,641). As we will see, the Citroën lock uses a tail-piece on the core that must be displaced longitudinally to the rear of the case to complete the linkage and operate the lock. An obvious advantage of the construction is that forcing the plug to turn will not operate the lock.

The Citroën lock appears in Figs. 7.42–7.46. The key is cruciform and reversible. There are five cuts on the top and bottom edges of the blade and four cuts on either side. Only two mutually perpendicular edges of the key are active when the key is inserted; the other two edges are required for reversibility.

The lock barrel comprises around 20 parts, not including the pins. There are a number of levels of nesting. The innermost component is a core with a slideable carriage (Fig. 7.45, right). The core is encircled by a plug (Fig. 7.43), which in turn

Figure 7.42: Citroën Simplex key and 9-pin barrel.

Figure 7.43: (Left) Plug with locking ball. (Right) Core with two rows of pins.

Figure 7.44: Driverless pins from Simplex lock.

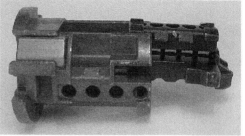

Figure 7.45: (Left) Carriage pins aligned by one side of key. (Right) Core and carriage with pins removed.

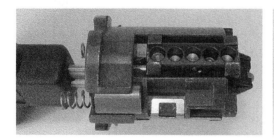

Figure 7.46: Operation of Simplex lock: carriage slides forward to release locking ball.

is housed in a barrel that includes the front trim of the lock (Fig. 7.42). Finally, the barrel is mounted inside a cylinder body to which a cam and linkage rod are attached (not shown).

The cylinder body merely houses the lock mechanism proper and is not essential to understanding the operating principle. Most of the action takes place at the interfaces between the core and plug and between the plug and barrel. A tail-piece is molded onto the rear of the plug that operates the cam of the lock. The relative movement between the plug and barrel is controlled by a pair of ball bearings mounted opposite each other in two bores at the rear of the plug. The balls lock the plug by engaging two short channels in the barrel. Unless the locking balls are retracted, the plug cannot turn relative to the cylinder body and barrel.

Zooming-in one level to the interface between the core and the plug, we find that it is the rear rim of the core that supports the two ball bearings. A third spring-loaded ball locates the core relative to the plug. When a key, any key, is inserted in the core and torque is applied, the locating ball pops out of its socket and allows the plug to turn. This does not operate the lock, however, because the rim of the core still acts to retain the locking balls in their channels in the barrel. What is required to release the locking balls is the forward movement of the carriage inside the core. The outcome of this action is twofold: first, it causes the locking balls to drop a few millimeters into a recess in the carriage; second, the rear edges of the carriage engage two slots in the end of the plug. So at the same time that the plug is freed to turn, it is coupled to the core, which can then be rotated along with the tail-piece that drives the locking cam.

It remains to be understood how the key operates the carriage, as illustrated in Fig. 7.46. The carriage is a molded component designed to slide on a set of rails at the rear of the core; it would slide right off the back were it not for two intervening sets of driverless pin-tumblers, shown in Fig. 7.44. The carriage has five vertically aligned pin chambers at 12 o'clock containing five miniature "top" pins. The chambers are capped by a steel cover that retains the driver springs.

Instead of top pins or drivers, the pins are ribbed, having a deep circumferential groove at various heights. The pins straddle a longitudinal channel in the carriage that intersects with a horizontal slot in the core at 12 o'clock. The pins are free to move vertically across this interface under the action of the key bittings, but cannot move longitudinally due to the chambering of the slot, unless their grooves line up with the channel. Some of the pins also have false-depth grooves.

The second set of four "side" pins is mounted in the core at 3 o'clock. Again, these are driverless and have one deep groove around their girth. The shanks of the pins cross a corrugated surface in the carriage on its lower right edge. When a key with the correct top- and side-bittings is inserted, the top pins are set such that their grooves are aligned along the channel in the carriage. This creates a path for the free movement of the top pins, which no longer block along the channel in the core. At the same time the grooves in the side pins have also formed a line that coincides with the corrugated edge of the carriage. It follows that the carriage is no longer bound by the pins and it slides backwards until its end stop contacts the bottom of the plug, engaging the latter and disengaging the locking balls from the barrel.

The system is ingenious and highly resistant to manipulation. With five sizes of top pin and three sizes of side pin the total number of system codes is $5^5 \times 3^4$ or 253,125. Like the Rivers lock in Chapter 2, it turns freely until a proper key is presented, and this gives it a degree of immunity from wrenching. On newer models from Citroën the lock is teamed with an electronic key-top transponder.

Chapter 8

Conclusion

> It is submitted, that the true principles of perfect security, strength, simplicity, and durability should be combined in every good lock. *John Chubb, c. 1850*

Let us briefly review the material that has been covered in the preceding chapters. We started with a historical perspective on the development of locks since the industrial revolution. This led into a discussion of the trade-off between the locksmith's responsibility to protect the public from dishonest people (by not divulging sensitive information) and the public's right to be fully informed about the product it is paying for. There followed a section on innovation in the lock industry, which dealt with the important issues of patenting, design registration, customer requirements, industry standards and lockpicking. Some administrative matters such as terminology, classification of locks, conventions, combinations, and grading of manipulation resistance were then discussed.

In Chapter 2, we surveyed the area of pin-tumbler locks, from their invention by Yale Senior and Junior in the latter half of the 19th century to the present day. We explained concepts such as the number of codes in a system, multiplex key profiles, MACS, and master-keying at a very basic level. We gave a cursory overview of lockpicking, impressioning and decoding. We then described measures taken to protect locks from manipulation and drilling. We explained the differences between passive and active profile pins. We also noted the current trend toward keys with active elements, which makes it next to impossible to produce unauthorized copies of keys. The following categories of pin-tumbler locks were identified: (1) inline; (2) inline with passive profile pins; (3) inline horizontal keyway; (4) twin inline; (5) inline with active profile pins; (6) cruciform; (7) multiple inline; (8) tubular or axial; (9) concentric pin; (10) rotating pin; (11) pin matrix; and (12) key-changeable. This was backed up by detailed descriptions of the operating principles of locks that exemplify each of these classes. We noted that the multiple inline dimple-key lock by Kaba AG has met with substantial commercial success.

Chapter 3 treated the subject of disc-tumbler or wafer locks. From their origins in the last quarter of the 19th century, these locks have generally been used as cam locks and switch locks in the industrial and automotive sectors. We considered both single- and double-throw wafer locks, as well as the number of codes and the concept of try-out keys. We noted some of the security features that provide resistance to picking and impressioning. The historically important Bramah lock as well as the more recent Bell lock with its side-track key were also discussed in this chapter. We adopted the following classification scheme for wafer locks: (1) conventional; (2) contoured; (3) three-sided; (4) inline push; (5) Bell; (6) axial; (7) reverse cut; (8) split-wafer; and (9) key-changeable. Detailed descriptions of locks of each of these types were subsequently presented.

Side-bar locks were the subject of Chapter 4. Although the inclusion of side-bars in cylinder locks is relatively recent, dating from the first quarter of the 20th century, the concept is closely connected with ancient designs such as the letter combination padlock and Scandinavian disc padlocks of the early 1700s. The side-bar lock category includes locks from the conventional types (pin-tumbler, wafer, and lever) that utilize a side-bar mechanism. We identified the following categories: (1) disc; (2) lever; (3) driverless pin; (4) wafer; and (5) dual-action. Next, we examined the operating principles of a number of high-security locks in these classes. Side-bar locks are the main product line of ASSA Abloy, one of the largest lock companies in the world today.

Chapter 5 was concerned with the most celebrated high-security lock—the lever lock. We traced its history from the late 18th-century Barron lock, through the more familiar form of the Chubb lock, to the double-bitted forms seen in Italy and Germany. We noted Kromer's important contribution in the development of double-bitted key locks. We recounted some of the events that have contributed to the colourful history of this class of locks. We also described master-keying in the context of lever locks, including the Butter's system and various other embodiments. Many of the early lever locks were exclusively used in safes, surviving today in museums and private collections. The lever locks we presented in the remainder of the chapter were arranged in the classes of: (1) conventional; (2) Italian double-throw; (3) lever locks with end-bitted keys; (4) radial lever locks; (5) cylindrical; (6) geared; (7) trap-door; (8) key-changeable; and (9) dual-control.

Chapter 6 was on the topic of magnetic locks, that is, locks using permanent magnets. These are quite distinct from electromagnetic locks, which use solenoids and/or card readers, and are not within the scope of the book. We explained that the development of magnetic locks depended on the availability, from the mid-1960s, of lightweight, high-flux-density permanent magnets. The early designs replaced conventional pin-tumblers with various arrangements of magnetic tumblers that were driven by the magnets in the key. Some of these designs teamed magnetic and conventional pin-tumblers. We also saw more recent examples based on magnetically oriented discs, and others that incorporate a matrix of small magnets.

Finally, in Chapter 7 we encountered a number of the more interesting automotive locks. We had a cursory look at the history of the motor car during the 20th century and its impact on the development and spread of wafer locks in particular. We saw that, apart from the Briggs & Stratton wafer side-bar lock introduced by General Motors in 1935, almost all vehicles used conventional locks with standard pin- or wafer-tumblers until the 1970s. Since then, quite a number of new designs have emerged, and we considered some of these. Specifically, we dealt with car locks utilizing twin inline pin-tumblers, the Abloy and Chubb locks used by Ford, various implementations of double-sided and split-wafer locks with two-track and four-track keys, and the Simplex lock used by Citroën.

Despite the numerous topics that we have dealt with, a number of important areas were omitted. As mentioned in the book's introduction, we did not cover keyless combination locks, time locks, or the general topic of safes. This is a vast and highly specialized area that is the preserve of only a relatively small fraction of the locksmithing and security profession. Information about this field, particularly detailed technical information, is generally unavailable to the public because of the risk of compromising the security of the institutions that rely on this technology. Notwithstanding this point, many of the keyed locks covered in this book have been and continue to be used on smaller safes and safe deposit boxes. For a discussion of combination locks and safes in general, the reader may wish to consult [13, 59, 122] and the references they contain.

Although we did mention in passing a number of nondestructive techniques for opening locks (picking, decoding, impressioning, etc.), the general area of opening locks nondestructively without access to the key was not discussed in any great detail. It seems prudent in a publicly available book to omit very detailed information of this sort. We chose instead to rate the manipulation resistance of individual locks and to note whether they included such features as drill and pry protection. The reader is referred to the references in Chapter 1, where several resources dealing with lock-picking, decoding, impressioning, opening tools, and bypass techniques were cited. Suffice to say that no lock is fully burglar proof; it is more a question of the amount of skill or force required and the time available for its application. Many specialized techniques developed in laboratories are of no practical value to burglars because they are either too difficult to learn, require expensive tools, or take too long to put into action.

As mentioned in the introductory chapter, the enthusiastic reader may wish to refer to the various manufacturers' Web sites, some of which contain historical or technical information on locks and locksmithing products. There are also a number of online forums, some moderated and others not, dedicated to locksmithing and related activities (e.g., Lock Picking 101). The FAQ from the alt.locksmithing newsgroup [60] is a good place to start searching for public-domain information on locks. In particular, the "Montmartre" Web site [7] contains a wealth of information on high-security locks and because it is an online resource, is well positioned to follow emerging technologies. Bearing in mind that much of the non–peer-reviewed information that one finds on the Web is unreliable, a search of the patent databases that have been made publicly available online is also recommended. This will acquaint the uninitiated reader with the higher level of detail required to specify a new lock design precisely. The material that is published through the patenting process has been reviewed by technical experts and suffers less from subjectivity and marketing "hype." The patent references in Appendix D provide a starting point for further study.

What of the future of mechanical locks? The 20th century saw the analog electronic computer appear and vanish because it was less flexible and more difficult to maintain than the digital computer: this led to a higher ratio of cost to performance, which signaled the death knell for this technology. The situation with locks, however, is somewhat less clear. For thousands of years, people have dwelt in houses and more recently apartments with wooden doors. At least for the foreseeable future this will continue to be the case. Therefore, there will always be a need for a mechanical "front-end" to a lock, that is, the part that actually secures the door to the frame. The idea that we will eventually either do away with doors or use electric force fields or some other barrier, as depicted in popular science fiction movies, seems fanciful to say the least. Mechanical locks will always be needed where there is no electricity supply (e.g., for perimeters and remote sites), or in applications where the power supply cannot be relied upon to ensure the locking function. In many situations there is still no substitute for the low cost, simplicity, and reliability of a mechanical cylinder lock or padlock.

On the other hand, in numerous applications we see an increasing prevalence of electronic locks teamed with electric strike plates and also electromagnetic door fasteners. The end-users of this technology are typically large industrial and institutional complexes, like universities, government offices, and hotels, although they also include smaller private concerns. Examples of these types of systems include digital locks, smart cards, magnetic stripe/swipe cards, noncontact RF proximity cards, and biometric access systems controlled by fingerprint, retinal, or voice recognition.

Digital combination locks have both mechanical and electronic embodiments. Both types have the advantage of being easily code-changed and not requiring a key. Digital or push button mechanical locks such as Simplex and Lockwood Digital are still in use, although we have not covered them in this book. Key-operated and fully electronic digital locks, where a code is entered by the user via a key pad, must also use an electric strike or motor drive. Electronic digital locks, such as those made by Mas Hamilton, La Gard, and Rosengrens, are steadily replacing mechanical keyless combination locks in bank safes and vaults.

While digital locks may provide a higher level of security in terms of the probability of someone accidentally stumbling on the correct code, there may be a risk that the system can respond to an override code keyed in by an unauthorized individual. It is impossible to tell whether such a code exists from an inspection of the lock since the access code exists only in the memory of a silicon chip inside the lock. The widespread use of electronic digital combination locks is also reducing the demand for persons skilled in the art of safe lock manipulation, possibly the most difficult aspect of locksmithing to master.

Factor	Pin-Tumbler Lock	Side-bar Lock	Keypad	Swipe Card	Proximity Card
Access Traceability	None	None	Logged	Logged	Logged
Installation	Easy	Easy	Harder	Harder	Harder
Key Portability	Good	Good	N/A	Fair	Fair
Key Copyability	Easy	Hard	Trivial	Easy	Hard
Key Replacement Cost	Low	High	N/A	Low	Medium
Maintenance	Medium	Medium	Medium	High	Low
Physical Strength	High	High	Low	Low	Low
Recombination	Medium	Medium	Easy	Easy	Easy
Requires Electricity	No	No	Yes	Yes	Yes
System Codes	Medium	High	Low	Unlimited	Unlimited
Lock Unit Cost	Low	Medium	Medium	Medium	High
Weather Resistance	Good	Fair	Fair	Poor	Good

Table 8.1: Engineering trade-offs associated with mechanical and electronic locks.

Table 8.1 sets out a simplified summary of the advantages and disadvantages of various kinds of mechanical and electronic locking systems.[1] This information may help the reader to appreciate the differences between these two generic technologies, although for any particular application a more detailed analysis should be carried out. There is no doubt that with the increasing demand for electronic access control systems, the unit cost of supply will decrease and there will be more avid competition with conventional mechanical locks. Many manufacturers are now hedging their bets by supplying both electronic and mechanical locking products.

In some applications (e.g., the automotive industry), a combination of both electronic and mechanical design aspects is more appropriate. All of the major car manufacturers have now incorporated key-top transponders in their late-model vehicles. The key operates the mechanical locks fitted to the doors, ignition, glove box, and trunk. The transponder also operates the external locks and in some cases may switch the alarm system on/off or enable/disable the car computer. Some vehicles feature fully electronic locking, having dispatched with the need for key-operated locks on the doors.

Electromechanical locks are also being produced for a number of access control applications. Examples of this technology include mortice locks with key-operated latchbolts and card-operated electric strikes or solenoid-operated mechanisms. An advantage of this type of system is that access times for certain cardholders can be restricted, while the mechanical lock provides an emergency override function. Electromechanical lock cylinders that incorporate a mechanical, key-operated plug and electronically controlled linkage or blocking mechanism are equally available. These "smart key" locks include electronics in the key head that transfers a coded

---

[1]In Table 8.1, keypad combinations are assumed to be limited to a length that is easily memorized, for example, four or five digits.

signal to the monitoring system in the lock via either a set of contact pads or wireless RF induction. Such systems are designed to eliminate unauthorized key duplication while offering greater flexibility in terms of access control.

Epilogue

We hope to have stirred your interest in the fascinating subject of high-security locks. There is certainly a lot more that can be said about mechanical locks and where the industry is headed. Naturally, many locks have not been included in this book due to lack of time or availability of samples.

The progress that has been made in this industry in the space of one generation is staggering. The leading lock-making companies invest considerably in ongoing R&D and are continuing to produce innovative products at a rate that is difficult to keep up with. On the other hand, the number of lock and security companies is continuing to diminish due to acquisitions of small- to medium-sized firms by global players such as ASSA Abloy and Kaba. It will be interesting to see the shape of the industry over the coming years and what products emerge from this corporate fusion.

Faced by this continual stream of new technology, it serves us well to remember the fundamental design principles of what constitutes a good lock, espoused over 150 years ago by John Chubb in his dissertation to the Institution of Civil Engineers in London: "security, strength, simplicity, and durability."

Bibliography

[1] Arnall, F. M. *The Padlock Collector*, 6th edition, 1998, Claremont, CA.

[2] Arnold Magnetic Technologies. Magnetics Technology Center Web site, http://www.arnoldmagnetics.com/mtc/index.htm.

[3] Assa Abloy. *Assa High Security Locks: Technical Specifications and Product Catalogue*, Assa Abloy Group, 2004.

[4] Association of Builders Hardware Manufacturers. "Best Practice Guide: Cylinders for Locks," BS EN 1303 (1998).

[5] Beck, R. A. *The Lock Collector*, Issue 8, July/August 2005, U.K.

[6] Beck, R. A. "Edwin Cotterill: His Middle Age and Lock Inventions (Part 2)," in *The Lock Collector*, Issue 10, January/March 2006, U.K.

[7] Becker, J. R. Lock pictures and references, available from http://www.chez.com/montmartre/, http://www.protections-vol.com/.

[8] Bengué, G. *Pour le Serrurier: La Serrure: Mise au Point, Dérèglements, Réparations*, Artisan-Pau, St Joseph, Tarbes, 1963.

[9] Berger, E. *Prunk-Kassetten: Europaïsche Meisterwerke aus acht Jahrhunderten*, Hans Schell Collection, Graz, Austria, Arnoldsche Art Publishers, 1998.

[10] Blaze, M. "Notes on Picking Pin Tumbler Locks," http://www.crypto.com/papers/notes/picking/, November 2003.

[11] Blaze, M. "Rights Amplification in Master-Keyed Mechanical Locks," *IEEE Security and Privacy Magazine*, pp. 24–32, March 2003.

[12] Blaze, M. "Notes on Schlage Everest Locks," http://www.crypto.com/photos/misc/everest/, April 2003.

[13] Blaze, M. "Safe Cracking for the Computer Scientist," Dept. Computer and Information Science, University of Pennsylvania, draft report, December 2004. Available from http://www.crypto.com/papers/safelocks.pdf.

[14] Bowman, D. L. "Servicing Medeco 3 Mortise Cylinders," *Locksmith Ledger*, Fort Atkinson, Wisconsin, pp. 26–38, October 2004.

[15] Bramah, J. "A Dissertation on the Construction of Locks," London, 1785.

[16] British Standards Institution. "Building Hardware—Cylinders for Locks—Requirements and Test Methods," BS EN 1303 (1998).

[17] British Standards Institution. Permission to reproduce extracts from BS EN 1300: 2004 and BS EN 1303: 1998 is granted by BSI. British Standards can be obtained from BSI Customer Services, 389 Chiswick High Road, London W4 4AL. Tel: +44 (0)20 8996 9001. E-mail: cservices@bsi-global.com.

[18] British Standards Institution. "Secure Storage Units—Requirements, Classification and Methods of Test for Resistance to Burglary—Part 2: Deposit Systems," BS EN 1143-2 (2001).

[19] British Standards Institution. "Secure Storage Units—Classification for High Security Locks According to Their Resistance to Unauthorized Opening," BS EN 1300 (2004).

[20] Bûgg, D. E., and Potter, D. P. K. *Burglary Protection and Insurance Surveys*, 4th edition, Stone & Cox, London, 1982.

[21] Butter, F. J. *An Encyclopaedia of Locks and Builders Hardware*, Josiah Parkes & Sons, Union Works, Willenhall, England, 1958, reprinted in 1968 and 1979.

[22] Chubb, J. "On the Contruction of Locks and Keys," *Proc. Inst. Civ. Eng.*, Vol. 9, 1850.

[23] Debito, A. "Generating the Foreigner Crime Wave," *Japan Times*, The Zeit Gist, October 4, 2002. Available from http://www.jref.com/forum/archive/index.php/t-834.html See also http://www.debito.org/TheCommunity/communityissues.html.

[24] Dickens, C. *The Pickwick Papers*, Chapter 53, "Comprising the final exit of Mr. Jingle and Job Trotter, with a great morning of business in Gray's Inn Square—Concluding with a double knock at Mr. Perker's door," Chapman and Hall, London, 1837.

[25] Diderot & d'Alembert. "Recueil de Planches sur les Sciences, Les Arts Libéraux, Les Arts Méchaniques, avec leur Éxplication: Serrurerie," 57 pages, 1762, in *Encyclopédie*, 1751–1772, Paris.

[26] Diederichsen, O. Photographs of German safe locks, http://www.tresoroeffnung.de, accessed July 25, 2007.

[27] Diederichsen, O. "Kromer Protector Locks," private correspondence, October 7, 2005.

[28] Duhamel du Monceau, H. L. *Descriptions des Arts et Métiers*, Académie Royale des Sciences, J. Desaint & C. Saillant, Paris, 1761–1782.

[29] Duhamel Du Monceau, H. L. *l'Art du Serrurier*, Paris, 1767.

[30] Eras, V. J. M. *Locks and Keys throughout the Ages*, published in Dutch 1941, reprinted in English by Lips, Dordrecht, 1957, and by Bailey Bros. and Swinfen, 1974, U.K.

[31] Erroll, D., Erroll, J., and Day, A. *American Genius: Nineteenth-Century Bank Locks and Time Locks*, Quantuck Lane, New York, 2006.

[32] European Certification Board—Security Systems. *Lock List ECB S R01*, Forschungs und Prüfgemeinschaft Geldschränke und Tresoranlagen e. V. Frankfurt, March 2005, Germany.

[33] Evans, J. "A Gazetteer of Lock and Key Makers," http://www.localhistory. scit.wlv.ac.uk/Museum/locks/gazetteer/gazhoh-izz.htm, 2002.

[34] Evans, J. "A Gazetteer of Lock and Key Makers," http://www.localhistory. scit.wlv.ac.uk/Museum/locks/gazetteer/chubbs.htm, 2002.

[35] Faúndez-Zanuy, M. "Biometric Security Technology," *IEEE Aerospace and Electronic Systems Magazine*, pp. 15–26, June 2006.

[36] Fey, H. "The Drumm Geminy Shield," July 2006. Available from http:// www.toool.nl/drumm-geminy.pdf.

[37] Fey, H. "Cutaways Part III," *Journal of Lock Collecting, A Publication of the American Lock Collectors Association* (R. Dix, Ed.), Vol. 36, No. 5, pp. 4–14, September 2005.

[38] Fey, H. "The Discs Make the Difference," *Journal of Lock Collecting, A Publication of the American Lock Collectors Association* (R. Dix, Ed.), Vol. 35, No. 6, pp. 4–14, November 2004.

[39] Fey, H. "Evolution of Abloy (Part 2)," *Journal of Lock Collecting, A Publication of the American Lock Collectors Association* (R. Dix, Ed.), Vol. 37, No. 2, pp. 4–14, March 2006.

[40] Fey, H. "High Security European Locks Part III," *Journal of Lock Collecting, A Publication of the American Lock Collectors Association* (R. Dix, Ed.), Vol. 36, No. 2, pp. 4–11, March 2005.

[41] Fichet-Bauche. *Histoire et Petites Histoires de la Serrurerie*, Paris-Versailles, 1976.

[42] Fincher, J. M. "Influence of the Bramah Lock on the Development of German Safe Locks (1800–1900)," in *Lock Collector*, R. A. Beck (Ed.), Issue 9, September/December 2005, U.K.

[43] Fincher, J. M. "The Chatwood Locks," in *The Chatwood Story*, compiled by A. C. Clare, Exeter, U.K., pp. 27–37, 1999.

[44] Fincher, J. M. "History of the Theodor Kromer Company," address to the Master Locksmiths Association Convention, De Vere Hotel, Coventry, U.K., August 25–30, 1983.

[45] Fincher, J. M. "Evolution of German Safe Locks (Part 2): Theodor Kromer," in *Lock Collector*, R. A. Beck (Ed.), Issue 11, April/June 2006, U.K.

[46] Fincher, J. M. Excerpts of correspondence between George H. Chubb and William H. Chubb, July 1877. Available from P. A. Prescott's site http://www.antique-locks.com/kromerchubb.htm, 2002.

[47] Friend, M. *The Encyclopaedia for Locksmiths*, Bertrams Print On Demand, available from *Authors OnLine*, March 2004, U.K.

[48] Gillespie, T. "High Security Locks: Schlage Primus," *Locksmith Ledger*, Fort Atkinson, Wisconsin, pp. 26–30, February 2002.

[49] Gunn, P. J. "Chubb—A Brief History," 2002, available from: http://www.chubbarchive.co.uk/.

[50] Hennessy, T. F. *Early Locks and Lockmakers of America*, Nickerson & Collins, Des Plaines, IL, 2nd edition, 1976.

[51] Hobbs, A. C. "On the Principles and Construction of Locks.," *Proc. Inst. Civ. Eng.* 1854.

[52] Hobbs, A. C. *On the Construction of Locks*, in *Construction of Locks and Safes cicra 1850*, C. Tomlinson (Ed.), Virtue and Co., London, 1868. Reprinted 1970, Redwood Press, Trowbridge, U.K.

[53] Hobbs, A. C. *Rudimentary Treatise on the Construction of Locks*, Weale, London, 1853.

[54] Hobbs, A. C. "Report from the Arbitrators to Whom the Bramah Lock Controversy Was Referred," September 1851.

[55] Hogg, G. *Safe Bind, Safe Find, the Story of Locks, Bolts and Bars*, Phoenix House, London, 1961.

[56] Holmes, F. S. "Bank Vault Construction and Equipment," *Architectural Forum*, reprinted in *The Lure of the Lock*, A. A. Hopkins, General Soc. of Mechanics and Tradesmen, Mechanics Institute, NY, 1954, pp. 220–230.

[57] Hopkins, A. A. *The Lure of the Lock*, General Soc. of Mechanics and Tradesmen, Mechanics Institute, NY, 1928, reprinted 1954, 1980, 1991.

[58] Hopkins, R. Notes from private correspondence with R. Chenovick concerning Miller padlocks, December 2005, U.K.

[59] Hunkin, T. "Illegal Engineering," lecture given at London Science Museum, 1997, http://www.timhunkin.com/94_illegal_engineering.htm.

[60] Ilacqua, J., Schaffer, H., et al. "Answers to Frequently Asked Questions," alt.locksmithing newsgroup, http://www.indra.com/archives/alt-locksmithing/, also available from http://www.faqs.org/faqs/locksmith-faq/index.html.

[61] Jain, A. K., Ross, A., and Pankanti, S. "Biometrics: A Tool for Information Security," *IEEE Transactions on Information Forensics and Security*, Vol. 1, No. 2, pp. 125–143, June 2006.

[62] Jain, A. K., Ross, A., and Prabhakar, S. "An Introduction to Biometric Recognition," *IEEE Transactions on Circuits and Systems for Video Technology*, Vol. 14, No. 1, pp. 4–20, January 2004.

[63] Johnson, G. "Classic Car Key Facts," *Locksmith Ledger*, Fort Atkinson, Wisconsin, pp. 52–55, March 2002.

[64] Jousse, M. *La Fidelle Ouverture de l'Art de serrurier: où l'on Void les Principaulx Préceptes, Desseings et Figures, Touchant les Expériences et Opérations Manuelles dudict Art*, Georges Griveau, La Flèche, France, 1627. English translation by Smith, C. S., and Sisco, A. G., *Technology and Culture*, vol. 2, pp. 131–145, 1961. Reprinted in 2004, Librairie des Arts et Métiers—Éditions Jacques Laget, Nogent-le-Roi, France.

[65] Kagawa, C. "Attacks Against the Mechanical Pin Tumbler Lock," GIAC Security Essentials Certification, SANS Institute, January 2004.

[66] Krühn, J. *Schliesszylinder: Entwicklungsgeschichte Technik Anwendung*, Gert Wohlfarth GmbH, Verlag Fachtechnik + Mercator-Verlag, Duisburg, 1996.

[67] Kyodo News International, Inc. "Lock Business Thrives in Japan, along with Rising Burglaries," Japan Policy & Politics, December 11, 2000. Available from http://www.findarticles.com/p/articles/mi_m0XPQ/is_2000_Dec_11/ai_68163430.

[68] Levine, J. "The ASSA Twin V-10," *Locksmith Ledger*, Fort Atkinson, Wisconsin, pp. 8–10, February 2002.

[69] Levine, J. "Yale High Security Cylinders," *Locksmith Ledger*, Des Plaines, Illinois, April 1991.

[70] Linder, N. (Ed.). Article on Christopher Polhem, *Nordisk Familjebok Konversationslexikon och Realencyklopedi*, Vol. 12, pp. 1490–1494, Gernandts Boktryckeri-Aktiebolag, Stockholm, 1888. Facsimile by Projekt Runeberg, Sweden, 2002, http://runeberg.org/nf/.

[71] Locksmith Ledger. Advertisement for BHI Huck pin. Des Plaines, Illinois, February 1976.

[72] Loss Prevention Certification Board. "Requirements and Testing Procedures for the LPCB Approval and Listing of Cylinders for Locks," Loss Prevention Standard LPS 1242, Issue 1, U.K., December 2003. Please check www.RedBookLive.com for the most up-to-date issue of this standard.

[73] Loss Prevention Certification Board. "Requirements and Testing Procedures for the LPCB Approval and Listing of Burglary Resistant Building Components, Strongpoints and Security Enclosures," Loss Prevention Standard LPS 1175, Issue 5.1, U.K., February 2004.

[74] Loss Prevention Certification Board. "Requirements and Testing Procedures for the LPCB Approval and Listing of Safe Storage Units—Part 1: Safes and Strong-rooms," Loss Prevention Standard LPS 1183, Issue 4.1, U.K., January 2004.

[75] Mandel, G. *Clefs*, Littostampa Instituto Grafico, Bergamo, 1990; French edition translated by C. Lavigne, EDDL, 2001, Paris.

[76] Mandelbrot, B. B. "The Fractal Geometry of Nature," W. H. Freeman, New York, 1985.

[77] Mangine, R. F. "Examination of Steering Columns and Ignition Locks," in *Forensic Investigation of Stolen-Recovered and Other Crime Related Vehicles*, Chapter 9, E. Stauffer and M. Bonfanti (Eds.), Academic Press, New York, 2006.

[78] Master Locksmiths Association. "Guidelines for Minimum Security Requirements for Domestic Property," 3rd edition, April 2005, U.K.

[79] Mauer GmbH Werk Kromer. Response to letter from J. M. Fincher concerning serial numbers of Kromer Protector locks, Umkirch, October 11, 2001.

[80] Mauer GmbH Werk Kromer. Response to letter from J. M. Fincher concerning Cawi and Kromer Protector locks, Umkirch, July 2, 2001.

[81] McBennett, M. (webmaster). "True Crimes," Newsletter of the Japan Zone Web site http://www.japan-zone.com/, Issue 5, July 25, 2001.

[82] McNeil, I. *Josepth Bramah, a Century of Invention 1749–1851*, David & Charles, Newton Abbot, U.K., 1968.

[83] Miller, H. C. "Mosler PK-5900 Series Safe Deposit Box Lock," brochure MO-002, The Harry C. Miller Lock Collection, 1977.

[84] Miller, H. C. "Sargent & Greenleaf 4400 Series Safe Deposit Box Lock," brochure SG-004, The Harry C. Miller Lock Collection, 1977.

[85] Miller, H. C. "Diebold 175 Series Safe Deposit Box Lock," brochure DI-001, The Harry C. Miller Lock Collection, 1977.

[86] Miller, H. C. "Mosler 5700 Series Safe Deposit Box Lock," brochure MO-001, The Harry C. Miller Lock Collection, 1977.

[87] Millington, J. "Lock Collecting—Patents," *Keyways*, Vol. 24, No. 4, pp. 40–42, February 2002.

[88] Millington, J. Private correspondence relating to the Code Lock, November 8, 2005.

[89] Millington, J. "Bibliography of Francis J. Butter," *Keyways*, Vol. 22, No. 3, pp. 28–30, December 1999.

[90] Miwa Lock Co. Ltd. Product advertisements in *Shuukan Asahi* (newspaper), Japan, pp. 76, 142–143, February 25, 2000.

[91] Monk, E. *Keys—Their History & Collection*, Shire Publications, Aylesbury, U.K., 1974.

[92] Murphy, K. & Kay, J. "Automatic Identification & Information Systems," in *Automatic Identification, an IFS Executive Briefing* (R. L. Chase, Ed.), Springer-Verlag, U.K., pp. 111–115, 1988.

[93] Myers, H. P. *Introductory Solid State Physics*, 2nd edition, Taylor & Francis, New York, pp. 373–375, 1997.

[94] Patent Office Designs Registry. United Kingdom. http://www.patent.gov.uk/design/.

[95] Phillips, B. *Professional Locksmithing Techniques*, 2nd edition, McGraw-Hill, New York, 1996.

[96] Prescott, P. A. "Hobbs and Co.," http://www.antique-locks.com/hobbsint.htm, 2002. See also http://www.antique-locks.com/viewtopic.php?t=161.

[97] Prescott, P. A. "Decoders for Ross Locks," http://www.antique-locks.com/decoders.htm.

[98] Prescott, P. A. "Chatwood Safe Company," http://www.antique-locks.com/chtemp1.htm, 2002.

[99] Price, G. *Treatise on Fire & Thief-Proof Depositories and Locks & Keys*, Simpkin Marshall & Co., London, 1856.

[100] Price, G. *A Treatise on Gunpowder-Proof Locks, Gunpowder-Proof Lock-Chambers, Drill-Proof Safes, &c.*, E. & F. Spon, London, 1860.

[101] Pulford, G. W. "Catalogue of High Security Locks v1.00," April 1994, available from ftp://ftp.indra.com/archives/alt-locksmithing/hiseclox.ps.Z.

[102] Rathjen, J. E. *Locksmithing: From Apprentice to Master*, TAB Books, New York, 1995.

[103] Roberts, J. *The National Locksmith Guide to Antique Padlocks*, National Publishing Company, Streamwood, IL, 1990.

[104] Robinson, A. "Improvements in or Appertaining to Locks for Safes and the Like," U.K. patent 179,970, May 1922.

[105] Robinson, R. L. *Complete Course in Professional Locksmithing*, Nelson-Hall, Chicago, 1983.

[106] Roper, C. A. *The Complete Book of Locks and Locksmithing*, TAB Books, Blue Ridge Summit, PA, 1983.

[107] Sargent & Greenleaf. "Key Operated Models 6804 and 6824—Operating Instructions."

[108] Schlage. "Cylinders, Keys and Key Control," Ingersoll Rand, 2002, U.S.

[109] Schlage. "Everest Answer Book," Ingersoll Rand, 2002, U.S.

[110] Silca^{TM}. 101 Key Catalogue 1993.

[111] Silca^{TM}. 201 Key Catalogue 1993.

[112] Silca^{TM}. 503 The Car Book, 1994.

[113] Sloane, E. A. *Complete Book of Locks, Keys, Burglar and Smoke Alarms and Other Security Devices*, 1977, William Morrow, New York.

[114] Stenner, B. *The Lockmakers. A Century of Trade Unionism in the Lock and Safe Trade 1889-1989*, Malthouse Press, Oxford, U.K., 1989.

[115] The Mathworks. http://www.mathworks.com/.

[116] Theodor Kromer KG. Translation of text from Kromer's 100-year booklet from 1968, Freiburg.

[117] Theodor Kromer KG. "Katalog über Sicherheitsschlösser und Sicherheitskombinationen für Geldschränke, Panzergewölbetüren, Tresoranlangen und Wertbehälter aller Art, sowie dazugehörende Beschläge," Ausgabe Nr. 13, Freiburg, Germany, c. 1950.

[118] Theodor Kromer KG. Translation of "A Short Report on the Opening of a Lock by Meister Bierhaus," Freiburg, December 22, 1954.

[119] Theodor Kromer KG. Translation of response to letter from J. M. Fincher concerning the picking of the RP37E Kromer Protector lock, Freiburg, September 21, 1976.

[120] Tobias, M. W. "Method and Apparatus for Decoding a Pin Tumbler Lock," U.S. patent 5,355,701, October 18, 1994.

[121] Tobias, M. W. *Locks, Safes and Security: A Handbook for Law Enforcement Personnel*, Charles C. Thomas, Springfield, IL, 1971.

[122] Tobias, M. W. *Locks, Safes and Security: An International Police Reference*, Charles C. Thomas, Springfield, IL, 2000.

[123] Tyler, B., et al. *Fractint*, fractal generating software (freeware), first written 1989, available from http://spanky.triumf.ca/www/fractint/fractint.html.

[124] Van Latum, E. "NTS Miwa locks," private correspondence, June 8, 1994.

[125] Vaudour, C. *Catalogue du Musée Le Secq des Tournelles. Fascicule II: Clefs et Serrures des Origines au Commencement de la Renaissance*, Rouen, France, September 1980.

[126] Volpe, F. P. and Volpe S. *Chipkarten—Grundlagen, Technik, Anwendungen*, Verlag Heinz Heise, Hannover, 1996.

[127] Walker, A. Articles on Vingcard and Marlok, published online in *Informatik: The Journal of Privileged Information*, Issue 5, October 31, 1992.

[128] Wanlass, M. private correspondence, June 9, 1995.

[129] Wanlass, M., and Hall, S. "Impressioning Manual For Amateur Locksmiths," http://www.gregmiller.net/locks/impress.html, (revised) 1997.

[130] Wels, B., and Gonggrijp, R. "Bumping Locks," The Organization of Open Lockpickers (TOOOL), January 2005, http://www.toool.nl/bumping.pdf.

[131] Wels, B., et al. References on bumping locks. The Organization of Open Lockpickers (TOOOL) Web site, http://www.toool.nl/index-eng.php.

[132] Wels, B., et al. "Certified Locks Get Knocked Out," originally published in *Consumentenbond Magazine*, April 2006, Netherlands. English version available from http://www.toool.nl/consumer-reports-nl.pdf.

[133] Wikipedia. The Free Encyclopedia, 2001–2006. Article on Christopher Polhem, http://www.reference.com/browse/wiki/Christopher_Polhem.

[134] Yale and Towne Manufacturing Company. "A Bit of History: The Trade-Mark 'YALE' and Its Origin," Catalog 22, Wynkoop Hallenbeck Crawford Co., New York, pp. 30–38, 1917.

[135] Yale, L., Jr. *A Dissertation on Locks and Lockpicking, and the Principles of Burglar Proofing Invented by Linus Yale, Jr.*, T. K. Collins and P. G. Collins, Philadelphia, 1856.

Figure and Table Credits

Photographs

Fig. 3.64 courtesy of O. Diederichsen.

Fig. 3.66 courtesy of O. Diederichsen.

Fig. 4.4 courtesy of R. Loschiavo.

Fig. 5.17 courtesy of R. Hopkins.

Fig. 5.38 (top) courtesy of P. Prescott.

Fig. 5.99 courtesy of J. M. Fincher.

Fig. 5.124 courtesy of O. Diederichsen.

Tables

Table 1.1 reproduced with permission of BSI [17].

Table 1.2 reproduced with permission of BSI [17].

Table 8.28 reproduced with permission of LPCB [72].

Patent Diagrams

All patent diagrams were obtained from the European Patent Office esp@cenet Web site at http://ep.espacenet.com. In this section, the publication date of the patent is indicated rather than the filing date. The initials EP stand for "European patent." Country codes are as defined in Table 1.4 in Chapter 1.

Fig. 1.4 from 1922 US patent 1,403,753.

Fig. 1.5 from 1943 US patent 2,309,677.

Fig. 1.6 from 1925 US patent 1,639,919.

Fig. 1.7 from 1928 US patent 1,667,223.

Fig. 1.8 from 1936 US patent 2,064,818.

Fig. 1.9 from 1956 US patent 2,763,027.

Fig. 1.10 from 1966 US patent 3,251,206.

Fig. 1.12 from 1992 US patent 5,172,578.

Fig. 2.8 from 1889 US patent 414,720.

Fig. 2.9 from 1965 US patent 3,206,958.

Fig. 2.19 from 1987 US patent 4,683,740.

Fig. 2.22 from 1942 US patent 2,283,489.

Fig. 2.23 from 1939 US patent 2,158,501.

Fig. 2.24 from 1973 US patent 3,762,193.

Fig. 2.25 from 1983 US patent 4,377,940.

Fig. 2.26 from 1968 US patent 3,418,833.

Fig. 2.27 from 1984 US patent 4,434,636.

Fig. 2.28 from 1983 US patent 4,377,082.

Fig. 2.29 from 1975 US patent 3,928,993.

Fig. 2.50 from 1930 US patent 1,770,864.

Fig. 2.68 from 1998 US patent 5,724,841.

Fig. 2.77 from 1968 US patent 3,393,542.

Fig. 2.88 from 1982 US patent 4,320,638.

Fig. 2.101 from 1973 Swiss patent 578,105.

Fig. 2.115 from 1963 UK patent 940,778.

Fig. 2.131 from 1995 US patent 5,457,974.

Fig. 2.144 from 1992 US patent 5,131,249.

Fig. 2.158 from 1973 US patent 3,738,136.

Fig. 2.159 from 1989 US patent 4,802,354.

Fig. 2.173 from 1993 FR patent 2,678,670.

Fig. 2.184 from 1921 US patent 1,390,222.

Fig. 2.188 from 1989 US patent 4,856,309.

Fig. 2.196 from 1980 US patent 4,208,894.

Fig. 2.197 from 1978 US patent 4,098,103.

Fig. 2.203 from 1979 US patent 4,149,394.

Fig. 3.10 from 1934 US patent 1,965,889.

Fig. 3.11 from 1936 US patent 2,039,126.

Fig. 3.12 from 1918 US patent patent 1,287,882.

Fig. 3.13 from 1990 US patent 4,966,021.

Fig. 3.20 from 1984 US patent patent 4,429,554.

Fig. 3.38 from 1966 US patent 3,263,461.

Fig. 3.45 from 1966 US patent 3,264,852.

Fig. 3.51 from 1964 US patent 3,237,436.

Fig. 3.72 from 1990 US patent 4,966,021.

Fig. 3.75 from 1978 US patent 4,069,694.

Fig. 4.2 from 1879 US patent 213,300.

Fig. 4.3 from 1874 US patent 156,113.

Fig. 4.5 from 1981 US patent 4,267,717.

Fig. 4.6 from 1901 US patent 688,070.

Fig. 4.8 from 1982 US patent 4,356,713.

Fig. 4.9 and Fig. 4.10 from 1970 US patent 3,499,302.

Fig. 4.13 from 1983 US patent 4,404,824.

Fig. 4.14 from 1968 US patent 3,367,156.

Fig. 4.20 from 1971 US patent 3,621,689 and from 1976 US patent 3,948,065.

Fig. 4.21 from 1978 US patent 4,083,212.

Fig. 4.22 from 1987 US patent 4,651,546.

Fig. 4.23 and Fig. 4.24 from 1978 US patent 4,109,495.

Fig. 4.38 from 1985 US patent 4,512,166.

Fig. 4.42 from 1977 US patent 4,044,578.

Fig. 4.47 from 1971 US patent 3,604,231.

Fig. 4.48 from 2002 US patent 6,490,898.

Fig. 4.56 from 2000 JP patent 2000 291300.

Fig. 4.60 from 1973 US patent 3,722,240.

Fig. 4.61 from 1994 US patent 5,375,444.

Fig. 4.66 from 1984 US patent 4,478,061.

Fig. 4.67 from 1986 US patent 4,603,565.

Fig. 4.74 from 2004 US patent 6,681,609.

Fig. 4.79 from 1996 US patent 5,517,840.

Fig. 4.80 from 1937 US patent 2,070,233.

Fig. 4.93 from 1990 US patent 4,977,767.

Fig. 4.94 from 2003 US patent 6,622,538 and from 2004 US patent 6,758,074.

Fig. 4.99 and Fig. 4.100 from 1987 US patent 4,635,455.

Fig. 4.101 from 1984 US patent 4,450,699.

Fig. 4.106 from 1983 US patent 4,393,673.

Fig. 4.110 from 1991 US patent 5,067,335.

Fig. 5.18 from 1937 DE patent 646,623.

Fig. 5.25 from 1924 UK patent 224,175.

Fig. 5.28 from 1951 UK patent 661,501.

Fig. 5.31 from 1901 US patent 666,697.

Fig. 5.32 from 1983 US patent 4,375,159.

Fig. 5.46 from 1968 US patent 3,402,581.

Fig. 5.48 from 1984 US patent 4,462,230.

Fig. 5.104 from 1980 US patent 4,196,606.

Fig. 5.111 and Fig. 5.112 from 1966 UK patent 1,030,921.

Fig. 5.127 from 1954 DE patent 911,220.

Fig. 5.133 from 1986 US patent 4,601,184.

Fig. 6.3 from 1890 US patent 428,247.

Fig. 6.4 from 1914 US patent 1,172,203.

Fig. 6.5 from 1939 US patent 2,177,996.

Fig. 6.6 from 1978 US patent 4,073,166.

Fig. 6.7 from 1969 US patent 3,444,711.

Fig. 6.8 from 1990 US patent 4,932,228.

Fig. 6.9 from 1970 US patent 3,512,382.

Fig. 6.10 from 1970 US patent 3,518,855.

Fig. 6.11 from 1974 US patent 3,855,827.

Fig. 6.12 from 1989 UK patent 2,214,226.

Fig. 6.13 from 1976 US patent 3,935,720.

Fig. 6.27 from 1978 US patent 4,073,166.

Fig. 6.36 from 1978 US patent 4,084,416.

Fig. 7.1 from 1920 US patent 1,345,014.

Fig. 7.2 from 1980 US patent 4,185,482.

Fig. 7.3 from 1942 US patent 2,279,592.

Fig. 7.4 from 1979 DE patent 7,203,658.

Fig. 7.19 from 1934 US patent 1,965,336.

Appendices

Appendix A1: Permutations and Combinations

This appendix is intended to introduce some basic concepts in permutations and combinations. These ideas are quite mathematical but of considerable importance for understanding key codes and master-keyed systems. Readers already acquainted with this material may wish to skip to the following section, which deals with permutations and locks and how the two relate to a new kind of fractal image.

The starting point for any discussion on permutations and combinations is a set $\mathcal{S}_n$ of n *distinct* objects, which we will assume are just counting numbers between 1 and n, thus $\mathcal{S}_n = \{1, 2, 3, \ldots, n\}$. By "disctinct," we mean to imply that there are no repeated entries in the set. (We later relax the assumption that the objects are distinct.) The first idea is that of a *rearrangement* of the set. For instance, if the set is $\mathcal{S}_3 = \{1, 2, 3\}$, then (1 2 3) is one rearrangement and (2 1 3) is another rearrangement of the elements.

The question arises of how many rearrangements are possible for a given set (including the original ordering). For small sets like $\mathcal{S}_3$ with $n = 3$ objects, the rearrangements can be written out explicitly or *enumerated*. Thus for the case in question we have (1 2 3), (1 3 2), (2 1 3), (2 3 1), (3 1 2), (3 2 1), from which we see that there are 6 rearrangements in total. For n greater than 5 or so, it becomes very tedious to enumerate all the rearrangements. Luckily, there is a simple analytical answer. It turns out that for $\mathcal{S}_n$, with n elements, there are $n \times (n-1) \times (n-2) \times \cdots \times 3 \times 2 \times 1$ possible rearrangements. This formula arises so frequently in the mathematics of sets that it is given a special name: $n!$, which is read "n factorial." Notice that it only makes sense when n is a positive whole number.

For example, when $n = 4$ there are $4! = 4.3.2.1 = 24$ rearrangements and for $n = 5$ there are $5! = 5.4.3.2.1 = 120$ (in mathematical notation, the $\times$ symbol is often replaced with a simple dot '.'). The factorial number grows *very* rapidly. For 10 objects there are $10! = 3,628,800$ rearrangements. What is the use of this formula? Suppose you had to recombinate a lock with five pins, all of different sizes, using only the pins in the lock (ignoring MACS and other constraints). How many ways could you do it before you started repeating codes? The answer is $5!$ or 120 ways.

So far we have considered sets of *distinct* objects. The next step in the process is to look at what happens if some of the objects are the same or *repeated*. To help us visualize this situation, consider a set of four objects that are distinct: $S_4 = \{1, 2, 3, 4\}$. We know that there are $4! = 24$ possible rearrangements, and we can list them according to Table A1.

Now suppose that two of the objects are the same, say, objects 1 and 2. Whenever we see a 1 or a 2 in the table, we can mark it with an "x" since we cannot distinguish them. If we do this, our table looks like Table A2.

What we notice about this new table is that for every row in it, we can find another row that is the same. So instead of having 24 different rows there are only 12 possible rearrangements or *permutations* (we revisit this idea in the following section). What would happen if the first three objects were the same, so that we only had two distinct objects (4 and x) out of the four? By setting $x = 3$ in Table A2 we soon find that the only different rows are (x x x 4), (x x 4 x), (x 4 x x), and (4 x x x). Thus with four objects, only one of which is *nonrepeated* ("4" and the other three the

1 2 3 4	3 1 2 4
1 2 4 3	3 1 4 2
1 3 2 4	3 2 1 4
1 3 4 2	3 2 4 1
1 4 2 3	3 4 1 2
1 4 3 2	3 4 2 1
2 1 3 4	4 1 2 3
2 1 4 3	4 1 3 2
2 3 1 4	4 2 1 3
2 3 4 1	4 2 3 1
2 4 1 3	4 3 1 2
2 4 3 1	4 3 2 1

Table A1: Permutations of the set $\{1, 2, 3, 4\}$.

x x 3 4	3 x x 4
x x 4 3	3 x 4 x
x 3 x 4	3 x x 4
x 3 4 x	3 x 4 x
x 4 x 3	3 4 x x
x 4 3 x	3 4 x x
x x 3 4	4 x x 3
x x 4 3	4 x 3 x
x 3 x 4	4 x x 3
x 3 4 x	4 x 3 x
x 4 x 3	4 3 x x
x 4 3 x	4 3 x x

Table A2: Permutations of the set $\{1, 2, 3, 4\}$ when $1 = 2 = x$.

same), we have only 4 permutations. On reflection we can explain why this is so. The original number of permutations is 24, but three of the objects are identical. Thus for every permutation, there are a number of other permutations that are indistinguishable from it. Suppose we look at the permutation (4 3 2 1). Since $3 = 2 = 1 = x$, this permutation cannot be distinguished from (4 3 1 2), (4 2 1 3), (4 2 3 1), (4 1 3 2), and (4 1 2 3), which are all equivalent to (4 x x x). We can see that the number of identical permutations starting with a "4" is 6 because there are $3! = 6$ ways of arranging the three remaining numbers 1, 2, and 3. The exact same argument applies regardless of where the "4" turns up in the permutations. Thus there are 6 permutations that are equivalent to (x 4 x x) and so on.

What we have shown is that the number of permutations of four objects when only $m = 1$ of them is nonrepeated is:

$$\frac{4!}{3!} = \frac{24}{6} = 4.$$

The same line of argument allows us to explain why we got 12 for the case of $m = 2$ nonrepeated objects out of 4:

$$\frac{4!}{2!} = \frac{24}{2} = 12.$$

Of course, if all the objects were the same, and hence none of them nonrepeated or $m = 0$, we would have only one permutation of the form (x x x x), which is explained as:

$$\frac{4!}{4!} = \frac{24}{24} = 1.$$

Also recall our formula for rearrangements of four distinct objects, which we can write as the $m = 3$ or $m = 4$ case since if three of the four are distinct then they all are:

$$\frac{4!}{1} = \frac{24}{1} = 24.$$

We can summarize all of the four above cases in a *single* formula, valid for $m = 0, 1, 2, 3$:

> The number of permutations of four objects when m of them are nonrepeated is $4!/(4 - m)!$

Mathematicians give this formula a special symbol, P_m^4, where the P stands for permutation. It is not difficult to show that, in general, if we have n objects, m of which are nonrepeated, then the number of permutations is:

$$P_m^n = \frac{n!}{(n - m)!}$$

understanding that n must be greater than or equal to m.

Let us apply this formula to find the number of different ways to recombinate a lock with five pins when some of the pins are the same size (and we cannot bring in any new pins). We know that if there are five pins, all of which are distinct, e.g., (1 2 3 4 5), then there are $P_5^5 = 5!/0! = 120$ permutations (notice that we define 0! to be 1 for convenience). Now suppose that two of the pins are the same size, e.g., (1 1 3 4 5). There are only three nonrepeated pins, so the number of permutations is:

$$P_3^5 = \frac{5!}{(5-3)!} = \frac{5!}{2!} = \frac{120}{2} = 60.$$

If three of the pins are the same size, e.g., (1 1 1 4 5), then only two are nonrepeated, we can only recombinate in:

$$P_2^5 = \frac{5!}{(5-2)!} = \frac{5!}{3!} = \frac{120}{6} = 20 \text{ ways.}$$

Lastly, if four of the pins are the same, e.g., (1 1 1 1 5), there are only:

$$P_1^5 = \frac{5!}{(5-1)!} = \frac{5!}{4!} = \frac{120}{24} = 5 \text{ permutations available.}$$

This example should convince the reader that our formula for permutations is correct. For completeness we include the mathematical definition of P_m^n, which can be written as:

> P_m^n is the number of ways of choosing m objects from n distinct objects without replacement when the *ordering is important*.

Another idea that comes in handy is that of a *combination*. In mathematics this is *not* the same as a permutation. The number of combinations is defined as:

> C_m^n = the number of ways of choosing m objects from n distinct objects without replacement when the ordering is *not* important.

The symbol C_m^n is sometimes written as $\binom{n}{m}$.

Thus the difference between a permutation and a combination is that, in the case of a combination, no distinction is made between, for instance, (1 1 3 4 5) and (1 1 5 4 3), since they both use the same *selection* of objects. There are clearly fewer combinations than permutations for the same values of m and n; in fact, one can show that:

$$C_m^n = \frac{P_m^n}{m!}$$

since there are $m!$ possible rearrangements of the m objects, all of which are considered to be equivalent (since order does not matter). Thus using the formula for P_m^n we can write:

$$C_m^n = \frac{n!}{(n-m)!\,m!}, \quad 0 \le m \le n.$$

Suppose that we had 10 sizes of pins numbered from 0 to 9 and we could only use 5 of them at a time. How many ways could we choose 5 from the 10? The answer is:

$$C_5^{10} = \frac{10!}{(10-5)!\,5!} = \frac{10!}{5!\,5!} = \frac{10.9.8.7.6}{5.4.3.2.1} = 252.$$

The list would look something like:

$$0\ 1\ 2\ 3\ 4$$
$$0\ 1\ 2\ 3\ 5$$
$$0\ 1\ 2\ 3\ 6$$
$$0\ 1\ 2\ 3\ 7$$
$$0\ 1\ 2\ 3\ 8$$
$$0\ 1\ 2\ 3\ 9$$
$$0\ 1\ 2\ 4\ 5$$
$$0\ 1\ 2\ 4\ 6$$
$$\cdots$$

ending in the combination (5 6 7 8 9).

It is sometimes important to distinguish between permutations and combinations, although in common speech the two terms are often used interchangeably. In locksmithing we usually mean permutation, since the order of pins in a lock clearly makes a difference.

Another frequently used concept that may need clarification is that of a bitting code for a key. This is the code that indicates what cuts are needed on the key to operate the lock when it is pinned according to the code. Throughout the book we have made use of the following simple idea: if we have a pin-tumbler lock with n pins, each of which may be any one of m different sizes, then the total number of system codes or permutations is m^n (read "m to the power of n"). To understand this, we note that in pin chamber 1 we can put any one of the m sizes, in pin chamber 2 we can also put any one of the m sizes, and so on. The net effect is that the number of possibilities is being *multiplied* as many times as there are pin chambers. Thus with n chambers we have:

$$\underbrace{m \times m \times \cdots \times m}_{n \text{ times}} = m^n.$$

We have chosen to refer to these possibilities as "codes" rather than "permutations" since not all the codes are permutations of each other. For instance, (1 3 6 4 9) and

(7 2 7 4 5) are not permutations of each other since they use different pin sizes, that is, one code cannot be *permuted* into the other by changing the order of its entries. We can now readily see that with 5 pins of 10 sizes there are theoretically $10^5 = 100,000$ codes. In practice, many of these must be ruled out due to MACS and other bitting constraints.

The final idea that we would like to present in this section relates to locks with two or more different sorts of independent features. For example, in a dual-action lock with six conventional pin-tumblers and three side-bar pins, how many possible codes are there? Suppose there are 10 sizes of regular pins and 4 different side-bar pin sizes. The total number of side-bar bittings is $4^3 = 64$, and there are of course $10^6 = 1,000,000$ theoretical codes for the top cuts. Since the two features are *independent*, that is, our choice of side-bar codes does not constrain the regular key code in any way, the total number of codes for the system is the *product* of the two previous numbers, or 64,000,000. In general, if we have a lock with M possible codes and add a feature that gives L possible independent codes, the total number of codes is increased to $L \times M$.

Appendix A2: Lock Permutations and Fractals

In this section we revisit the idea of a permutation when some of the elements are the same. We ask the reader's forbearance as we venture further into the land of permutations (there are some nice pictures coming!). Recall the example from the preceding section, which took all permutations of the set $\{1, 2, 3, 4\}$ with the proviso that objects 1 and 2 were the same (marked by an "x"). When we eliminate the duplicate rows from Table A2, making sure that it is sorted in increasing order, we obtain Table A3, which, as we remarked earlier, retains only 12 of the original 24 rows. For convenience we have also numbered the rows in this table.

1	x x 3 4
2	x x 4 3
3	x 3 x 4
4	x 3 4 x
5	x 4 x 3
6	x 4 3 x
7	3 x x 4
8	3 x 4 x
9	3 4 x x
10	4 x x 3
11	4 x 3 x
12	4 3 x x

Table A3: Permutations of the set $\{1, 2, 3, 4\}$ when $1 = 2 = x$ and copies of rows are removed.

Now suppose we were only interested in the position of the two identical objects in the permutation. Hence, we would no longer make a distinction between, say, row 1 = (x x 3 4) and row 2 = (x x 4 3) because both of these contain an "x" in positions 1 and 2. We would then be led to consider which rows of the table are equivalent in the sense that they have an "x" in the same place. It is quite easy to see that in this scheme, the following pairs of rows are equivalent: (1, 2), (3, 5), (4, 6), (7, 10), (8, 11), and (9, 12). If we eliminate any row that is equivalent to any other row, we are left with a reduced table with only 6 rows as shown in Table A4.

Suppose now that we wished to visualize the process of going from Table A3 to Table A4. We would be led to consider a *matrix* of size 12 × 12 with each entry (i, j) being given the value 1 if row i was equivalent to row j, and zero otherwise. (By convention the entry (i, j) is at row i and column j in the matrix, where the numbering starts from 1.) Thus our matrix in this case contains 144 entries, all of which are zero except for entries (1, 2), (2, 1), (3, 5), (5, 3), (4, 6), (6, 4), (7, 10), (10, 7), (8, 11), (11, 8), (9, 12), and (12, 9). If we then represent each nonzero entry of this matrix by a dot, we would obtain the graph shown in Fig. A1, which we refer to as a *permutation equivalence matrix*.

By now the reader is probably wondering about the use of all this, to which the answer is: to move to higher levels of complexity and see what happens. The matrix for the previous example for a lock with four pins, two of which are the same, seems rather innocuous. Things get a bit more interesting, however, as soon as we consider locks with five or more pins corresponding to permutations with five or more entries.

An elegant example is furnished by a lock with six pins, with two pins being the same. The permutations for the lock are of the form (x x 3 4 5 6), where we have put 1 = 2 = x. We know that for a lock with six pins, all of different sizes, there are 6! = 720 permutations. When two pins are of equal size, this number is reduced to 360. To obtain the permutation equivalence matrix, we ask the following question: of the 360 permutations, which ones are equivalent in the sense that the two pins marked "x" are in the same positions? For instance, the permutations (x x 3 4 5 6) and (x x 5 6 3 4) would be equivalent according to this definition since they both have an x in positions 1 and 2. It is useful at this point to resort to a computer program

x x 3 4

x 3 x 4

x 3 4 x

3 x x 4

3 x 4 x

3 4 x x

Table A4: Permutations of the set $\{1, 2, 3, 4\}$ when $1 = 2 = x$ and only the position of the x's is important.

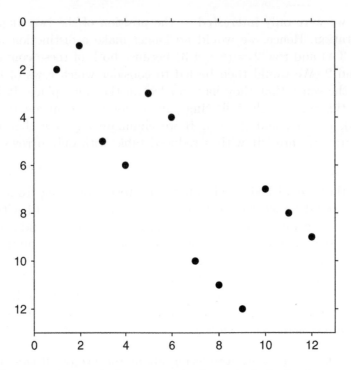

Figure A1: Permutation equivalence matrix for a 4-pin lock when two pins are the same.

to obtain an answer to this question, since there are hundreds of permutations to consider. The answer, which is a 360×360 matrix of binary (0 or 1) entries, can be plotted as in the previous example to give the permutation equivalence matrix appearing in Fig. A2. This plot, as well as the other plots presented in this section were obtained using the Matlab[TM] computer software package [115].

The result displays some interesting properties: the pattern appears to repeat at different *scales*, becoming more complex as it gets bigger. This property is similar to one found in fractals like the now famous Mandelbrot set [76], shown in Fig. A3, except that the similarity pertains to larger scales; that is, we see similar patterns as we zoom out. The permutation equivalence matrix is also only defined on a set of whole numbers (the indexes of a matrix), whereas conventional fractals are defined over the set of complex or imaginary numbers (i.e., numbers of the form $a + ib$ where a and b are real numbers and $i = \sqrt{-1}$).

It is straightforward to generalize this idea to permutations where more than a single group of entries have the same value, as occurs in a lock code with several sets of pins of the same sizes. For instance, in locks with seven pins where there are three pairs of identical pins, the permutations are of the form (x x y y z z 7) where $1 = 2 = x$, $3 = 4 = y$ and $5 = 6 = z$. The permutation equivalence matrix for this case is given in Fig. A4, displaying fractal qualities on a different base pattern.

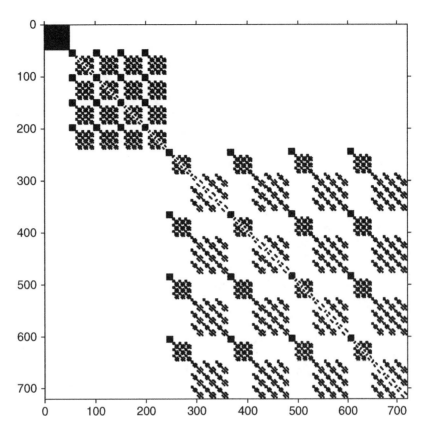

Figure A2: Permutation equivalence matrix for a 6-pin lock when two pins are the same.

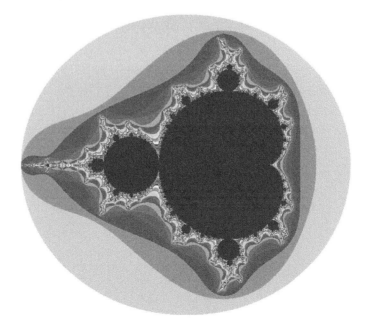

Figure A3: Fractal image from the Mandelbrot set (obtained using Fractint software package [123]).

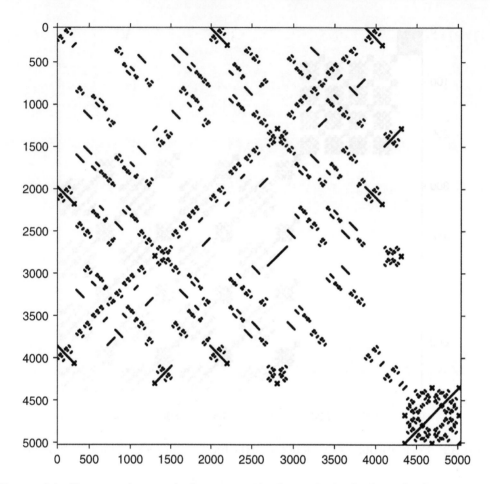

Figure A4: Permutation equivalence matrix for a 7-pin lock with three pairs of identical pins.

Appendix B: Translations of Lock Vocabulary

This Appendix deals with translations of lock-related terminology in French, Italian, and German. It is hoped that the list can be used as a starting point for English-speakers who wish to read European patent or other documents pertaining to security locks. Alternatively, people whose native language is not English may find the lists useful in reading this book or other material related to locksmithing.

Specifically, we consider a list of around 240 English words that arise frequently in the text of the book and provide, in the following tables, technical translations into French, Italian, and German. English verbs are denoted by "(to ∼)."

Although considerable care has been taken in finding the appropriate translation for each word, there is always a possibility of error. Naturally, translations also depend on context, and this is why in some cases several alternative words have been provided. In some cases there may not be a word-for-word equivalent, in which case an equivalent expression has been provided.

For accessibility, we have replaced the German letter ß by "ss."

ENGLISH	FRENCH	ITALIAN	GERMAN
access control	contrôle d'accès	controllo dell'accesso	Zugriffskontrolle
align (to ~)	aligner	allineare	ausfluchten, abstimmen, ausrichten
all	tout	tutto	alle(s)
alloy	alliage	lega	Legierung
angle	angle	angolo	Winkel
antipicking/antitamper	anti-tâtage	antieffrazione, antimanomissione	pickgeschützt
arm	bras	braccio	Arm
armored	blindé	blindato	gepanzert
assemble (to ~)	assembler, monter	assemblare	assemblieren
attract	attirer	attrarre, attirare	anziehen
authorized	autorisé	autorizzato	autorisiert
backward	en arrière	all'indietro	rückwärts, nach hinten
ball (bearing)	bille	sfera (cuscinetto a), sferetta	Kugellager
barrel	stator, canon, barrillet	canna, involucro	Trommel, Zylinder(fuss)
belly	ventre	ventre	Bauch, Kontur
bevel	biseau	smussatura	Schräge
bit (of lever key)	panneton	mappa (di chiave a leva)	Bart, Schlüsselbart
bitting	coupe, taille	dentatura	Stufung
blacksmithing	ferronnerie	lavoro di maniscalco, del fabbro	schmiedearbeit
blade	lame	lama	Klinge
block (to ~)	bloquer, coincer	bloccare, immobilizzare	blockieren
body, housing	corps, logement	corpo, alloggio	Körper, Gehäuse
bolt	pène, verrou	catenaccio, chiavistello	Riegel
bolt-step	barbe	gradino del chiavistello	Riegelstufe
bore (to ~)	vriller, forer	trapanare, forare	(aus)bohren

Table B1: Translations of English lock vocabulary into French, Italian, and German.

ENGLISH	FRENCH	ITALIAN	GERMAN
bottom pin	goupille inférieure	spina inferiore	Kernstift
bow of key	anneau de clé	anello	Schlüsselreide
brass	laiton	ottone	Messing
break (to ~)	casser, fracturer	rompere, spezzare	brechen
bring to shear line	mettre au passage	portare alla linea di taglio	auf die Scherlinie bringen
burglary	cambriolage, effraction, vol	furto con scasso	Einbruch
bypass (to ~)	contourner (les sûretés)	bypassare	umgehen
cam	came, mentonnet	camma	Schliessbart, Nocke
cap	capuchon	coperchio, calotta	Verschlusskappe
card	carte	scheda	Karte
case	boîtier	scatola	Gehaüse, Koffer
chamber	chambre, logement	camera, foro di alloggiamento	Kammer
change (to ~)	changer	cambiare	wechseln
channel	canal	canale	Kanal
class/grade	classe	classe, grado	Klasse
clip	griffe, pince	clip, graffetta di fissaggio, fermaglio	Klammer
clockwise	dans le sens des aiguilles d'une montre	in senso orario	im Uhrzeigersinn
close (to ~)	fermer	chiudere	schliessen
code, coding	code, enchiffrement	codice, cifratura	Code, Kode
combination	combinaison	combinazione	Kombination
component	composant(e)	componente, elemento	Bauteil
copy (to ~)	copier	copiare	kopieren
core	noyau	nucleo	Kern
counterclockwise	dans le sens inverse des aiguilles d'une montre	in senso antiorario	entgegen dem Uhrzeigersinn
cover	couvercle	copertina	Abdeckung

Table B2: Translations of English lock vocabulary into French, Italian, and German (cont.).

ENGLISH	FRENCH	ITALIAN	GERMAN
cut	coupe, taille	taglio	Einschnitt
cut (to ~)	couper, tailler	tagliare	schneiden
cylinder	cylindre	cilindro	Zylinder
deadlock	condamner, verrouiller	chiudere a chiave chiudere a doppia mandata	Riegelschloss
decode (to ~)	décoder	decodificare	dekodieren, decodieren
decoder-picking	crochetage-tâtage	manomissione del decodificatore	Decoderpicken
decrease/reduce (to ~)	réduire	diminuire, ridurre	vermindern, reduzieren
deep	profond	profondo	tief
defeat (to ~)	vaincre	vincere, annullare	besiegen
depth	profondeur	profondità	Tiefe
design	conception	disegno, progetto	Entwurf
design (to ~)	concevoir	progettare	entwerfen
destroy (to ~)	détruire	distruggere	vernichten
diameter	diamètre	diametro	Durchmesser
differ, variation	variation	divergenza, variazione	Wechsel, Variation, Veränderung
differ (to ~)	(faire) varier	differire	abweichen
difference	différence	differenza	Unterschied
dimple key	clé à trous, clé micro-points	chiave a fossetta	Bohrmuldenschlüssel
disc/disk	disque	disco	Scheibe
dismantle (to ~)	démonter	smantellare, smontare	auseinanderbauen
displace (to ~)	déplacer, décaler	dislocare, spostare	verschieben, verlagern
door	porte	porta	Tür
double-bitted	à double panneton	a doppia mappa	Doppelbartschlüssel, Tresorschlüssel
down	en bas	giù	abwärts, hinab
drill (to ~)	percer	trapanare, forare, perforare	bohren

Table B3: Translations of English lock vocabulary into French, Italian, and German (cont.).

ENGLISH	FRENCH	ITALIAN	GERMAN
drilling	perçage	trapanatura, foratura	bohren
duplicate (to ~)	doubler, reproduire	duplicare, riprodurre	duplizieren, vervielfältigen
end	bout	fondo, estremità	Ende
even	pair	pari	gerade
factory	usine, fabrique	fabbrica	Fabrik
false	faux	falso	falsch
feeling, probing	tâtage, palpage	sensazione, sondaggio	Gefühl, untersuchen
file (to ~)	limer	limare	(ab)feilen
finger	doigt	dito	Finger
fixed	fixe	fisso	fest
flange	flasque, collerette	flangia	Flansch
flat	plat	piatto	flach
foot, base	pied, socle, base	piede, base	Fuss, Sockel
force (to ~)	forcer	forzare	zwingen, forcieren
forward	en avant	in avanti	vorwärts
front	(en) avant	anteriore	Vorderseite
function, use	fonction, utilisation	funzione, utilizzo	Funktion, Verwendung
gate (lever)	passage, cage	cancello, passaggio	Öffnung, Tourstiftkanal
gear	engrenage, roue dentée	ingranaggio	zahnrad
grind (to ~)	fraiser	molare	abschleifen
groove	rainure	scanalatura	Rille
guard	protège-	protezione	Schutz
half	demi, moitié	metà, mezzo	Hälfte, halb
handle/knob	poignée, clanche	maniglia, manopola	Knauf
hardened	trempé	temprato	gehärtet
head	tête	testa	Kopf

Table B4: Translations of English lock vocabulary into French, Italian, and German (cont.).

ENGLISH	FRENCH	ITALIAN	GERMAN
height	hauteur	altezza	Höhe
high	haut, supérieur	alto	hoch
hinge	gond, charnière	cerniera	Scharnier
hole	trou	foro	Loch, Bohrung
impact	impact, choc	impatto	Stoss, Aufprall
impressioning	taille de clés par empreintes/à la trace	impressione	Impressionstechnik
increase (to ~)	augmenter	aumentare	erhöhen
insert (to ~)	insérer, introduire, intercaler	inserire	einsetzen
interface	interface	interfaccia	Schnittstelle
iron	fer	ferro	Eisen
key	clé	chiave	Schlüssel
key blank	ébauche de clé	chiave grezza	Schlüsselrohling
key cutting	taille de clés	taglio di chiave, cianfrinatura	Schlüssel schneiden
keyhole	trou de (la) serrure	buco della serratura	Schlüsselloch
keyway	entrée de clé	canale della chiave	Zylinderkernöffnung
	passage de la clé		Schlüsselführung
latch	demi-tour	chiavistello, saliscendi	Riegel
left	(à) gauche	sinistra	links
length	longueur	lunghezza	Länge
level	niveau	livello	Ebene, Niveau
lever	gorge	mandato, nottolino	Hebel
lift, raise (to ~)	soulever	sollevare	heben
linkage	accouplement, entretoise	connessione, raccordo	Verbindung
lock	serrure	serratura	schloss
lock (to ~)	verrouiller, fermer à clé	serrare	schliessen, sperren
lock-pick	crochet, tâteur, palpeur	attrezzo per manomettere la serratura	Pickwerkzeug

Table B5: Translations of English lock vocabulary into French, Italian, and German (cont.).

ENGLISH	FRENCH	ITALIAN	GERMAN
lockpicking	crochetage	raccolto, manomissione della serratura	picken
locksmith	serrurier	fabbro	Schlüsseldienst, Schlosser
long	long(ue), grand	lungo	lang
low	bas, inférieur	basso, inferiore	niedrig
lower (to ~)	abaisser	abbassare	herunterlassen
machine	machine	macchina	Maschine
machine (to ~)	tourner	tornire, lavorare a macchina	bearbeiten
magnet	aimant	magnete	Magnet
manipulate (to ~)	manipuler	manipolare	manipulieren
manipulation	manipulation	manipolazione	Manipulation
manufacturer	fabricant	produttore, fabbricante	Hersteller
master key	clé maître, clé passe-partout	chiave maestra	Hauptschlüssel, Zentralschlüssel
master-keyed	à clés hiérarchisées sur organigramme	a chiave maestra a chiave gerarchica	Schliessanlagen bestiftet
mechanism	mécanisme	meccanismo	Mechanismus
mill (to ~)	fraiser	fresare	fräsen
mortice lock	serrure à encastrer/larder	serratura a mortasa	Zusatzschloss
move (to ~)	déplacer, bouger	spostare	bewegen
movable	amovible	mobile	beweglich
mushroom pin	goupille anti-tâtage	perno a fungo, perno antiscasso	Pilzkopfstift
none	aucun	nessuno	keine
notch	cran, encoche	tacca, intaglio	Kerbe, Nute
number	nombre, numéro	numero	Nummer
odd	impair	dispari	ungerade
offset	décalage	offset	Versatz
one, any	un(e), quelconque	uno/una, qualsiasi	ein

Table B6: Translations of English lock vocabulary into French, Italian, and German (cont.).

ENGLISH	FRENCH	ITALIAN	GERMAN
open (to ~)	ouvrir	aprire	öffnen
operate (to ~)	opérer, fonctionner	azionare, operare	operieren, betätigen
owner	propriétaire	proprietario, titolare	Inhaber, Besitzer
padlock	cadenas	lucchetto	Hangschloss, Vorhängeschloss
part, piece, portion	partie, pièce	parte, pezzo, porzione	Teil, Stück, Portion
patent	brevet	brevetto	Patent
pick (to ~)	crocheter	scassinare, manomettere	picken
pin	goupille	perno, spina	Stift
pipe	tuyau	tubo	Rohr
pivot	pivot	fulcro, cardine	Drehzapfen, Drehpunkt
plate	écusson, plaque	piastra	Platte
plug	rotor, (barrillet)	cilindretto	Kern
plug (retainer)	bouchon, tampon de retenue	pastiglia di chiusura	Verschlussstift
position	position	posizione	Position
profile, section	profil	profilo, sezione	Profil, Abschnitt
profile cylinder	cylindre profilé	cilindro a profilo/da infilare	Profilzylinder
prong	dent, broche	dente	Zacke
protection	protection	protezione	Schutz
protrude (to ~)	saillir	sporgere	herausragen
pull (to ~)	tirer	tirare	ziehen
push (to ~)	pousser	spingere	drücken
push-key	clé à pompe	chiave a pompa	Pumpenschlüssel, Stechschlüssel
reading	lecture	lettura	Messung
rear	(en) arrière	posteriore	hinter
recess	chambrage, encastrement, niche	rientranza, cavità	nische

Table B7: Translations of English lock vocabulary into French, Italian, and German (cont.).

ENGLISH	FRENCH	ITALIAN	GERMAN
register, engage (to ~)	enregistrer, engrener	registrare, impegnare	registrieren
rekey (to ~)	changer la clé	cambiare la chiave	neu bestiften
repel	repousser	respingere	abstossen
resistance	résistance	resistenza	Widerstandskraft
reversible key	clé réversible	chiave reversibile	Wendeschlüssel
right	(à) droit(e)	destro	rechtes, rechts
rim lock	serrure en applique	serratura da applicare	Felgenschloss
ring	anneau	anello	Ring
rocker	bascule	bilanciere	Kipphebel
rod	tringle, verge	verga, asta	Stange
roller	rouleau	rullo	Rolle
round	rond	rotondo	rund
rule/constraint	règle, contrainte	regola, vincolo	Regel, Auflage
safe/vault	coffre-fort, chambre forte	cassaforte, volta	Geldschrank, Tresor
same	même	stesso	selbe, gleich
saw (to ~)	scier	segare	sägen
screw (to ~)	visser	avvitare	schrauben
security	sécurité	sicurezza	Sicherheit
security feature (blocking device)	sûreté, organe de blocage	meccanismo di ritenuta meccanismo di blocaggio/sicurezza	Sicherheitseinrichtung Sperrelement
shackle (of padlock)	anse	maniglia del lucchetto	Bügel
shallow	peu profond	poco profondo, basso	seicht
shape	forme	forma	Form
shear line	ligne de césure	linea/piano di taglio	Scherlinie
shift (to ~)	décaler	spostare	versetzen
short	court, petit	corto	kurz, klein

Table B8: Translations of English lock vocabulary into French, Italian, and German (cont.).

ENGLISH	FRENCH	ITALIAN	GERMAN
shoulder	épaule, embase	fermo, spalla(mento)	Schulter
side-bar	barre latérale	barra laterale	Sperrleiste
side/edge	côté, bord	lato, bordo	Seite, Ecke
size	taille, longueur	dimensione, misura	Grösse
skeleton key	passe-partout, rossignol	passepartout, chiave universale	Dietrich
sleeve	chemise	manicotto	Ärmel
slider	coulisseau	barra di scorrimento	Schieber
slope	pente	pendenza, pendio	Gefälle
space	espace, intervalle	spazio, intervallo	Platz
spool	bobine	bobina	Pilzkopfstift, Spulenstift
spring	ressort	molla	Feder
standard	norme	norma	Norm
steel	acier	acciaio	Stahl
stem	tige	gambo, canna	Halm, Schaft
stop	arrêt, boutoir	arresto	Anschlag, Stop
stop (to ~)	arrêter	arrestare	halten, stoppen
strength	force	forza	Stärke, Kraft
stump	ergot, pilier	tacchetto	Stummel, Stift
supplier/dealer	fournisseur, point de vente	fornitore, punto vendita, rivenditore	Lieferant, Händler
symmetric/double-entry key (for lever lock)	clé bénarde	chiave a mappa simmetrica	symmetrischer Schlüssel
system	système	sistema, impianto	System
tail-piece	entraîneur	appendice	Endstück
tampering	tâtage	manomissione	einmischend, herumpfuschend
tensioner	entraîneur	tensionatore	Spanner
thickness	épaisseur	spessore	Dicke

Table B9: Translations of English lock vocabulary into French, Italian, and German (cont.).

ENGLISH	FRENCH	ITALIAN	GERMAN
tilt, tip (to ~)	mettre en biais, tilter	inclinare	kippen
time	temps	tempo	Zeit
tip	pointe	punta	Spitze
tool	outil	attrezzo, strumento, arnese	Werkzeug
tooth	dent	dente	Zahn
top pin/driver	goupille supérieure	controperno, spina superiore	Gehaüsestift
torque	couple	coppia	Drehung, Drehmoment
true	vrai	vero	wahr
try-out key	trousseau de Saint Pierre	chiave di prova	Probierschlüssel
	clés résumés, rossignols		
tubular key	clé tubulaire	chiave tubolare	Rundschlüssel, Tubularschlüssel
tubular lock-pick	parapluie	grimaldello tubolare	Tubularpick
tumbler	organe mobile	nottolino, dispositivo mobile	taumeln
	garniture	chiavetta mobile	
turn, rotate (to ~)	tourner	girare, ruotare	drehen, rotieren
up	en haut, supérieur	su, sopra	auf, hinauf
wafer	paillette, barrette	wafer, disco	Scheibe
wall	mur, paroi	parete	Wand, Mauer
ward(ing)	garde, garniture, ève, nervure	scontro	Schutz
wheel	roue	ruota	Rad
width	largeur	larghezza	Breite, Weite
window	fenêtre	finestra	Fenster

Table B10: Translations of English lock vocabulary into French, Italian, and German (cont.).

Appendix C: Terminology and Abbreviations

Terminology

U.S.	U.K.	Context
plug	core	cylinder lock
cylinder	barrel	cylinder lock
saddle, sweep	belly	lever lock
trap	pocket	lever lock
bolt post	bolt stump	lever lock
slot, fence	gate	lever lock
flag	bit	lever lock
wafer	disc	wafer lock
driver	top pin	pin-tumbler lock
guard key	guardian key	safe deposit lock

Table C1: Equivalence of U.S. and U.K. terminolgy.

Term	Equivalences	Context
head	bow	key
blade	stem, shaft, shank, neck	key
cut	bitting	key
position	space	key
milling	track	key
side-track	wave, internal-cut	key
stop	shoulder	key
differ	combination, change	keying
profile	broaching, section, warding	keyway
ward	rib	keyway
chamber	bore	cylinder lock
cylinder	barrel, housing, stator	cylinder lock
plug	core, rotor	cylinder lock
shell	cage, drum, armature	cylinder lock
bottom pin	key pin	pin-tumblers
top pin	upper pin, driver pin	pin-tumblers
stack	pack	tumblers

Table C2: Roughly equivalent terms as used in the text.

Abbreviations

#	number
″	inches
A2P	Assurance Prévention Protection
ANSI	American National Standards Institute
AS	Australian Standard
BHMA	Builders Hardware Manufacturers Association
BRE	Building Research Establishment
BS	British Standard
BSI	British Standards Institution
CCW	counterclockwise
CK	change key
CNC	computer numerical control
CNPP	Centre National de Prévention et de Protection
CW	clockwise
DIN	Deutsches Institut für Normung
DLP	DiskLock Pro (Abloy)
EN	European Norm
EP	European Patent
FAQ	frequently asked questions (file)
GM	General Motors
GMK	grand master key
IC	interchangeable core
LPCB	Loss Prevention Certification Board
LPS	Loss Prevention Standard
MLA	Master Locksmiths' Association
MK	master key, master-keyed, master-keying
N/A	not applicable
NG	New Generation (BiLock)
QCC	Quick Change Cylinder (BiLock)
R&D	research and development
RF	radio frequency
RU	resistance unit
S&G	Sargent and Greenleaf
SBM	side-bit milling
SFIC	small-format interchangeable core
TMK	top master key
UL	Underwriters Laboratories
US	United States
UK	United Kingdom
VATS	Vehicle antitheft system
VdS	Vertrauen durch Sicherheit
WO	World Intellectual Property Organization

Table C3: Abbreviations used in the text. See also country codes in Table 1.4.

Appendix D: Lock Patents

This Appendix summarizes the patent documents that have been used or cited in the text. It is in no way intended to be exhaustive in terms of the available public-domain information. Patents have been classed as relating to either magnetic or nonmagnetic (mechanical, key-operated) locks. For each patent, we give the filing date, country of filing, serial number, author and/or assignee's company, relevance, and a one-line description of the content of the patent. Country codes are defined in Table 1.4 in Chapter 1. Relevance (for nonmagnetic lock patents only) is as defined in the table caption.

Filed	Patent no.	Author/Company	Relevance	Description
1832	UK 0,006,350	Parsons	L	Balance lever lock
1844	US 0,003,630	Yale, senior	P	Quadruplex radial pin-tumbler lock
1857	US 0,018,169	Yale, senior	P	5-pin-tumbler sliding-body padlock
1865	US 0,048,475	Yale, junior	P	Pin-tumbler cylinder lock
1868	US 0,076,066	Felter	W	Plate-wafer lock
1870	US 0,099,013	Shepardson	A,W	Antipicking wafer lock
1874	US 0,156,113	Ahrend (Romer & Co.)	S	Scandinavian padlock with one-sided key
1875	US 0,167,088	Shepardson	A,S,W	Wafer side-bar lock
1879	US 0,213,300	Romer & Company	S	Improved Scandinavian padlock
1889	US 0,414,720	O'Keefe	P	Master ring cylinder
1891	US 0,457,753	Taylor (Yale & Towne)	P	Paracentric keyway, beveled pin with tab
1892	CH 0,004,935	Theodor Kromer	L,W	Protector lock with one-part sliders
1892	DE 71,766	Max Zahn	L,W	Doppel Bramah-Chubb lock
1901	US 0,688,070	Denn	S	6-disc Abloy-type lock
1904	US 0,758,026	Taylor (Yale & Towne)	P	Pin-tumbler lock curtain
1907	DE 0,214,691	Panzer A.G.	L	Quadruple-bitted key cylinder lock
1909	US 1,136,067	Watson	L	Interchangeable key multiple pivot lever lock
1912	UK 0,027,511	Eras	P	Drill protection for cylinder
1913	US 1,224,021	Renaux	P	Early 3-pin dimple-key padlock
1916	DE 295,060	Fraigneux	L,S	Liega twin side-bar safe lock
1918	US 1,287,882	Christoph	W	Multiple action plate-wafer lock
1919	US 1,514,318	Henriksson	S	Abloy 10-disc lock
1920	US 1,390,222	Wise	P	Coaxial tube lock
1922	FR 0,552,963	Bézard	W	Axial wafer lock
1922	US 1,403,753	Epstein	P	Pick gun
1924	UK 0,224,175	Eras, Lips	L	Safe deposit lock with twin levers
1925	US 1,639,919	Baron	P	Vibratory pick
1927	US 1,784,444	Homolle (Bournisien Beau et Companie)	L,W	AVA-type plate-wafer lock

Table D1: Nonmagnetic key-operated lock patents. Legend: P = pin-tumbler, W = wafer, L = lever, S = side-bar, A = automotive.

Filed	Patent no.	Author/Company	Relevance	Description
1928	FR 0,659,113	Coffres-Forts Bauche	L	Bauche Monopole
1928	US 1,667,223	Simpson	P	Rapping (999) key
1928	US 1,739,964	Hainline	P	Antipicking sleeve
1928	US 1,770,864	Rivers	P	Rivers lock
1929	US 1,863,525	Long	L	Variable-stump lever + pin-tumbler lock
1928	US 1,819,853	Von Mehren	P	Code lock
1929	US 1,899,739	Von Mehren	P	Key for Code lock
1930	US 1,860,712	Gutman	P	Antipicking chamber shape
1931	DE 0,560,425	Theodor Kromer GmbH	L	Kromer Proector with split-levers
1931	US 1,917,302	Hill (Sargent & Greenleaf)	W	Recombinatable wafer lock
1932	US 2,043,205	Thompson	P	Mushroom pin with undercut cylinder groove
1933	US 1,965,889	Fitzgerald	W	Serrated-edge wafer-tumbler
1933	US 2,049,742	Lowe (Yale & Towne)	A,W	Sliding core wafer lock
1933	DE 0,595,834	Theodor Kromer GmbH	L	Protector key with angled bittings
1933	DE 0,642,131	Bode	L	Tangential 4-bitted key with complex cuts
1933	DE 0,646,623	Bode	L	Tangential 4-way lever cylinder lock
1933	US 2,070,233	Liss (Briggs & Stratton)	P,S	Dual side-bar 5-pin lock
1934	UK 0,421,715	Bauer	P	(4+4)-pin Kaba lock
1934	US 1,965,336	Fitz Gerald (Briggs & Stratton)	A,W,S	GM 5-wafer side-bar lock
1934	US 2,030,836	Full, Muntner	W	Dudley side-track key split-wafer lock
1934	US 2,039,126	Svoboda	W	Multiple-action Bell-type wafer lock
1934	US 2,064,818	Buday	P	Two-part comb pick
1935	US 2,059,376	Lombardo	P	Tubular lock-pick
1938	US 2,158,501	Gutman (Independent Lock Co.)	P	Impact & pick-resistant pin-tumbler
1938	US 2,182,588	Jacobi (Briggs & Stratton)	S,W	Tension-resistant wafer side-bar lock
1938	US 2,279,592	Machinist (Dudley Lock Corp.)	W	Picking tool for 4-sided wafer lock

Table D2: Nonmagnetic key-operated lock patents (cont.). Legend: P = pin-tumbler, W = wafer, L = lever, S = side-bar, A = automotive.

Filed	Patent no.	Author/Company	Relevance	Description
1938	US 2,424,514	Sterner	P	Improved version of Code lock
1939	US 2,309,677	Segal	P	Pick gun
1940	US 2,283,489	Crousore	P	Grooved upper and lower pins
1948	US 2,578,211	Spain (Yale & Towne)	S	Spring-biased Abloy disc lock
1949	UK 0,661,501	Butter	L	Edge-gated lever lock
1950	UK 0,678,123	(Fichet-Beau & Cie)	W,S	Fichet 484 10-lever twin side-bar lock
1951	DE 0,911,220	Bode-Panzer	L	CAWI lock
1953	UK 0,737,547	Saarento	L,W	AVA 7-slider lock
1954	DE 1,027,552	Sellin & Münzer	L	Protector with curtain disc tabs
1955	DE 1,061,226	Sellin (Kromer)	L	Recessed cuts in Protector key bit
1955	US 2,763,027	Tampke	P,W	Comb-foil impressioning tool
1959	UK 0,940,778	Pearson	P	Cruciform locks
1959	US 3,035,433	Testa	S,W	Dudley wafer lock with side-bar
1963	UK 1,030,921	Stanton, Brett	L,W	Chubb AVA 10-slider lock
1963	US 3,181,320	Bauer	P	Deadlocking antipicking cylinder
1963	US 3,206,958	Best	P	Interchangeable-core cylinder
1963	US 3,234,768	Russell, Jennie, Armstrong	P	Construction-keyed cylinder
1963	US 3,251,206	Gruber	P	Tubular lock-pick
1963	US 3,263,461	Tartaglia	W	Recombinatable Bell lock
1964	US 3,237,436	Williams	W	Supra Title 6-slider lock
1964	FR 1,425,311	(Fichet)	S,W	Fichet-Bauche 666 7-wafer side-bar lock
1964	US 3,264,852	Gysin	W	SEA four-track wafer lock
1964	US 3,509,749	Regan, Arzig (Duncan Ind.)	W	Recombinatable Bell lock
1965	DE 1,553,294	Bauer	P	Kaba 20
1965	US 3,267,706	Kerr	P	Tamper-proof axial cylinder
1965	US 3,393,542	Crepinsek	P	4 inline DOM-type cylinder

Table D3: Nonmagnetic key-operated lock patents (cont.). Legend: P = pin-tumbler, W = wafer, L = lever, S = side-bar, A = automotive.

Filed	Patent no.	Author/Company	Relevance	Description
1966	US 3,303,677	Bauer (Sargent & Co.)	P	Sargent KESO dimple-key lock
1966	US 3,367,156	Johnstone (Gen. Motors Corp.)	A,S,W	Pivoting wafer side-bar lock
1966	US 3,402,581	Schweier, Thompson (Bell Labs)	L	Floating-cam lever lock
1966	US 3,418,833	Kerr (Chicago Lock Co.)	W	Passive profile pin wafer lock
1967	US 3,411,331	Schlage	P	8-pin axial cylinder (see Pollux)
1967	US 3,499,302	Spain, Oliver, Powell	P,S	Medeco cam and dual-action locks
1968	US 3,499,303	Spain	P,S	Medeco grooved-pin side-bar
1968	US 3,603,123	Best (Best Lock Corp.)	P	Interchangeable-core cylinder
1969	US 3,621,689	Koskinen, Solitanner	S	Doubly-gated Abloy lock
1970	UK 1,374,288	(CISA)	L	Italian detector lever lock
1970	US 3,604,231	Buschi	L,S	Early form Mottura 5-bar pump lock
1971	DE 7,203,658	Neiman	A	Two-track key
1971	US 3,722,240	Spain, Oliver	S	Medeco cam lock
1971	US 3,762,193	Hucknall	P	BHI Huck pick-proof pin
1972	US 3,731,507	Wolter (Josef Voss KG)	P	DOM iX 10-pin cylinder + pin design
1972	US 3,738,136	Falk (Fort Lock Corp.)	P	Master-keying of axial cylinder
1972	US 3,789,638	Roberts, Cohn, Ward	S	Early Abloy DLP key and discs
1972	UK 1,408,340	Vachette	P	Vachette 2000 active profile pin cylinder
1974	DE 2,433,918	Wolter (Josef Voss KG)	P	DOM iX-10 profile wards + split key
1974	US 3,928,993	Epstein	P	Chain-of-balls force-transmission cylinder
1974	US 3,987,654	Iaccino, Idoni	S	Decoder for Medeco cam lock
1975	UK 1,543,940	Sieg	P	Bilateral key cylinder lock
1975	US 3,948,065	Martikainen (Oy Wartsila AB)	S	Antipicking features for Abloy disc lock
1976	IL 0,050,984	Bahry, Dolev	P	Mul-T-Lock
1976	US 4,044,578	Guiraud (Fichet-Bauche)	S	Enhanced Fichet-Bauche 484 cylinder
1976	US 4,069,694	Raymond, Millett	W	Winfield recombinatable bicentric wafer lock

Table D4: Nonmagnetic key-operated lock patents (cont.). Legend: P = pin-tumbler, W = wafer, L = lever, S = side-bar, A = automotive.

Filed	Patent no.	Author/Company	Relevance	Description
1976	US 4,069,696	Steinbach (Chicago Lock Co.)	P	Restricted-keyway axial cylinder
1977	US 4,083,212	Proefrock (Sargent & Greenleaf)	S	Half pipe-key Abloy-type lock
1977	US 4,098,103	Raskevicius	P	Early form Emhart twist-and-lift pin cylinder
1977	US 4,099,396	Kerr (Chicago Lock Co.)	P	Serrated-pin axial cylinder
1977	US 4,109,495	Roberts	S	Early form Abloy DLP key and discs
1977	US 4,142,389	Bahry, Dolev	P	Mul-T-Lock
1978	FR 2,415,185	Frank	P,W	JPM 6-pin axial cylinder
1978	US 4,149,394	Sornes (Trioving)	P	Early form 5 × 5 matrix Vingcard
1978	US 4,185,482	Nail	A	Decoding tool for GM side-bar lock
1978	US 4,187,705	Guiraud (Fichet-Bauche)	L	Geared push-lever safe lock
1978	US 4,196,606	Guiraud (Fichet-Bauche)	L	Twin-cut Fichet-Bauche Monopole
1978	US 4,208,894	Surko	P	Emhart interlocking pin cylinder
1978	US 4,296,618	Guiraud (Fichet-Bauche)	L,S	Reinforcement of Fichet-Bauche 484 lock
1979	US 4,267,717	Martikainen (Oy Wartsila AB)	S	Turn limiting of Abloy 10-disc lock
1980	FR 2,491,531	Ferrari, Pollastri	P	ISEO R6
1980	US 4,320,638	Dunphy, Newman (Ogden Ind.)	P	Lockwood V7
1980	US 4,343,166	Hofmann (Bauer Kaba AG)	P	Kaba Quattro 4 inline cylinder
1980	US 4,356,713	Widen (GKN Stenman AB)	P,S	ASSA Twin
1980	US 4,377,082	Wolter (DOM)	P	DOM iX-10 floating-ball cylinder
1980	US 4,377,940	Hucknall	P,S	Impression and pick-resistant pin design
1980	US 4,434,636	Prunbauer (EVVA)	P,S	Active-profile pin design
1981	US 4,375,159	Bechtiger (Sargent & Greenleaf)	L	Double-bitted key wheel-pack lock
1981	US 4,385,510	Harper	A,W	AVA automotive lock
1981	US 4,404,824	Hennessy (Lori Lock Corp.)	S	6-Slider side-bar lock with dimple key
1981	US 4,429,554	Litvin, Scherz	P	(5+4)-pin bilateral key cylinder
1981	US 4,450,699	Genakis	P,S	Rotating female-pin side-bar lock

Table D5: Nonmagnetic key-operated lock patents (cont.). Legend: P = pin-tumbler, W = wafer, L = lever, S = side-bar, A = automotive.

Filed	Patent no.	Author/Company	Relevance	Description
1981	US 4,446,709	Steinbach (Chicago Lock Co.)	P,S	Tubar 8-pin axial lock
1982	US 4,462,230	Evans (Sargent & Greenleaf)	L	Key-changeable safe deposit lock
1982	US 4,478,061	Preddey	S	BiLock 12-pin 2-side-bar cylinder
1982	US 4,512,166	Dunphy, Newman (Ogden Ind.)	S	DOM Diamant-type disc-tumbler lock
1983	US 4,603,565	Strassmeir	P,S	3-Bladed BiLock
1984	US 4,601,184	Doinel	L	Fichet-Bauche 787
1985	AU 0,550,647	Ogden Ind.	S	Similar to DOM Diamant
1985	UK 2,158,870	Hakkarainen (Oy Wartsila AB)	S	Abloy DLP with transponder
1985	US 4,635,455	Oliver (Medeco Secu. Locks)	S	Medeco Biaxial cylinder
1985	US 4,667,495	Girard, Gaell	P	Vachette Radial with mobile element
1985	US 4,683,740	Errani (CISA)	P	ABUS TS 5000/CISA TSP cylinder
1986	US 4,651,546	Evans (Sargent & Greenleaf)	P,S	Abloy-type pin-tumbler disc lock
1986	US 4,787,225	Häuser, Stefanescu	P	DOM D 10-pin bilateral key lock
1987	US 4,756,177	Widen	P,S	Schlage Primus design
1987	US 4,802,354	Johnson (Fort Lock Co.)	P	Pick-resistant axial cylinder
1987	US 4,815,307	Widen	P,S	Schlage Primus design
1987	US 4,856,309	Eizen (Mul-T-Lock)	P	Mul-T-Lock with wavy shear line
1987	US 4,932,229	Genakis	P,S	Rotating interlocking pin side-bar lock
1988	US 4,836,000	Hirvi (AB Fas Lasfabrik)	L	Read-resistant lever lock keyway cam
1988	US 4,966,021	Boag	W	Mechanically reprogrammable 8-wafer lock
1988	US 5,067,335	Widen	P,S	ASSA Twin V-10
1989	US 4,977,767	Prunbauer (EVVA)	S,W	EVVA 3KS
1990	US 5,101,648	Kuster (Bauer Kaba)	P	Profile-control pins for Kaba lock
1990	US 5,131,249	Baden, Hinz	P	BKS Janus
1991	FR 2,678,670	Bonnard, Millier, Decosse	P	Pollux Interactive
1991	US 5,172,578	Bitzios	P	Sputnik decoder-pick for pin-tumbler lock
1991	US 5,375,444	Smith	S	Shield 2-key twin side-bar cylinder

Table D6: Nonmagnetic key-operated lock patents (cont.). Legend: P = pin-tumbler, W = wafer, L = lever, S = side-bar, A = automotive.

Filed	Patent no.	Author/Company	Relevance	Description
1991	WO 93/09317	Prunbauer	P	EVVA DPS/DPX
1992	US 5,224,365	Dobbs	A	Sputnik decoder for GM side-bar lock
1992	US 5,517,840	Häggström	S	ASSA Desmo
1992	US 5,582,050	Häggström	S	ASSA Desmo
1993	UK 2,271,807	Jones, Shields	P	Kaba Vario code change lock
1993	US 5,325,691	Embry (HSL Marketing)	A	Bypass tensioner for picking GM lock
1993	US 5,355,701	Tobias	P	Pressure-sensitive resistive decoder for Vingcard
1993	US 5,640,865	Widen	P,S	ASSA Twin Combi
1994	US 5,438,857	Kleinhaeny (Bauer Kaba)	P	KESO 1000 S pin design
1994	US 5,457,974	Keller	P	KESO Omega
1994	US 5,613,389	Häuser	S	DOM Diamant 10-disc cylinder
1995	DE 195,15,129	Krühn	P	IKON Multiprofile
1995	US 5,560,234	Keith James Ross	L	Ross 600 vertical lift lever lock
1995	US 5,724,841	Botteon (Silca)	P	Mobile key for "mezzanine" horizontal keyway
1995	US 5,791,181	Sperber, Sperber	L,S	Changeable Mottura push-key lock
1995	US 5,956,986	Vonlanthen	W,S	SEA twin side-bar bar-wafer lock
1996	US 5,797,287	Prunbauer	P	EVVA DPS/DPX
1997	US 5,839,308	Eizen, Markbreit	P	Mul-T-Lock interactive
1998	EP 0,903,455	Errani (CISA)	L	Antipicking Italian lever lock
1998	US 6,725,696	Blight, Esser (Lockwood Aust.)	S	Similar to DOM Diamant
1999	JP 2000 291300	Ikuo (Miwa Lock KK)	W,S	Modification to Miwa U9 side-bar lock
1999	US 6,681,609	Preddey (Australian Lock Co.)	S	New Generation BiLock
2000	US 6,477,876	Kim	P,S	Scorpion CX-5
2000	US 6,490,898	Mottura	P,S	Champions model 48
2002	US 6,622,538	Prunbauer, Amon (EVVA)	W,S	Axial side-bar wafer lock
2003	US 6,758,074	Prunbauer (EVVA)	S	Disc side-bar lock with round "3KS" key
2003	EP 1,375,790	Mottura	L	Recombinatable Italian lever lock

Table D7: Nonmagnetic key-operated lock patents (cont.). Legend: P = pin-tumbler, W = wafer, L = lever, S = side-bar, A = automotive.

Filed	Patent no.	Author/Company	Description
1890	US 0,428,247	Fenner	Very early magnetic pivoting pin lock
1914	US 1,172,203	Fuller	Air gap attractor key
1925	US 1,669,115	Anakin	Folded attractor key with blocking balls
1937	US 2,121,301	Ractliffe	Hybrid pin-tumbler lock with cotter magnets
1938	US 2,177,996	Raymond (Eagle Lock Co.)	Balls in channels driven by magnetic key
1959	US 3,056,276	Allander	Magnetic tilting pin principle
1963	US 3,234,767	Allander	Rolling ball in groove dragged by magnetic force
1964	US 3,271,983	Schlage	Matrix magnetic tumbler
1965	US 3,416,336	Felson (Liquidonics)	Twin inline magnetic tumbler lock
1966	US 3,444,711	Sedley	Early-form matrix magnetic card
1966	DE 1,553,365	Heimann	Rotating magnetic pin multiple inline lock
1966	US 3,566,637	Hallmann	Round key axial/radial magnetic lock
1967	US 3,393,541	Wake	Miwa EC type magnetic tumbler lock
1967	US 3,421,348	Hallmann	Round key axial magnetic lock
1967	US 3,518,855	Wake	Miwa EC type magnetic tumbler lock
1968	US 3,512,382	Check, Mauro, Valentinetti	(Liquidonics) Miracle Magnetic hybrid lock
1968	US 3,633,393	Hisatsune	Rotating magnetic 4-tumbler lock
1969	US 3,552,159	Craig	Rotating magnet push-key padlock
1969	US 3,581,030	Sedley	Magnetic matrix card and reed switch array
1970	US 3,657,907	Böving	Tilting magnetic pin padlock
1970	US 3,665,740	Goal Co.	Two rows of 4 magnetic tumblers
1971	US 3,705,277	Sedley	Magnetic version of Vingcard
1972	US 3,742,739	Hickman	Rotating magnetic tumbler padlock
1972	US 3,834,197	Sedley	Early-form CorKey
1973	US 3,837,194	Fish (Unican)	Hybrid 7 pin + 7 magnet lock
1973	US 3,837,195	Pelto	MagLok-type axial magnetic lock
1973	US 3,855,827	Hallman	Recessed rotating magnetic pin lock

Table D8: Magnetic key-operated lock patents.

Filed	Patent no.	Author/Company	Description
1973	US 3,857,262	Sidiropolous	Tilting magnetic pin padlock
1974	US 3,935,720	Böving	Magnetic rotors controlling side-bar radially
1975	US 3,948,068	Stackhouse (Am. Locker Co.)	MagLok 8-position tilting pin lock
1975	US 3,995,460	Sedley	Early-form CorKey
1976	NL 7,604,725	LIPS	Hybrid multiple inline cylinder with magnetic pins
1976	US 4,084,416	Prunbauer (EVVA)	Magnetic 6-rotor lock with inline pins
1976	US 4,133,194	Sedley	Later-form CorKey
1977	US 4,073,166	Clark	Magnetic lock-pick for Genii/Sima padlock
1979	US 4,220,021	Burger	EVVA 8-rotor 2 side-bar rotor control
1979	US 4,229,958	Burger	EVVA 8-rotor 2 side-bar profile control
1979	US 4,229,959	Easley	Decoder-pick for Miwa 14 magnet lock
1979	US 4,312,198	Sedley	CorKey
1980	EP 0,024,242	Sedley	CorKey with change carriage
1980	US 4,333,327	Wake/Miwa Lock Co.	Miwa 8 magnetic tumbler lock
1980	US 4,398,404	Wake/Miwa R.K.K.	Hybrid 8-magnet + 4-pin lock
1984	UK 2,151,295	Rogers, Robinson (Chubb)	4-rotor magnetic tumbler lock
1985	US 4,686,841	EVVA	Magnetic rotor damping
1986	US 4,676,083	Sedley, Eisermann	Magnetic card-in-slot lock
1987	AT 0,387,065B	Prunbaum (EVVA)	Early-form magnetic-rotor cylinder
1988	US 5,074,135	Eisermann	(Schulte-Schlagbaum) Magnetic Card-in-slot lock
1988	US 4,932,228	Eisermann	Magnetic card-in-slot lock with override
1988	US 5,072,604	Eisermann	Magnetic card-in-slot successor key sequencing
1988	DE 3,901,483	Elzett	Rotating magnetic pin twin side-bar Parnis lock
1990	US 5,074,136	Kim, Kwak	Concentric pin with magnetic inner tumbler
1990	US 5,267,459	Sedley	Modified code changer and change key for CorKey
1992	EP 0,571,311	Mondragon	Horizontal keyway lock with magnetic pins

Table D9: Magnetic key-operated lock patents (cont.).

Appendix E: Brief History of the Bramah Lock

Joseph Bramah (1748–1814) was a Yorkshire cabinet-maker who became a celebrated inventor and engineer [82]. During his lifetime, Bramah made numerous innovative contributions to society. Among his list of patents we find a design for a water closet (1778), the first hydraulic press (1795), a machine for filtering and storing beer (1797), and a printing machine for numbering banknotes (1806). His long-lasting contribution to locksmithing was made in 1784 with the patenting of his celebrated axial wafer lock, a key for which is displayed in Fig. E1. His achievements in this area are all the more remarkable since at the time almost all other locks in circulation were of the simple warded type or single-acting/single-lever locks. The only other English lock providing a reasonable level of security was a two-lever "double-action" lock invented by Robert Barron six years earlier in 1778. This lock, a precursor to the Chubb lever lock, had stumps on the levers and gates in the bolt (see Fig. 5.10 in Chapter 5). It was, however, not difficult to impression due to the differing sizes of the bellies of the levers.

Bramah's lock was so far ahead of the competition that he was unable to produce enough to satisfy the demand. In order to overcome the supply problem, he enlisted the help of an outstanding young mechanic named Henry Maudslay. During the term of his service, which lasted for nine years, he and Bramah conceived and implemented a number of machine tools without parallel at the time. These included a slotting machine for producing the lock barrel, a quick-grip vise for clamping the work, a nibbling machine for bitting the keys, a winding machine for producing the coil springs, and a compound slide tool for the lathe. In an era when factories relied overwhelmingly on manual labor, these new machine tools greatly contributed to the accuracy and speed of production, as well as to the concept of interchangeable component parts.

So confident was Bramah of the infallibility of his lock that he wrote in his 1785 dissertation [15], which formed the specification for his patent:

Figure E1: Key from an old 7-slider Bramah lock.

> I have contrived a security, which no instrument but its proper key can reach; and which may be so applied, as not only to defy the art and ingenuity of the most skilful workman, but to render the utmost force ineffectual, and thereby to secure what is most valued as well from dishonest servants as from the midnight ruffian.

Despite these advances in process mechanization, the Bramah lock was not a lock that was affordable to the working class, and this is reflected in the literature of the times. For example, Charles Dickens in his first book *The Pickwick Papers* [24] describes a meeting between Mr. Pickwick and a clerk from a legal firm in Gray's Inn Square located in the City of London:

> Mr. Lowten extracted the plug from the door-key; having opened the door, replugged and repocketed his Bramah, and picked up the letters which the postman had dropped through the box, he ushered Mr. Pickwick into the office.

As it turned out, Bramah's original design was susceptible to manipulation of the sliders with a pair of fine steel forceps. The advertising of this fact did not help bolster public confidence in the product.

With the benefit of hindsight, Bramah may have been a bit hasty in declaring his lock to be infallible. Nonetheless, it was still a vast improvement in security compared with warded and Barron-type lever locks. The subsequent introduction of false notches or gates into the sliders by W. Russell, one of Bramah's team, rendered the lock unpickable in any practical sense due to the length of time required to determine the correct depths for each of the sliders.

Around 1790, to reinforce the point still further, Bramah had his engineer Maudslay construct a 4″ padlock with no fewer than 18 sliders in a $1\frac{1}{2}$″ diameter barrel for display in their Piccadilly shop window (see Fig. E2). The engraving on the lock issued a public challenge to any would-be lockpicker, reading:

> The Artist who can make an Instrument that will pick or open this lock, shall receive 200 guineas the moment it is produced.

The lock remained unpicked in the shop front window for the remainder of Bramah's lifetime.

It was not until 1851, the year of the Great Exhibition in Crystal Palace, Hyde Park, London, that a fresh attempt was made by Alfred C. Hobbs of the Day & Newell Lock Company, Boston. Only two days prior, Hobbs had demonstrated his expertise by opening a 6-lever Chubb detector lock attached to a strong-room door in only 25 minutes. Even with Hobbs's considerable prowess, the manipulation of the Bramah lock was to occupy him for a full month, although he worked on the lock

Figure E2: Bramah's challenge padlock remained unpicked for around 60 years.

for only 44 hours in total over this period, and further delays were experienced as a result of the Bramah Company's protestations at Hobbs's methods. The ensuing debate over whether or not Hobbs had adhered to the rules of the challenge became known as the Great Lock Controversy and is documented in Hobbs's book of 1853 [51—53] and in the report of the committee overseeing the challenge [54]. Suffice to say that the method would not be well suited to use "in the field."

Notwithstanding the contentions of the Bramah Company, the 200 guineas reward was paid to Hobbs on the 9th of September 1851, which he promptly exhibited at the Day & Newell stand at the Great Exhibition. Bramah's design, despite succumbing on this occasion to Hobbs's methods, has withstood the test of time and is still used for security locks and safes in the United Kingdom and elsewhere. Bramah's original challenge padlock, pictured in Fig. E2, can be seen today in the London Science Museum, where Maudslay's screw-cutting lathe also resides. The Bramah principle found its way into many other lock designs, particularly in France and Germany, some of which were covered in the preceding chapters.

For a more detailed account of this interesting episode in the history of lock development, the reader is urged to consult Price's 1856 book [99], which devotes a chapter of 50 pages to the Lock Controversy at the time of the Great Exhibition and a further chapter of 77 pages to challenges engendered by the Lock Controversy up to 1856 (one of which was decided in court). Accounts of the tools and techniques that Hobbs made use of are contained in Price's book and also in McNeil's biography of Joseph Bramah [82]. Hogg's book [55] relates the Bramah/Hobbs encounter in an entertaining style. John Chubb's paper [22] is also a good source of information, containing, for example, a description of Owen's clockwork decoder for a Bramah 6-slider padlock: the device was powered by a weight and line and regularly attained the correct combination in half an hour to three hours.

Appendix F: Key Code Computations

This Appendix contains listings of a number of programs useful for determining key series with realistic constraints, such as MACS. All programs are written in the Matlab^{TM} language and are compatible with Matlab versions 5 and 6. The output from the programs is a list or matrix of cut depths, which can be interpreted as direct key codes. The cut positions are numbered from bow to tip.

The first three programs are for pin-tumbler locks. This is followed by a program for GM-type wafer locks. The last two programs are for lever locks with symmetric keys. On an algorithmic note, all of the programs are nonrecursive; that is, they do not involve recursive function calls. It is possible to implement the code generation algorithms in a direct, recursive form so that they run more efficiently.

For 5-pin locks with 10 cut depths the two functions "check_macs.m" and "perms_macs.m" can be used to generate the code list. The step of computing codes subject to MACS is done first in check_macs.m. The first time the program is run, a file called perms5.mat with all possible codes is generated; this file is reloaded on subsequent runs. The output is stored in another file called perms5_macs[n].mat, where n is the value of MACS used. The user can view the results, which contain nonsense codes like (1 1 1 1 1). The second step takes the MACS-compliant results and prunes off codes that do not satisfy other constraints, such as "up to three pins of the same size." The result of this step is a variable called all_perms containing the key series as a matrix. The number of codes is also displayed. The program code is easily modified for 6-pin locks, although the run time can be quite long for large values of MACS. For MK systems with a progression step of 2, the program "differ_by_2.m" can be applied to the output of perms_macs.m to generate a unique sequence of codes that differ by two or more increments in at least one position.

For 6-wafer locks of the General Motors variety, the program "perms_macs_gm.m" should be used. This program combines all the code-checking steps, including generation of the full list for five wafer sizes; checking codes against the MACS = 3 constraint; and checking that codes satisfy the parity and other constraints, for example, "sum of sizes must be even."

For 5-lever locks the functions "check_macs_lever.m" and "perms_macs_lever.m" are provided. The constraints are typical of Chubb-type mortice locks for two-sided operation with 8 lever sizes and a symmetric 7-cut key. The user must specify the MACS for both programs (for consistency, the same value should be used for both). The output matrix holds all usable 7-cut codes, although these are determined by the cuts in the first four positions due to symmetry. It is straightforward to modify the code for a different number of lever sizes or to cater for an asymmetric MACS constraint.

CHECK_MACS

```
% function all_perms=check_macs(MACS)
%
% Compute all feasible permutations satisfying constraints
% by exhaustive enumeration and elimination of codes that
% do not satisfy the MACS constraint.
% Cylinder assumed to have 5 pins.
%
% Input
% M - MACS (maximum adjacent cut specification)
%
% Example
% all_perms=check_macs(5);
%
% Copyright G. Pulford 2005
% Licence is granted to use, modify and distribute this code for non-profit
% purposes provided that my name is referenced as the original author
% in any modified versions and in any supporting documentation.

function all_perms=check_macs(MACS)
P=5;
if exist('perms5.mat','file')
    load perms5.mat
else
    disp('running perms5...')
    all_perms=perms5;
    disp('done')
end
N=size(all_perms,1); % total number of permutations
feas_perms=[];
% Eliminate infeasible codes
for i=1:N
    feas=1;
```

```
    % check MACS first
    for j=2:P
        if abs(all_perms(i,j)-all_perms(i,j-1))>MACS
            feas=0;
            break
        end
    end
    if feas
        feas_perms=[feas_perms,i];
    end
end
all_perms=all_perms(feas_perms,:);
size(all_perms,1)

function all_perms=perms5
% generate all codes for 10 sizes and 5 pins
% data saved to perms5.mat
P=5;
all_perms=zeros(10^P,P);
r=1;
for i=0:9
    for j=0:9
        for k=0:9
            for l=0:9
                for m=0:9
                    all_perms(r,:)=[i,j,k,l,m];
                    r=r+1;
                end
            end
        end
    end
end
save perms5.mat all_perms
```

PERMS_MACS

```
% function all_perms=perms_macs(MACS)
%
% Compute all feasible permutations satisfying constraints
% by elimination of offending codes. Cylinder assumed to have 5 pins.
% Code permutations that satisfy the MACS constraint are generated by
% check_macs(MACS) and retrieved from perms5_macs[MACS].mat for checking
% against the other constraints.
%
% Input
% M - MACS (maximum adjacent cut specification)
%
```

```
% Example
% all_perms=perms_macs(5);
%
% Copyright G. Pulford 2005
% Licence is granted to use, modify and distribute this code for non-profit
% purposes provided that my name is referenced as the original author
% in any modified versions and in any supporting documentation.

function all_perms=perms_macs(MACS)
P=5;
filename=['perms5_macs', num2str(MACS),'.mat']
if exist(filename,'file')
    disp('loading from perms5_macs file')
    cmd=['load ', filename];
    eval(cmd)
else
    disp('running check_macs...')
    cmd=['all_perms=check_macs(', num2str(MACS), ');']
    eval(cmd);
    disp('done')
end
N=size(all_perms,1); % total number of permutations
disp(['Number of codes satisfying MACS is: ', num2str(N)])
feas_perms=[];
% Eliminate infeasible codes
for i=1:N
    feas=1;
    %  MACS is already satisfied for these codes
    if all_perms(i,1)==8
        feas=0;
    end
    if all_perms(i,1)==9
        feas=0;
    end
    if feas
        % Only two adjacent pins of same size allowed
        for j=3:P
            if all_perms(i,j-2)==all_perms(i,j-1)
                if all_perms(i,j-1)==all_perms(i,j)
                    feas=0;
                end
            end
        end
    end
    if feas
        % Up to three pins can be the same size
        perm=all_perms(i,:);
        perm=sort(perm); % ascending order
        for j=4:P
            if perm(j-3)==perm(j-2)
```

```
                    if perm(j-2)==perm(j-1)
                        if perm(j-1)==perm(j)
                            feas=0;
                        end
                    end
                end
            end
        end
        if feas
            % must have at least 3 pins of different sizes
            perm=unique(perm);
            if length(perm)<3
                feas=0;
            end
        end
        if feas
            feas_perms=[feas_perms,i];
        end
end
all_perms=all_perms(feas_perms,:);
disp(['Codes satisfying all constraints:   ', num2str(size(all_perms,1))])
```

DIFFER_BY_2

```
% function all_perms=differ_by_2(all_perms);
%
% Given matrix of code vectors (code book), determine all codes that differ
% by only 1 unit in any component from another code and eliminate them.
% Resulting code book will have no codes that differ by only 1 unit.
% For instance the code 78751 differs by 1 unit from the following
% 10 codes:
% 68751, 77751, 78651, 78741, 78750
% 88751, 79751, 78851, 78761, 78752
%
% Notes:
% Length 5 code assumed.
% Code book should already satisfy other constraints (MACS, etc).
%
% Requires: code2dec_LUT()
%
% Copyright G. Pulford, 2005
% Licence is granted to use, modify and distribute this code for non-profit
% purposes provided that my name is referenced as the original author
% in any modified versions and in any supporting documentation.

function all_perms=differ_by_2(all_perms)
fwd=input('loop forwards? yes=1 (default=0): ');
```

```
% if set to 1, loop goes from 1 -> 5, else it goes from 5 -> 1
if isempty(fwd), fwd=0; end
if fwd
    disp('loop order: forwards')     first=1;
    last=5;
    inc=1;
else
    disp('loop order: backwards')
    first=5;
    last=1;
    inc=-1;
end
for n=first:inc:last
    disp(['position: ', num2str(n)])
    N=size(all_perms,1); % number of code vectors in code book
    disp(['code vectors: ', num2str(N)])
    all_perms_dec=zeros(N,1);
    for i=1:N
        all_perms_dec(i)=code2dec_LUT(all_perms(i,:),n);
    end
    [all_perms_dec,isrt]=sort(all_perms_dec);
    disp('code2dec loop done')
    feas=[isrt(1)];
    itest=1;
    for i=2:N % eliminate all codes differing by < 2 in n-th position
        if abs(all_perms_dec(itest)-all_perms_dec(i))>=2
            feas=[feas,isrt(i)];
            itest=i;
        end
    end
    disp('test loop done')
    feas=sort(feas);
    all_perms=all_perms(feas,:);
end
N=size(all_perms,1);
disp(['code vectors: ', num2str(N)])

% function d = code2dec_LUT(v,LSF)
% Converts code vector v of up to length 5 to decimal number.
% All entries of code vector should be in [0..9] (base 10).
% LSF in [1..5] is the index of the least-significant-figure
% e.g. given v= [3 0 6 4 1] LSF=5 => 1 is units
% LSF=4 => 4 is units, 1 is 10,000's
% LSF=3 => 6 is units, 4 is 10,000's, 1 is 1000's etc.
% Ignores leading zeros.
%
% Example
% d = code2dec_LUT([3 0 6 4 1],4)
%
```

```
% Copyright G. Pulford, 2005
% Licence is granted to use, modify and distribute this code for non-profit
% purposes provided that my name is referenced as the original author
% in any modified versions and in any supporting documentation.

function d = code2dec_LUT(v,LSF)
switch LSF
    case 1
        pow10=[10,100,1000,10000,1];
    case 2
        pow10=[100,1000,10000,1,10];
    case 3
        pow10=[1000,10000,1,10,100];
    case 4
        pow10=[10000,1,10,100,1000];
    case 5
        pow10=[1,10,100,1000,10000]; % Note: pow10(i)=10^(i-1) for i=1..5
end
N=length(v);
d=0;
for i=1:N
    vi=v(i);
    if vi~=0
        pow10Ni=pow10(N-i+1);
        d=d+vi*pow10Ni;
    end
end
```

PERMS_MACS_GM

```
% function all_perms=perms_macs_gm
%
% Compute all feasible permutations satisfying constraints for GM Holden
% by elimination of offending codes. Cylinder assumed to have 6 wafers.
% MACS (maximum adjacent cut specification) is assumed to be 3.
% Code permutations that satisfy the MACS constraint are generated or
% retrieved from perms6_gm_macs[MACS].mat for checking
% against the other constraints:
% MACS=3, no more than 3 adjacent same, perm must sum to even number.
%
% Example
% all_perms=perms_macs_gm;
%
% Copyright G. Pulford 2005
% Licence is granted to use, modify and distribute this code for non-profit
% purposes provided that my name is referenced as the original author
% in any modified versions and in any supporting documentation.
```

```matlab
function all_perms=perms_macs_gm
MACS=3
P=6;
filename=['perms6_gm', '.mat']
if exist(filename,'file')
    disp('loading from perms6_gm file')
    cmd=['load ', filename];
    eval(cmd)
else
    disp('generating perms for 5 sizes, 6 wafers...')
    all_perms=zeros(5^P,P);
    r=1;
    for i=1:5
        for j=1:5
            for k=1:5
                for l=1:5
                    for m=1:5
                        for n=1:5
                            all_perms(r,:)=[i,j,k,l,m,n];
                            r=r+1;
                        end
                    end
                end
            end
        end
    end
    save perms6_gm.mat all_perms
    disp('done')
end
disp('checking MACS')
N=size(all_perms,1); % total number of permutations
disp(['Total number of permutations: ', num2str(N)])
feas_perms=[];
% Eliminate infeasible codes
for i=1:N
    feas=1;
    % check MACS first
    for j=2:P
        if abs(all_perms(i,j)-all_perms(i,j-1))>MACS
            feas=0;
            break
        end
    end
    if feas
        feas_perms=[feas_perms,i];
    end
end
all_perms=all_perms(feas_perms,:);
filename=['perms6_gm_macs', num2str(MACS),'.mat']
cmd=['save ', filename, ' all_perms'];
```

```
eval(cmd)
N=size(all_perms,1); % total number of permutations
disp(['Number of codes satisfying MACS is: ', num2str(N)])
feas_perms=[];
% Eliminate infeasible codes
for i=1:N
    feas=1;
    %  MACS is already satisfied for these codes
    if feas
        % Only three adjacent wafers of same size allowed
        perm=all_perms(i,:);
        p=perm(3);
        if perm(4)==p
            if perm(4)==p
                % middle 2 sizes are same
                if perm(1)==p & perm(2)==p
                    % have four same in a row
                    feas=0;
                end
                if perm(2)==p & perm(5)==p
                    % have four same in a row
                    feas=0;
                end
                if perm(5)==p & perm(6)==p
                    % have four same in a row
                    feas=0;
                end
            end
        end
    end
    if feas
        % sum of sizes must be even
         if mod(sum(perm),2)~=0
            feas=0;
        end
    end
    if feas
        feas_perms=[feas_perms,i];
    end
end
all_perms=all_perms(feas_perms,:);
disp(['Codes satisfying all constraints:   ', num2str(size(all_perms,1))])
```

CHECK_MACS_LEVER

```
% function all_perms=check_macs_lever(MACS)
%
% Compute all feasible permutations satisfying constraints
```

```
% by exhaustive enumeration and elimination of codes that
% do not satisfy the MACS constraint for a 5 lever two-sided
% mortice lock with 8 lever sizes and a symmetric 7-cut key.
%
% Input
% M - MACS (maximum adjacent cut specification)
%
% Example
% all_perms=check_macs_lever(4);
%
% Copyright G. Pulford 2005
% Licence is granted to use, modify and distribute this code for non-profit
% purposes provided that my name is referenced as the original author
% in any modified versions and in any supporting documentation.

function all_perms=check_macs_lever(MACS)
P=4; % number of independent bitting positions
L=8; % number of cut depths (if changed, delete old perms5_lever.mat file)
if exist('perms5_lever.mat','file')
    load perms5_lever.mat
else
    disp('running perms5...')
    all_perms=perms5_lever(L);
    disp('done')
end
N=size(all_perms,1); % total number of permutations
feas_perms=[];
% Eliminate infeasible codes
for i=1:N
    feas=1;
    % check MACS first
    for j=2:P
        if abs(all_perms(i,j)-all_perms(i,j-1))>MACS
            feas=0;
            break
        end
    end
    if feas
        feas_perms=[feas_perms,i];
    end
end
all_perms=all_perms(feas_perms,:);
size(all_perms,1)
filename=['perms5_lever_macs', num2str(MACS),'.mat']
cmd=['save ', filename, ' all_perms'];
eval(cmd)

function all_perms=perms5_lever(L)
% Generate all codes for L sizes and 5 levers for a two-sided mortice locks
% with a symmetric 7-cut key. We only need to consider the first 4 cuts,
```

```
% since the remaining 3 cuts are determined by symmetry.
% For example: 1464 -> 1464641
% Data saved to perms5_lever.mat
P=4;
all_perms=zeros(L^P,P);
r=1;
for i=1:L
    for j=1:L
        for k=1:L
            for l=1:L
                all_perms(r,:)=[i,j,k,l];
                r=r+1;
            end
        end
    end
end
save perms5_lever.mat all_perms
```

PERMS_MACS_LEVER

```
% function all_perms=perms_macs_lever(MACS)
%
% Compute all feasible permutations satisfying constraints
% by elimination of offending codes for a 5-lever two-sided
% mortice lock with 8 lever sizes and a symmetric 7-cut key.
% Code permutations that satisfy the MACS constraint are generated by
% check_macs_lever(MACS) and retrieved from perms5_lever_macs[MACS].mat
% for checking against the other constraints.
%
% Input
% M - MACS (maximum adjacent cut specification)
%
% Example
% all_perms=perms_macs_lever(4);
%
% Copyright G. Pulford 2005
% Licence is granted to use, modify and distribute this code for non-profit
% purposes provided that my name is referenced as the original author
% in any modified versions and in any supporting documentation.

function all_perms=perms_macs_lever(MACS)
P=4; % number of independent bitting positions
filename=['perms5_lever_macs', num2str(MACS),'.mat']
if exist(filename,'file')
    disp('loading from perms5_lever_macs file')
    cmd=['load ', filename];
    eval(cmd)
else
```

```
        disp('running check_macs...')
        cmd=['all_perms=check_macs_lever(', num2str(MACS), ');']
        eval(cmd);
        disp('done')
end
N=size(all_perms,1); % total number of permutations
disp(['Number of codes satisfying MACS is: ', num2str(N)])
feas_perms=[];
% Eliminate infeasible codes
for i=1:N
    feas=1;
    % MACS is already satisfied for these codes
    % Exclude if three same out of four
    perm=all_perms(i,:);
    temp=sort(perm); % ascending order
    test=temp(2);
    if temp(3)==test
        if temp(1)==test
            feas=0;
        elseif temp(4)==test
            feas=0;
        end
    end
    % must have at least 2 different sizes
    % this is automatically satisfied if we get to here

    % must have at least one adjacent cut difference of 2 or more
    if feas
        if abs(perm(2)-perm(1))<=1
            if abs(perm(3)-perm(2))<=1
                if abs(perm(4)-perm(3))<=1
                    feas=0;
                end
            end
        end
    end

    if feas
        feas_perms=[feas_perms,i];
    end
end
all_perms=all_perms(feas_perms,:);
N=size(all_perms,1);
disp(['Codes satisfying all constraints:   ', num2str(N)])
all_perms=[all_perms, fliplr(all_perms(:,1:3))];
```

Appendix G: Security Gradings for Cylinder Locks

Notes

1. Drill resistance times are in minutes.

2. The units used for attack resistance by chisel are the number of blows.

3. The units for resistance to twisting are number of twists.

4. The notation used for extraction resistance is the applied force in kilo-Newtons (kN) followed by the time allowed in minutes.

5. Extraction and torque may be applied to either the plug or the cylinder. Tolerances for torque have been omitted.

6. Key registration applies to the manufacturer. Keys for grade H locks must be registered in the LPS 1224 database.

7. Restricted key cutting means that only the manufacturer may cut keys. Semirestricted key cutting means manufacturers and approved stockists may cut keys.

8. Drill type has been omitted.

9. Operation torque for all grades is 1.5 Nm.

Requirement/Grade	A	B	C	D	E	F	G	H
Min Effective Differs	100	300	15000	30000	100000	250000	500000	1000000
Min Movable Detainers	2	3	5	6	6	7	8	9
Max Cuts Same (%)	100	70	60	50	50	50	50	40
Max Adjacent Cuts Same	2	2	2	2	2	0	0	0
Direct Coding of Keys	Yes	Yes	No	No	No	No	No	No
Drill Resistance Time (net)	N/A	N/A	N/A	3	5	5	5	10
Drill Resistance Time (total)	N/A	N/A	N/A	5	10	20	20	30
Attack Resistance—Chisel	N/A	N/A	N/A	30	40	50	50	50
Resistance to Twisting	N/A	N/A	N/A	20	30	40	50	50
Resistance—Extraction	N/A	N/A	N/A	15(3)	15(5)	17(5)	17(5)	17(10)
Resistance—Torque (Nm)	2.5	5	15	20	30	40	50	50
Patented Key Design	N/A	N/A	N/A	N/A	Yes	Yes	Yes	Yes
Factory Key Registration	N/A	N/A	N/A	Yes	Yes	Yes	Yes	N/A
Key Cutting—Restricted	N/A	N/A	N/A	N/A	N/A	Yes	Yes	Yes
Key Cutting—semirestricted	N/A	N/A	N/A	Yes	Yes	N/A	N/A	N/A
Cylinder part of MK system	Yes	Yes	Yes	Yes	Yes	Yes	No	No

Table G1: Security grading for locks specified under Loss Prevention Standard 1242 (UK) [72]. Reproduced with permission of LPCB.

Index

Printed and bound by CPI Group (UK) Ltd, Croydon, CR0 4YY

03/10/2024

01040335-0017